METHODS IN MOLECULAR BIOLOGY™

Series Editor
John M. Walker
School of Life Sciences
University of Hertfordshire
Hatfield, Hertfordshire, AL10 9AB, UK

For further volumes:
http://www.springer.com/series/7651

Protein Nanotechnology

Protocols, Instrumentation, and Applications

Second Edition

Edited by

Juliet A. Gerrard

Biomolecular Interaction Centre and School of Biological Sciences, University of Canterbury, MacDiarmid Institute for Advanced Materials and Nanotechnology, Riddet Institute, Christchurch, New Zealand; Callaghan Innovation Research Limited, Lower Hutt, New Zealand

Editor
Juliet A. Gerrard
Biomolecular Interaction Centre
and School of Biological Sciences
University of Canterbury
MacDiarmid Institute for Advanced Materials
and Nanotechnology
Riddet Institute
Christchurch, New Zealand

Callaghan Innovation Research Limited
Lower Hutt
New Zealand

ISSN 1064-3745 ISSN 1940-6029 (electronic)
ISBN 978-1-62703-353-4 ISBN 978-1-62703-354-1 (eBook)
DOI 10.1007/978-1-62703-354-1
Springer New York Heidelberg Dordrecht London

Library of Congress Control Number: 2013932703

Printed on acid-free paper

Humana Press is a brand of Springer
Springer is part of Springer Science+Business Media (www.springer.com)

Dedication

To Lee

Preface

Since the first edition of this book, the intersection of protein science and nanotechnology has become an exciting frontier in interdisciplinary sciences. Proteins exist naturally on the nanoscale, which, coupled with our increasing ability to control their form and function, makes them obvious candidates to play a lead role in nanotechnology. Increasingly this potential is being recognized and realized, and we hope that this book plays a role in inspiring more research in this burgeoning field.

After an introductory chapter, this volume has three parts. The first, "*Old proteins, new tricks*," highlights examples of proteins that are generally well understood in the biological world, which are being harnessed for a wide range of applications, some more developed than others. We begin with the silk proteins, and other fibrous proteins, followed by chapters that highlight the increasingly recognized potential of amyloid fibrils, hydrophobins, and S-layer proteins for bionanotechnological applications. In the next part "*New proteins*," we take a look at engineering proteins for specific ends, how we might characterize these, and the sorts of uses to which these new proteins might be put. In the last part "*Tools of the trade*," we give an overview of the sorts of tools that are now readily available to manipulate the structure and function of proteins, both rationally and using methods inspired by evolution. We also take a look at some instrumental methods that are important for studying protein nanostructures as they assemble. Together, the aim is to provide an overview of this multifaceted field and a useful guide to those who wish to contribute to it.

Christchurch, New Zealand ***Juliet A. Gerrard***

Acknowledgments

This book has had a particularly difficult gestation, having being written during the course of a series of major earthquakes. So first of all, I would like to acknowledge the patience of those diligent authors who had their chapters in early and then had to wait too long while I found the time to chase the tardier. Thanks to all authors for their contributions, support, and enthusiasm to make this project work. I hope you are all pleased with the result.

On a personal note, I would like to thank the University of Canterbury for its generous sabbatical leave provisions, without which this book would never have made it into print. I also owe a great deal to Peter, for the usual list of forbearances. Thank you.

Contents

Contributors

STEVEN ARCIDIACONO • *U.S. Army Natick Soldier Research, Development & Engineering Center, Natick, MA, USA*

JOHN A. CARVER • *School of Chemistry & Physics, The University of Adelaide, Adelaide, Australia*

DANIEL CHRIST • *Garvan Institute of Medical Research, Sydney, Australia; Faculty of Medicine, University of New South Wales, Sydney, Australia*

ROY M. DANIEL • *Thermophile Research Unit, Department of Biological Sciences, University of Waikato, Hamilton, New Zealand*

SANTANU DEB-CHOUDHURY • *AgResearch Limited, Lincoln Research Centre, Lincoln, Christchurch, New Zealand*

SEAN R.A. DEVENISH • *Department of Biochemistry, University of Cambridge, Cambridge, UK*

LAURA J. DOMIGAN • *Biomolecular Interaction Centre and School of Biological Sciences, MacDiarmid Institute for Advanced Materials and Nanotechnology, University of Canterbury, Christchurch, New Zealand; Biomedical Engineering and Mechanical Engineering Departments, Tufts University, Medford, MA, USA*

JOLON M. DYER • *AgResearch Limited, Lincoln Research Centre, Lincoln, Christchurch, New Zealand*

HEATH ECROYD • *School of Biological Sciences, University of Wollongong, Wollongong, Australia*

OLGA ESTEBAN • *CIBER de Bioingeniería, Biomateriales y Nanomedicina (CIBER-BBN), Zaragoza, Spain; Institute for Bioengineering of Catalonia (IBEC), Barcelona, Spain*

CONAN J. FEE • *Biomolecular Interaction Centre, University of Canterbury, Christchurch, New Zealand; Department of Chemical & Process Engineering, University of Canterbury, Christchurch, New Zealand*

MARTIN FISCHLECHNER • *Department of Biochemistry, University of Cambridge, Cambridge, UK*

MEGAN GARVEY • *Max Planck Research Unit for Enzymology of Protein Folding, Halle (Saale), Saxony-Anhalt, Germany; School of Chemistry & Physics, The University of Adelaide, Adelaide, Australia*

JULIET A. GERRARD • *Biomolecular Interaction Centre and School of Biological Sciences, University of Canterbury, MacDiarmid Institute for Advanced Materials and Nanotechnology, Riddet Institute, Christchurch, New Zealand; Callaghan Innovation Research Limited, Lower Hutt, New Zealand*

SALLY L. GRAS • *Bio21 Molecular Science and Biotechnology Institute, The University of Melbourne, Melbourne, Australia; Department of Chemical and Biomolecular Engineering, The University of Melbourne, Melbourne, Australia*

PAULINA HANSON-MANFUL • *Institute of Natural Sciences, Massey University, Auckland, New Zealand*

FLORIAN HOLLFELDER • *Department of Biochemistry, University of Cambridge, Cambridge, UK*
MIRIAM KALTENBACH • *Department of Biochemistry, University of Cambridge, Cambridge, UK*
DAVID L. KAPLAN • *Biomedical Engineering and Mechanical Engineering Departments, Tufts University, Medford, MA, USA*
EMMANOUIL KASOTAKIS • *Department of Materials Science and Technology, and Institute for Electronic Structure and Laser, Foundation for Research and Technology-Hellas, (IESL-FORTH), University of Crete, Vassilika Vouton, Heraklion, Crete, Greece*
Institute for Electronic Structure and Laser, Foundation for Research and Technology-Hellas, (IESL-FORTH), Heraklion, Crete, Greece
CHARLES K. LEE • *Thermophile Research Unit, Department of Biological Sciences, University of Waikato, Hamilton, New Zealand*
GARY LEISK • *Biomedical Engineering and Mechanical Engineering Departments, Tufts University, Medford, MA, USA*
BRIDGET C. MABBUTT • *Department of Chemistry and Biomolecular Sciences, Macquarie University, Sydney, Australia*
ANTON P.J. MIDDELBERG • *Centre for Biomolecular Engineering, Australian Institute for Bioengineering and Nanotechnology and School of Chemical Engineering, The University of Queensland, Brisbane, Australia*
ANNA MITRAKI • *Department of Materials Science and Technology, and Institute for Electronic Structure and Laser, Foundation for Research and Technology-Hellas, (IESL-FORTH), University of Crete, Vassilika Vouton, Heraklion, Crete, Greece*
COLIN R. MONK • *Thermophile Research Unit, Department of Biological Sciences, University of Waikato, Hamilton, New Zealand*
VANESSA K. MORRIS • *School of Molecular Bioscience and Discipline of Pharmacology, University of Sydney, Sydney, Australia*
FIORENZO OMENETTO • *Biomedical Engineering and Mechanical Engineering Departments, Tufts University, Medford, MA, USA*
WAYNE M. PATRICK • *Department of Biochemistry, University of Otago, Dunedin, New Zealand; Institute of Natural Sciences, Massey University, Auckland, New Zealand*
JEFFREY E. PLOWMAN • *AgResearch Limited, Lincoln, Christchurch, New Zealand*
RUCSANDA C. PREDA • *Biomedical Engineering and Mechanical Engineering Departments, Tufts University, Medford, MA, USA*
ELIZABETH B. SAWYER • *Bio21 Molecular Science and Biotechnology Institute, The University of Melbourne, Melbourne, Australia; Department of Chemical and Biomolecular Engineering, The University of Melbourne, Melbourne, Australia*
BERNHARD SCHUSTER • *Department of NanoBiotechnology, University of Natural Resources and Life Sciences, Vienna, Austria*
UWE B. SLEYTR • *Department of NanoBiotechnology, University of Natural Resources and Life Sciences, Vienna, Austria*
JASON W. SOARES • *U.S. Army Natick Soldier Research, Development & Engineering Center, Natick, MA, USA*
MEGHNA SOBTI • *Structural and Computational Biology Division, Victor Chang Cardiac Research Institute, Sydney, Australia*

DMITRY V. SOKOLOV • *Institute of Fundamental Sciences, Massey University, Palmerston North, New Zealand*
DANIELA STOCK • *Faculty of Medicine, University of New South Wales, Sydney, Australia; The Victor Chang Cardiac Research Institute, Sydney, NSW, Australia*
MARGARET SUNDE • *Discipline of Pharmacology, University of Sydney, Sydney, Australia*
DAVID C. THORN • *School of Chemistry & Physics, The University of Adelaide, Adelaide, Australia*
ELIZABETH A. WELSH • *U.S. Army Natick Soldier Research, Development & Engineering Center, Natick, MA, USA*
CHUN-XIA ZHAO • *Centre for Biomolecular Engineering, Australian Institute for Bioengineering and Nanotechnology and School of Chemical Engineering, The University of Queensland, Brisbane, Australia*

Chapter 1

Protein Nanotechnology: What Is It?

Juliet A. Gerrard

Abstract

Protein nanotechnology is an emerging field that is still defining itself. It embraces the intersection of protein science, which exists naturally at the nanoscale, and the burgeoning field of nanotechnology. In this opening chapter, a select review is given of some of the exciting nanostructures that have already been created using proteins, and the sorts of applications that protein engineers are reaching towards in the nanotechnology space. This provides an introduction to the rest of the volume, which provides inspirational case studies, along with tips and tools to manipulate proteins into new forms and architectures, beyond Nature's original intentions.

Key words Protein nanotechnology, Self-assembly, Supramolecular, Tecton

1 Introduction

Proteins naturally exist in the nanoscale, so it comes as no surprise that protein nanotechnology is an emerging field, capturing both the excitement of recent advances in nanotechnology and harnessing our exponentially expanding knowledge of protein science. This happy marriage of Nature's intricate nanoscale machines and the drive to explore the technological benefits of the nanoworld leads to an exciting mix of disciplines coming together in the protein nanotechnology space. Proteins are finding utility in a host of nanotechnological applications, and likewise nanodevices are perfectly placed to interact with the biological world.

Proteins have long been recognized as the most versatile of the biological building blocks, but this versatility comes with the cost of complexity, so they are also hard to control in a predictive manner. In contrast, nucleic acids have been used to create complex 3D structures, using the predictive quality of Watson-Crick base pairing and ease of chemical synthesis, with considerable success (1). Intricate structures and nanomechanical devices have been built from DNA, including exciting shape-shifting structures such as gears and walkers, which respond to small molecules (2). However,

Juliet A. Gerrard (ed.), *Protein Nanotechnology: Protocols, Instrumentation, and Applications*, Methods in Molecular Biology, vol. 996, DOI 10.1007/978-1-62703-354-1_1,

some challenges remain. Functionalization of nucleotide-based nanostructures is still a challenge (3) and it has also proved difficult to control these structure in three dimensions (4).

Peptides are increasingly investigated as building blocks (or tectons) for nanostructures (5) and exciting applications are being explored for their use in creating functional biological materials (6); Chapters 10 and 11). Successes include the generation of fibrous materials from peptides that bridge the nano- to mesoscales (7). Routine chemical synthesis of peptides means that non-biogenic amino acids are readily incorporated, and structures assembled from peptides are often much more stable than protein structures. However, peptides lack the wealth of functionality exhibited by proteins, and are expensive to synthesize at large scale (8). This, coupled with increasingly routine techniques to create proteins with "designer properties" (Chapters 14–16), leads researchers in increasing numbers to explore proteins as tectons for building nanostructures.

Proteins represent the next frontier. Their potential for the creation of designer materials and intricate nanoscale machines is unquestioned, since biology is filled with examples of proteins acting in these roles, as will be described in the chapters of this book. Proteins offer enormous diversity in three-dimensional architecture and endless scope for modification and functionalization. Their complex architectures include rings, tubes, and cages that can in principle be used as components of nanomachines (9). This promise has been heralded for the last decade or more (10–13) but the challenge of designing and controlling these interactions in vitro to engender useful functionality is a large one. In contrast to the simpler DNA or peptide tectons, designed self-assembly of proteins has been hampered by chemical heterogeneity and the large size of molecular surfaces required for protein-protein interactions (14). However, as protein engineering comes of age, so do ambitions to design sophisticated multicomponent protein systems, which increasingly attract the eye of synthetic biology (15).

This chapter gives a select review of some of the exciting nanostructures that have already been created using proteins, and the sorts of applications that protein engineers are reaching towards in the nanotechnology space. The intention is to whet the reader's appetite for the chapters that follow, which provide tips and tools to manipulate proteins into new forms and architectures, beyond Nature's original intentions, along with detailed case studies.

2 Inspiration from Nature

Nature abounds with examples of hierarchical self-assembly (16) and nowhere is this more impressive than in the realm of proteins. These highly precise and regulated associations give rise to a vast range of structures that range from tough materials such as the silks (Chapters 2 and 3) to molecular machines in cells, such as ATP synthase (Chapter 12) and ribosomes that are capable of performing highly sophisticated functions (17). Protein assemblies control cellular morphology, position proteins within or outside cells, transport cargo, and provide compartmentalization within cells (8). This impressive repertoire inspires the de novo design of protein structures in vitro, based on the principles of self-assembly learnt from biology (18).

The intersection between nanoscience and protein science is well illustrated using collagen as a case study. Collagen is the most abundant protein in the body and beautifully illustrates the way that biology harnesses bottom-up assembly to construct materials that appear exquisitely designed for their function. Assembly typically begins with three amino acid chains self-assembling to form a triple helix, the properties of which are defined by the repeating sequence of amino acids in the individual polypeptide chains. A repeating sequence of proline-hydroxyproline-glycine engenders the specific properties of the collagen triple helix, which in turn engages in multi-hierarchical self-assembly to create nanofibrous strands. These strands continue to self-assemble both linearly and laterally to form the basis of a tough, strong proteinaceous material in vivo. Our understanding of this process has inspired many scientists to mimic this process in vitro, with a recent study by O'Leary et al. (19) illustrating how a small peptide, based on the characteristic proline-hydroxyproline-glycine repeating unit and incorporating salt-bridges between strands, will create a "sticky end-assisted" assembly, a particularly elegant example. Figure 1 illustrates schematically how these designer peptides successfully mimic all of the hierarchical stages of self-assembly seen in the native protein (19).

Elastin is another protein that is understood in great detail and has motivated much research into how the structure of the protein engenders its biological role in providing the elastic properties to the flexible tissues in which it dominates. Like collagen, elastin has become the subject of biomimetic research in which model peptides and proteins are used to create elastin-like protein assemblies in vitro, based on soluble precursors (tropoelastin and related proteins) that escape the constraints of the natural, insoluble, material. Understanding the interactions that dominate in the natural material has led to the design of a sophisticated range of designer proteins that have been used for a range of biomaterials with useful elasticity and cell interactive qualities (20).

The silk proteins (Chapters 2 and 3) are perhaps the best-understood protein in terms of how the structure and function relate. Kaplan and others have seized upon this wonderfully versatile material to explore a world of potential bio-inspired high-tech applications. Based on a controlled ability to turn silk proteins into gels, films, fibers and sponges, applications as diverse as photonics, electronics, optical

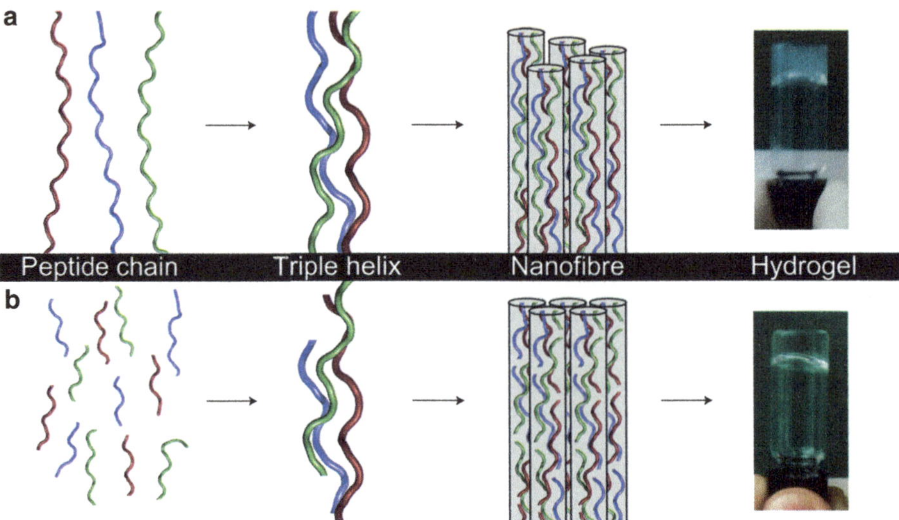

Fig. 1 (**a**) Self-assembly of collagen type I compared to that of collagen mimetic peptides: type I collagen assembly in which the peptide chains (shown in *red*, *blue*, and *green*) consist of 1,000 amino acids, the triple helices are 100 nm in length and the blunt-ended nanofibers (shown in *grey*) assemble via the staggered lateral packing of the triple helices. The hydrogel pictured is from a rat-tail collagen sample. (**b**) Scheme for the self-assembly of collagen mimetic peptides in which the peptides consist of 36 amino acids (shown in *red*, *blue*, and *green*), the triple helix is staggered with a length of 10 nm and the nanofibers (shown in *grey*) result from triple helical elongation as well as from lateral packing. The hydrogel pictured is the designed peptide $(\text{Pro-Lys-Gly})_4(\text{Pro-Hyp-Gly})_4(\text{Asp-Hyp-Gly})_4$ (Reproduced from O'Leary et al. (2012) with permission (19))

fibers, biomaterials for bones and ligaments, adhesives, microfluidics, and medical devices are being pioneered (21).

More generally, coiled coil proteins, which are a ubiquitous protein fold in vivo (22), have attracted attention as motifs for generating self-assembled protein materials. The basic "rules" of coiled coil assembly have been elucidated, allowing the design of new materials with precisely engineered structures with potential applications in drug delivery, regenerative medicine, and biosensing. Unlike the more specific examples of collagen, silk, and elastin, coiled coil proteins are not only involved in structural proteins but also gene regulation, intracellular vesicle, and viral fusion processes, potentially expanding the range of eventual uses to which these proteins might be put (23). With an increased understanding of sequence-structure rules for coiled coil proteins, the successful generation of designed materials that can compete with naturally derived biomaterials and synthetic polymer chemistry will be enabled (7). Amyloid fibrils are another generic protein fold that has attracted considerable attention as potential protein nanostructures for use in nanobiotechnology, with considerable recent headway being made (11, 24, 25). This will be explored in detail in later chapters of this volume (Chapters 5–7).

One of the first protein nanostructures to be studied was the cage protein, ferritin, which in biology stores iron in a microcrystalline nontoxic form and controls its release. The protein cage controls the size of the nanoparticulate iron, a fact that has been exploited by using the cage in vitro to instead grow controlled nanoparticles of cobalt and iron oxides (26, 27). Other cage structures, based on virus scaffolds, have also been adopted as very versatile containers and nanostructures, e.g. as enzyme nanocarriers. Virus capsid protein self-assembly is driven by precise supramolecular combinations of protein monomers, which have made them attractive building blocks. Thus virus-like particles are being used as nanoscaffolds for enzyme selection, enzyme confinement and patterning, phage therapy, raw material processing, and single-molecule enzyme kinetics studies (28). These particles afford several advantages, including their nanometer size range, physical and chemical stability, ease of production in large quantities, and ability to be conjugated to other molecules on their outer surface. It is likely that these proteins will find many applications as protein nanotechnology comes of age.

Beyond the immediate horizons for protein nanotechnology, many researchers are beginning to see the enormous inherent potential in building molecular machines out of proteins. Molecular motors such as ATP synthase and the bacterial flagellum motor are now understood in exquisite detail (Chapter 12). This has inspired a long term goal for bionanotechnology: to build nanodevices that carry out tasks of our designing. Components for such devices are beginning to appear, including proteins that can be triggered and switched, and existing motor proteins have been modified to perform specific tasks (29). We are entering an exciting time in the construction of proteinaceous molecular machines.

3 Beyond Nature's Repertoire: Designer Proteins

In the last decade, our understanding of how to manipulate the structure of proteins to create artificial constructs with predictable properties has increased exponentially. Clarke and Regan (30) present an overview of the state of the art of protein engineering and design in a very useful special issue. In later chapters of this volume, useful guides to create new proteins are outlined, covering basic tips and tools (Chapters 14 and 15), directed evolution (Chapter 15), and thermophilic proteins (Chapter 13). The advent of the Rosetta programmes (31) has brought the promise of computer designed proteins to fruition, with an ever increasing number of success stories in the literature (32). With clever combinations of rational design and smart use of directed evolution technologies (Chapters 14 and 15) it is increasingly possible to create a protein of desired structure and function, with good stability in nonbiological environments.

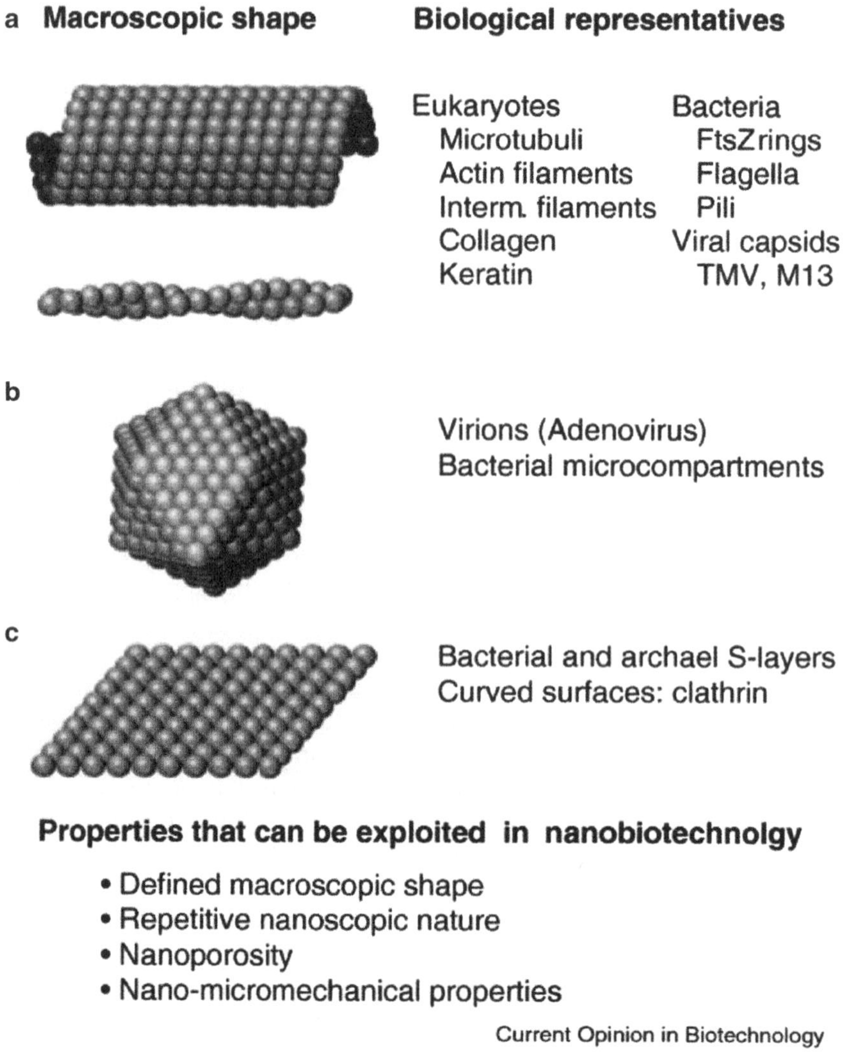

Fig. 2 Protein assemblies: morphological shape, biological representatives, and properties that can be exploited in nanobiotechnology. (**a**) Rods and cylinders, (**b**) icosahedrons, and (**c**) sheets and curved surfaces (Reproduced with permission from ref. 8)

Howorka (8) gives an excellent overview of the rational engineering of natural protein assemblies for nanobiotechnology. He highlights the design principles inherent in naturally occurring multimeric proteins and how these might be exploited for the manufacture of bottom-up structures made of protein, with potential applications in biomaterial design, vaccine development, biocatalysis, materials science, and synthetic biology. The shape of natural protein assembly architectures (Fig. 2) provides a useful start point for exploring the potential of their use in vitro. For example, rods and cylinders offer potential for formation of gels and films, as well as components of motors or nanodevices associated with transport

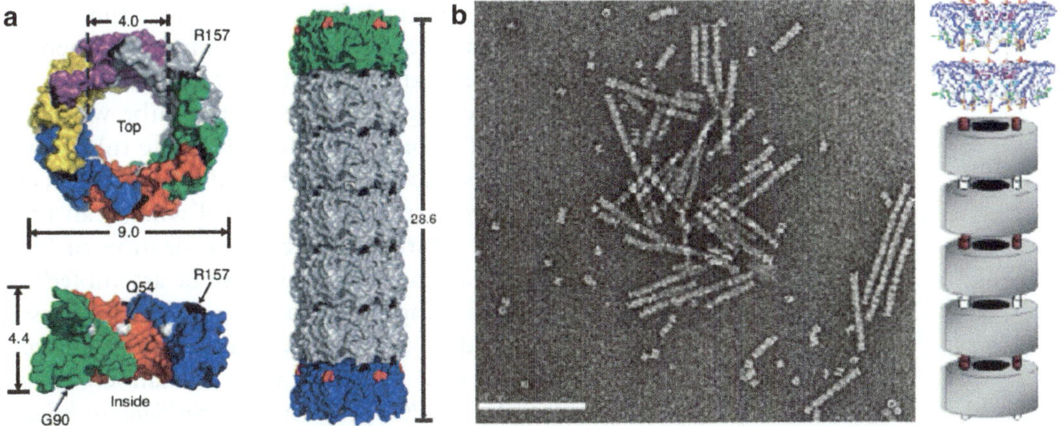

Fig. 3 (**a**) Hcp1 protein genetically modified to assemble into tubules reproduced with permission from ref. 42; (**b**) TEM of mutant TRAP protein tubes in presence of a weak reducing agent (Reproduced with permission from ref. 35)

and motility. Hollow assemblies afford encapsulation, compartmentalization, and protection from the environment, perhaps with triggered release, as well as possibilities for surface display of useful functionalities. Planar assemblies (such as S-layers, Chapter 9) suggest applications in protection, molecular filtration, and immobilization of useful functionalities, such as enzymes.

There are some well-characterized proteins that form assemblies that have been explored in a nanotechnological context. The bacterial tRNA attenuation protein (TRAP) is heat stable and reasonably tolerant to mutations and has been engineered to assemble as a twelve-membered ring, with a pore diameter of approximately 2 nm (33). This protein was used as a nanodot-binding module by replacing an unconserved arginine residue in the inner pore with a cysteine residue, thus artificially introducing unique thiol moieties in the pore of protein, which enabled selective binding of gold nanoparticles in the inner pore. Further functionalization with a titanium-binding peptide enabled immobilization of these protein rings on an inorganic surface (34). The TRAP protein ring was also modified to generate self-assembled nanotubes approximately 1 μm in length through disulfide and hydrophobic interactions (Fig. 3) (35).

Stable protein 1 (SP1) is another ring-shaped protein, a dodecamer, which is highly thermostable and protease resistant (36). SP1 nanostructures have been functionalized with glucose oxidase and then assembled as a nanotube with the enzyme coating the outer surface (37). It also forms regular arrays and interacts with gold nanoparticles, successfully forming gold nanowires (38). These gold nanoparticle conjugates have both been linked to glucose oxidase and layered on an electrode surface by means of dithiol-bridging units (39) and utilized as a component of basic logic circuit (40).

HcpI protein is another ring-shaped structure, this time a hexamer, with an outer diameter of 9 nm and an inner diameter of 4.0 nm (Fig. 3) (41). By introducing cysteine residues on the top and bottom surface of the ring, the protein nanotube was stabilized by disulfide bonds. To control the length of the tube, it was capped with HcpI mutants that had an attached epitope, thus creating an enclosed elongated cavity the inside of which can function as a nanocapsule (42). Biophysical characterization of HcpI protein showed that its quaternary structure can be assembled and disassembled without disrupting the secondary structure by using detergents, thereby providing control over its assembly process. This can be further exploited for fabrication of nanocontainers (43). In another example, HSP60 protein from *Sulfolobus shibatae* has a barrel-shaped structure which has been exploited as a nanocontainer using reactive cysteine residues to attach gold and CdSe-ZnS nanoparticles (44).

Moving beyond the use of building blocks that emerge directly from our understanding of the natural world, increasing efforts are being invested in designing de novo protein-protein associations to create new nanoarchitectures from protein. The analysis of natural interfaces between proteins has informed the formulation of some generic rules that govern these associations, which rely on a consideration of the symmetry of the assembly to control multiplicity and inform mutation strategies. As early as 2001, Padilla et al. (45) used an elegant symmetry-based design to construct a series of self-assembling nanohedra including cages, filaments, layers, and porous materials, with potential use as nanomaterials. Specifically, a general strategy was outlined in which protein A, which naturally forms self-assembling oligomer A_n, was fused to protein B, which naturally forms a distinct self-assembling oligomer B_m, to create a fusion protein A–B that self-assembled into bespoke nanohedral particles $(A–B)_p$ (see Fig. 4). The authors claim that this moves researchers a step closer to engineering self-assembling nanomaterials. Other examples of using symmetry considerations to creation of novel assemblies have been achieved with considerable success (46) and provide a useful set of guidelines for future researchers using different protein systems. More recently, designed cages have been engineered with a defined 16 nm cavity (47).

Tsai et al. (2, 48) present a bolder strategy which obviates the need to start with an ideal natural building block and instead propose starting with the nanostructure shape, then selecting candidate proteins and mapping them onto the selected shape. Starting with a desired shape, rather than a particular protein, widens the scope of imaginable structures substantially, and has been validated for the protein nanotube. The key to success is the mapping of the protein onto the nanostructure, followed by a global optimization process (2). The ever-growing protein data bank provides a vast range of shapes, surfaces, and chemical properties, which increases the chances of this strategy succeeding (48).

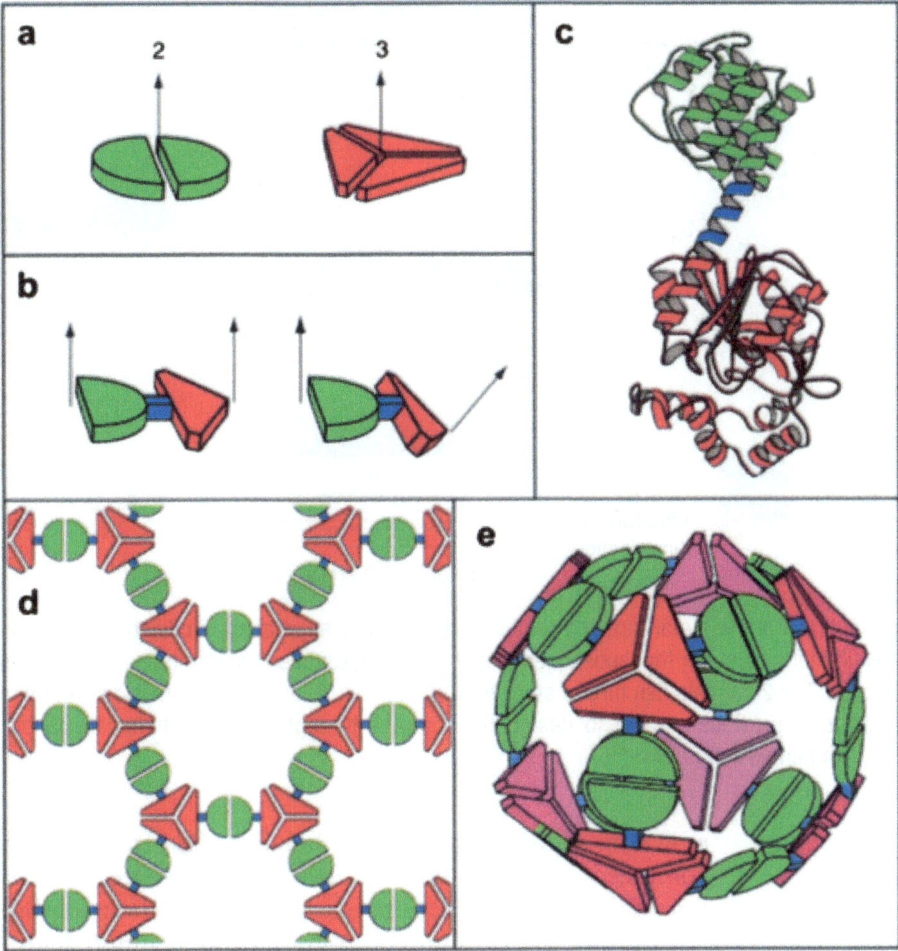

Fig. 4 A general strategy for designing fusion proteins that assemble into symmetric nanostructures. (**a**) The green semicircle represents a natural dimeric protein (i.e., a protein that associates with one other copy of itself), whereas the red shape represents a trimeric protein. The symmetry axes of the natural oligomers are shown. (**b**) The two natural proteins are combined by genetic methods into a single fusion protein. Each of the original natural proteins serves as an "oligomerization domain" in the designed fusion protein. Two different hypothetical fusion proteins are shown to illustrate that the oligomerization domains can be joined rigidly in different geometries. (**c**) A ribbon diagram of a fusion protein showing one method for joining two oligomerization domains (*red* and *green*) in a relatively rigid fashion. One of the natural oligomerization domains must end in an α-helical conformation, and the other must begin in an α-helical conformation. The two are then linked by a short stretch of amino acids (*blue*) that have a strong tendency to adopt an α-helical conformation. Thus, the two oligomerization domains are joined physically in a predictable orientation. (**d**) A designed fusion protein self-assembles into a particular kind of nanostructure that depends on the geometry of the symmetry axes belonging to its component oligomerization domains. A molecular layer arises from an arrangement like that in (**b**) (*Left*). (**e**) A cubic cage arises from an arrangement like the one in (**b**) (*Right*) (Reproduced with permission from ref. 45)

A pragmatic approach is to combine any of the design principles above with selection strategies to create useful proteins. This has been demonstrated in the remodelling of a protein-peptide-binding interface using Rosetta to interrogate possible sequences and then a

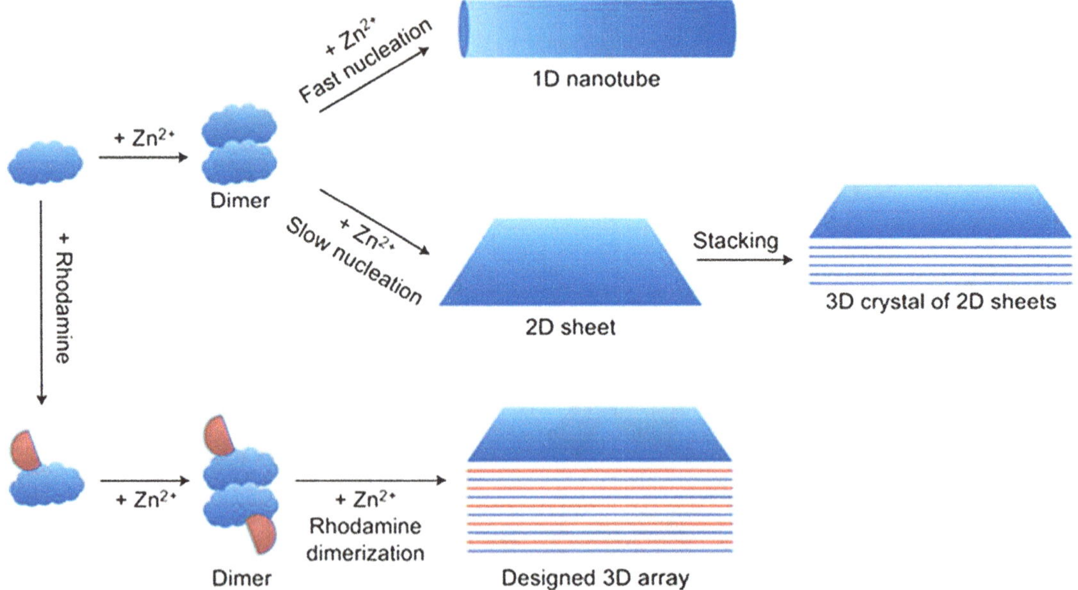

Fig. 5 Representation of the metal-directed protein self-assembly approach to building arrays. A monomeric protein (*blue*) first forms a dimer that subsequently self-assembles into either nanotubes or sheets, depending on the conditions. The 2D sheets can further stack into 3D crystals. Adding rhodamine side groups (*red*) to the monomeric proteins provides a route to designed 3D arrays through inter-sheet rhodamine dimerization (Reproduced with permission from ref. 51)

screening approach to optimize binding (49). It is likely that these combined strategies become increasingly adopted as researchers learn the benefits of each approach and how to best combine them.

Beyond the natural repertoire of protein amino acids, there is increasing interest in adding a degree of chemical control to protein assembly in vitro. Such approaches are reviewed by Fegan et al. (50) who reflect on the role of protein assembly in biological structures and build on these insights to suggest tools to use in "the 1–100 nm niche" which is too large to fill with synthetic organic chemistry but too small for the techniques of microfabrication. They speculate that filling this niche will enable the creation of protein nanobots, advanced protein therapeutics, and next-generation electronic devices. Examples are appearing in the literature that use multivalent ligands, chemically induced dimerization, and metal binding to expand the range of functionality available for building protein nanostructures. For example, a recently published study employs metal ions to produce a very elegant self-assembling system (14, 51).

As illustrated in Fig. 5, the protein incorporates both a dimerization site and metal coordination sites with different affinities, enabling exquisite control of the assembly of the protein into different morphological forms. Small variations in the ratio of protein to metal ion, or to the pH, allowed fine control of these structures, due to a high degree of kinetic control and reversibility of assembly. This encourages formation of the more thermodynamically stable extended arrays, in preference to less ordered associations.

4 Towards Applications

Following the huge interest in peptide- and protein-based materials over the last decade (52), the use of proteins for nanotechnological applications is fast becoming a reality. Silk proteins, in particular, are already being remodelled to create self-assembled structures of broad utility (Chapter 2; (21)), other proteins are being used for the fabrication of smart biomaterials (53, 54) and vaccines are a long way along the path from science to product (8). Assembled protein nanostructures have been recently described as "temptingly close to application in energy, biotechnology, and nanomedicine" (55) as the tools for the "nano toolkit" start to fall into place. Protein engineering is being recognized as offering "powerful solutions to the challenge posed by the creation of well-defined, multifunctional materials that guide cell and tissue behavior" (56). For example, tetratricopeptide repeat (TPR) proteins have been used to create pre-designed arrays for the assembly of stimuli-responsive gels (57).

Combined advances in experimental, computational, and theoretical methods to understand protein materials increasingly open up potential applications in materials design and nanotechnology (58). Dynamic proteinaceous materials, such as hydrogels, are being heralded as the solution to a broad range of material-related challenges, including robotic actuation, controlled drug release, and adaptive optics (59). Biomaterials designed as catalysts and designer materials are also developing quickly, although these face competition from more traditional organic and inorganic approaches (8). However, as we begin to see the impact of nanomaterials in areas such as biomedical imaging, drug delivery, biosensing, and nanocomposites, methods to effectively interface proteins with nanomaterials are emerging. There is interest both in how nanomaterials might influence the structure and therefore function of proteins, and conversely how proteins might be employed, as they are in Nature, to control the assembly of nanomaterials (60). In material science, the precise nanoscale structure of protein assemblies can help to generate metallic, inorganic, or organic materials of predesigned dimensions and favorable magnetic, electronic, or photonic properties. For metallic materials, the spherical, planar, and rod-like assemblies serve as templates to create nanoparticles, arrays of nanoparticles, or nanowires, respectively (8).

The range of applications accessible to the protein nanotechnologist is enhanced if proteins are used as a nanoscaffold that can be decorated with useful functionality. There are many examples of

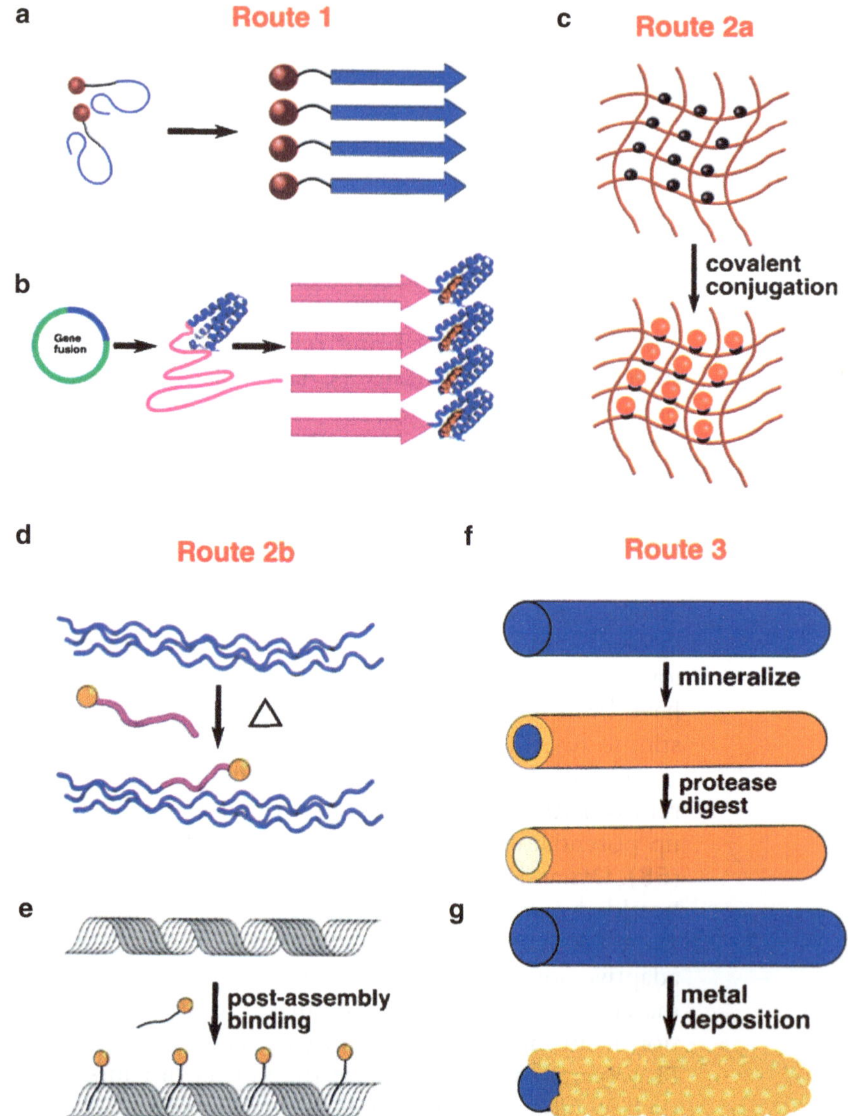

Fig. 6 Routes to decorating self-assembling systems. (**a**) Co-assembly of covalently linked structural and functional moieties. (**b**) A functional protein can be (genetically) fused to a self-assembling peptide/protein. (**c**) Post-assembly covalent decoration of scaffolds using techniques such as click chemistry. (**d**) Structural mimics or (**e**) noncovalent interactions can be exploited for functionalization of preformed scaffolds. Peptides/protein fibers can be used to template hollow tubes (**f**), and conducting nanowires (**g**) (Reproduced from ref. 61 with permission)

this emerging in the literature, and general strategies are outlined by Woolfson and Mahmoud (61). These are illustrated in Fig. 6.

Proteins can be produced on an industrial scale from bacterial expression systems relatively cheaply. They have a number of qualities that lend themselves to being useful materials for a range of applications: they are synthesized under ambient conditions, without toxic by-products (9) and their self-assembly, unlike other methods of nanofabrication, does not require a clean room. The future for protein nanotechnology seems bright.

5 Conclusion

Proteins are starting to find a niche as starting materials for the production of exciting nanostructured materials. Progress has been swift over the last few years, as will be illustrated in the chapters of this book. Designed nanodevices made from protein are on the horizon, but challenges remain before they become a reality. Hopefully this book inspires more research in this burgeoning field and brings this frontier closer to reality.

References

1. Seeman NC (2010) Nanomaterials based on DNA. Annu Rev Biochem 79:65–87
2. Tsai CJ, Zhang J, Aleman C, Nussinov R (2006) Structure by design: from single proteins and their building blocks to nanostructures. Trends Biotechnol 24:449–454
3. Jaeger L, Chworos A (2006) The architectonics of programmable RNA and DNA nanostructures. Curr Opin Struct Biol 16:531–543
4. Rothemund PWK (2006) Folding DNA to create nanoscale shapes and patterns. Nature 440:297–302
5. Gazit E (2007) Self-assembled peptide nanostructures: the design of molecular building blocks and their technological utilization. Chem Soc Rev 36:1263–1269
6. Matson JB, Zha RH, Stupp SI (2011) Peptide self-assembly for crafting functional biological materials. Curr Opin Solid State Mater Sci 15:225–235
7. Woolfson DN, Ryadnov MG (2006) Peptide-based fibrous biomaterials: something old, new and borrowed. Curr Opin Chem Biol 10:559–567
8. Howorka S (2011) Rationally engineering natural protein assemblies in nanobiotechnology. Curr Opin Biotechnol 22:485–491
9. Heddle JG (2008) Protein cages, rings and tubes: useful components of future nanodevices? Nanotechnol Sci Appl 1:67–78
10. Zhang SG (2002) Emerging biological materials through molecular self-assembly. Biotechnol Adv 20:321–339
11. Waterhouse SH, Gerrard JA (2004) Amyloid fibrils in bionanotechnology. Aust J Chem 57:519–523
12. Ferrari M (2005) Cancer nanotechnology: opportunities and challenges. Nat Rev Cancer 5:161–171
13. Ellis-Behnke RG, Liang YX, You SW, Tay DK, Zhang SG, So KF, Schneider GE (2006) Nano neuro knitting: peptide nanofiber scaffold for brain repair and axon regeneration with functional return of vision. Proc Natl Acad Sci U S A 103:5054–5059
14. Brodin JD, Ambroggio XI, Tang C, Parent KN, Baker TS, Tezcan FA (2012) Metal-directed, chemically tunable assembly of one-, two- and three-dimensional crystalline protein arrays. Nat Chem 4:375–382
15. Grunberg R, Serrano L (2010) Strategies for protein synthetic biology. Nucl Acids Res 38:2663–2675
16. Whitesides GM, Grzybowski B (2002) Self-assembly at all scales. Science 295:2418–2421
17. Nakamoto RK, Baylis Scanlon JA, Al-Shawi MK (2008) The rotary mechanism of the ATP synthase. Arch Biochem Biophys 476:43–50
18. Channon K, Bromley EHC, Woolfson DN (2008) Synthetic biology through biomolecular design and engineering. Curr Opin Struct Biol 18:491–498
19. O'Leary LER, Fallas JA, Bakota EL, Kang MK, Hartergerink JD (2011) Multi-hierarchical self-assembly of a collagen mimetic peptide from triple helix to nanofibre and hydrogel. Nat Chem 3:821–828
20. Almine JF, Bax DV, Mithieux SM, Nivison-Smith L, Rnjak J, Waterhouse A, Wise SG, Weiss AS (2010) Elastin-based materials. Chem Soc Rev 39:3371–3379
21. Omenetto FG, Kaplan DL (2010) New opportunities for an ancient material. Science 329:528–531
22. Mason JM, Arndt KM (2004) Coiled coil domains: stability, specificity, and biological implications. Chembiochem 5:170–176

23. Apostolovic B, Danial M, Harm-Anton Klok H-A (2010) Coiled coils: attractive protein folding motifs for the fabrication of self-assembled, responsive and bioactive materials. Chem Soc Rev 39:3541–3575
24. Gras SL, Tickler AK, Squires AM, Devlin GL, Horton MA, Dobson CM, MacPhee CE (2008) Functionalised amyloid fibrils for roles in cell adhesion. Biomaterials 29:1553–1562
25. Cherny I, Gazit E (2008) Amyloids: not only pathological reagents but also ordered nanomaterials. Angew Chem 47:4062–4069
26. Allen M, Willits D, Young M, Douglas T (2003) Constrained synthesis of cobalt oxide nanomaterials in the 12-subunit protein cage from *Listeria innocua*. Inorg Chem 42:6300–6305
27. Uchida M, Flenniken ML, Allen M, Willits DA, Crowley BE, Brumfield S, Willis AF, Jackiw L, Jutila M, Young MJ (2006) Targeting of cancer cells with ferrimagnetic ferritin cage nanoparticles. J Am Chem Soc 128:16626–16633
28. Cardinale D, Carette N, Michon T (2012) Virus scaffolds as enzyme nano-carriers. Trends Biotechnol 30:369–376
29. Astier Y, Bayley H, Howorka S (2005) Protein components for nanodevices. Curr Opin Chem Biol 9:576–584
30. Clarke J, Regan L (2010) Protein engineering and design: from first principles to new technologies. Curr Opin Struct Biol 20:480–481
31. Das R, Baker D (2008) Macromolecular modeling with Rosetta. Annu Rev Biochem 77:363–382
32. Kaufmann KW, Lemmon GH, DeLuca SL, Sheehan JH, Meiler J (2010) Practically useful: what the ROSETTA protein modeling suite can do for you. Biochemistry 49:2987–2998
33. Heddle JG, Yokoyama T, Yamashita I, Park SY, Tame JRH (2006) Rounding up: engineering 12-membered rings from the cyclic 11-Mer TRAP. Structure 14:925–933
34. Heddle JG, Fujiwara I, Yamadaki H, Yoshii S, Nishio K, Addy C, Yamashita I, Tame JRH (2007) Using the ring shaped protein TRAP to capture and confine gold nanodots on a surface. Small 3:1950–1956
35. Miranda FF, Iwasaki K, Akashi S, Sumitomo K, Kobayashi M, Yamashita I, Tame JRH, Heddle JG (2009) A self assembled protein nanotube with high aspect ratio. Small 5:2077–2084
36. Wang WX, Dgany O, Wolf SG, Levy I, Algom R, Pouny Y, Wolf A, Marton I, Altman A, Shoseyov O (2006) Aspen SP1, an exceptional thermal, protease and detergent resistant self assembled nano particle. Biotechnol Bioeng 95:161–168
37. Heyman A, Levy I, Altman A, Shoseyov O (2007) SP1 as a novel scaffold building block for self-assembly nanofabrication of submicron enzymatic structures. Nano Lett 7:1575–1579
38. Medalsy I, Dgany O, Sowwan M, Cohen H, Yukashevska A, Wolf SG, Wolf A, Koster A, Almog O, Marton I (2008) SP1 protein-based nanostructures and arrays. Nano Lett 8:473–477
39. Frasconi M, Heyman A, Medalsy I, Porath D, Mazzei F, Shoseyov O (2011) Wiring of redox enzymes on three dimensional self-assembled molecular scaffold. Langmuir 27: 12606–12613
40. Medalsy I, Klein M, Heyman A, Shoseyov O, Remacle F, Levine RD, Porath D (2010) Logic implementations using a single nanoparticle-protein hybrid. Nat Nanotechnol 5:451–457
41. Mougous JD, Cuff ME, Raunser S, Shen A, Zhou M, Gifford CA, Goodman AL, Joachimiak G, Ordoñez CL, Lory S (2006) A virulence locus of *Pseudomonas aeruginosa* encodes a protein secretion apparatus. Science 312:1526
42. Ballister ER, Lai AH, Zuckermann RN, Cheng Y, Mougous JD (2008) *In vitro* self-assembly of tailorable nanotubes from a simple protein building block. Proc Natl Acad Sci U S A 105:3733–3738
43. Schreiber A, Zaitseva E, Thomann Y, Thomann R, Dengjel J, Hanselmann R, Schiller SM (2011) Protein yoctowell nanoarchitectures: assembly of donut shaped protein containers and nanofibres. Soft Matter 7:2875–2878
44. McMillan RA, Paavola CD, Howard J, Chan SL, Zaluzec NJ, Trent JD (2002) Ordered nanoparticle arrays formed on engineered chaperonin protein templates. Nat Mater 1:247–252
45. Padilla JE, Colovos C, Yeates TO (2001) Nanohedra: using symmetry to design self assembling protein cages, layers, crystals, and filaments. Proc Natl Acad Sci U S A 98:2217–2221
46. Grueninger D, Treiber N, Ziegler MOP, Koetter JWA, Schulze MS, Schulz GE (2008) Designed protein-protein association. Science 319:206–209
47. Lai Y-T, Cascio D, Yeates TO (2012) Structure of a 16 nm cage designed by using protein oligomers. Science 336:1129
48. Tsai C-J, Zheng J, Zanuy D, Haspel N, Wolfson H, Alema C, Nussinov R (2007) Principles of nanostructure design with protein building blocks. Prot Struct Funct Bioinform 68:1–12
49. Grove TZ, Hands M, Regan L (2010) Creating novel proteins by combining design and selection. Prot Eng Design Select 23:449–455
50. Fegan A, White B, Carlson JCT, Wagner CR (2010) Chemically controlled protein assembly: techniques and applications. Chem Rev 110:3315–3336
51. Sinclair JC (2012) Self-assembly: proteins on parade. Nat Chem 4:346–347

52. Woolfson DN (2010) Building fibrous biomaterials from α-helical and collagen-like coiled-coil peptides. Biopolymers 94:118–127
53. Wagner DE, Philips CL, Ali WM, Nybakken GE, Crawford ED, Schwab AD, Smith WF, Fairman R (2005) Towards the development of peptide nanofilaments and nanoropes as smart materials. Proc Natl Acad Sci U S A 102:12656–12661
54. Kohli P, Martin CR (2005) Smart nanotubes for biotechnology. Curr Pharm Biotechnol 6:35–41, Curr Opin Biotech 17:562–568
55. Ulijn RJ, Woolfson DN (2010) Peptide and protein based materials in 2010: from design and structure to function and application. Chem Soc Rev 39:3349–3350
56. Maskarinec SA, Tirrell DA (2005) Protein engineering approaches to biomaterials design. Curr Opin Biotechnol 16:422–426
57. Grove TZ, Forster J, Pimienta G, Dufresne E, Regan L (2012) A modular approach to the design of protein-based smart gels. Biopolymers 97:508–517
58. Buehler MJ, Yung YC (2009) Deformation and failure of protein materials in physiologically extreme conditions and disease. Nat Mater 8:175–188
59. Shaikh Mohammed J, Murphy WL (2009) Bioinspired design of dynamic materials. Adv Mater 21:2361–2374
60. Asuri P, Bale SS, Karajanagi SS, Kane RS (2006) The protein–nanomaterial interface. Curr Opin Biotechnol 9:562–568
61. Woolfson DN, Mahmoud ZN (2010) More than just bare scaffolds: towards multi-component and decorated fibrous biomaterials. Chem Soc Rev 39:3464–3479

Part I

Old Proteins, New Tricks

Chapter 2

Bioengineered Silk Proteins to Control Cell and Tissue Functions

Rucsanda C. Preda, Gary Leisk, Fiorenzo Omenetto, and David L. Kaplan

Abstract

Silks are defined as protein polymers that are spun into fibers by some lepidoptera larvae such as silkworms, spiders, scorpions, mites, and flies. Silk proteins are usually produced within specialized glands in these animals after biosynthesis in epithelial cells that line the glands, followed by secretion into the lumen of the gland prior to spinning into fibers.

The most comprehensively characterized silks are from the domesticated silkworm (*Bombyx mori*) and from some spiders (*Nephila clavipes* and *Araneus diadematus*). Silkworm silk has been used commercially as biomedical sutures for decades and in textile production for centuries. Because of their impressive mechanical properties, silk proteins provide an important set of material options in the fields of controlled drug release, and for biomaterials and scaffolds for tissue engineering. Silkworm silk from *B. mori* consists primarily of two protein components, fibroin, the structural protein of silk fibers, and sericins, the water-soluble glue-like proteins that bind the fibroin fibers together. Silk fibroin consists of heavy and light chain polypeptides linked by a disulfide bond. Fibroin is the protein of interest for biomedical materials and it has to be purified/extracted from the silkworm cocoon by removal of the sericin. Characteristics of silks, including biodegradability, biocompatibility, controllable degradation rates, and versatility to generate different material formats from gels to fibers and sponges, have attracted interest in the field of biomaterials. Cell culture and tissue formation using silk-based biomaterials have been pursued, where appropriate cell adhesion, proliferation, and differentiation on or in silk biomaterials support the regeneration of tissues. The relative ease with which silk proteins can be processed into a variety of material morphologies, versatile chemical functionalization options, processing in water or solvent, and the related biological features of biocompatibility and enzymatic degradability make these proteins interesting candidates for biomedical applications.

Key words Biomaterials, Silk, Protein, Silkworm silk, Biodegradability, Hydrogel, Film, Scaffold, Sponge, Regenerative medicine, Fibroin, Spider

1 Introduction

Regenerated silk solutions have been used to generate a variety of biomaterial forms, such as gels, sponges, and films, for medical applications (1–4). The mechanical properties and rates of degradation of these silk biomaterials relate to the mode of processing

Juliet A. Gerrard (ed.), *Protein Nanotechnology: Protocols, Instrumentation, and Applications*, Methods in Molecular Biology, vol. 996, DOI 10.1007/978-1-62703-354-1_2, © Springer Science+Business Media New York 2013

and the corresponding content of beta-sheet crystallinity (5, 6). In addition, many types of cells have been grown on different silk biomaterials including epithelial (7), endothelial (8), osteoblasts, glial, keratinocytes (9), fibroblasts (10), and human bone marrow stromal cells (11–14) to demonstrate a range of biological outcomes. Silk biomaterials are biocompatible and have been successfully used in wound healing (15, 16) and in tissue engineering of bone (12, 17–20), cartilage (11, 14, 21), tendon (22), ligament (23, 24), vascular (8, 23–27), fat (13), kidney (28), breast (7, 29), cervical (30), urethral, and other tissues.

This chapter consists of eight protocols. The first method, silk purification, is the procedure employed to obtain silk fibroin solution from silkworm cocoons that is then used as the starting material for preparing silk-based biomaterials described in the other seven methods/protocols. The remaining seven protocols involve the use of the regenerated silk solutions to produce different formats of silk biomaterials.

2 Materials

2.1 Silk Purification

1. Silkworm cocoons, dried (heat treated).
2. Scissors.
3. Pyrex glass beaker, 2,000 mL capacity (Kimble Chase Kimble, Fisher cat. # 14000 2000).
4. Ultrapure water (water with a resistivity of 18.2 MΩ-cm) (see Note 1).
5. Lab scale (Denver Instrument Analytical Balance, Fisher cat. # 01-914-03).
6. Sodium carbonate, granular (Fisher, ACROS Organics, or Sigma-Aldrich, cat. # 451614).
7. Metallic spatula.
8. Hot plate (Fisher cat. # 11-100-49H).
9. Plastic beaker, 2,000 mL capacity (Fisher cat. # 02-591-33).
10. Heat-resistant gloves.
11. Lithium bromide, anhydrous (Fisher or Sigma-Aldrich, cat. # 213225).
12. Graduated cylinder.
13. Two glass beakers (i.e., 20 or 30 mL capacity).
14. Magnetic stir bar.
15. Dry heat oven (set at 60°C).
16. Dialysis cassette 3,500 MWCO, 3–12 mL capacity (Fisher cat. # PI-66110, Thermo Scientific cat. # 66110), float buoy, elastic band.

17. Syringes, 20 mL (Fisher cat. # 14-823-2B, BD Medical No. 309661).
18. Single-use needles 18 G (Fisher cat. # 14-826-5G, BD Medical No. 305195).
19. Plastic beaker, 1,000 mL capacity (see Note 2).
20. Conical tubes, sterile, 50 mL (BD Medical, Fisher cat. # 14-432-22).
21. Centrifuge with fixed angle rotor.
22. Freezer, –80°C.
23. Lyophilizer (freeze-dryer).

2.2 Silk Hydrogels

1. Aqueous silk fibroin solution, sterile or non-sterile (see Notes 3 and 4).
2. Conical tubes, 15 mL.
3. Eppendorf tubes, 1.5 mL.
4. Branson 450 Sonifier (Branson Ultrasonics Co., Danbury, CT): the model 450 power supply, converter (Part No. 101-135-022), externally threaded disruptor horn (Part No. 101-147-037), and 1/8″ (3.175 mm) diameter tapered microtip (Part No. 101-148-062).
5. Well plate (6-, 12-, or 24-well plate).
6. Petri dish (35 mm dia × 10 mm H).
7. Syringe (1, 3, or 5 mL).

2.3 Silk Porous Sponges: Aqueous-Based Silk Scaffolds

1. Aqueous silk solution, concentration 6–8% (w/v) (see Note 3).
2. Sodium chloride (granular particles), 100–1,000 μM particle size (e.g., Fisher cat. # S640-3, or BP 358-212) (see Note 5).
3. Teflon containers/vials with closures, 7 mL (Savillex, cat. # 200-007-10 and cat. # 600-024-01 for closure), or polyethylene vials with hinged cap (Fisher, cat. # 03-338-1E).
4. Ultrapure water.
5. Stir plate (Fisher, cat. # 11-510-49SHQ).
6. Sieves with nominal opening from 106 μM to 1 mm (e.g., Fisher, cat. # 04-881-10Z).
7. Scale.
8. 5 mL pipettes.
9. Disposable scalpel (Fisher, cat. # 08-927-5D).
10. Disposable biopsy punch (e.g., 5 mm diameter—Fisher, cat. # NC9642816).

2.4 Silk Porous Sponges: Solvent-Based Silk Scaffolds

1. Lyophilized silk fibroin (see Subheading 3.1.6).
2. HFIP (1,1,1,3,3,3-hexafluoro-2-propanol) (Fisher, cat. # AC147541000).

3. Sodium chloride (granular particles), 100–1,000 μM particle size (e.g., Fisher cat. # S640-3, or BP 358-212).
4. Teflon containers/vials with closures, 7 mL (Savillex, cat. # 200-007-10 and cat. # 600-024-01 for closure), or polyethylene vials with hinged cap, 10 mL (Fisher, cat. # 03-338-1E).
5. Glass vial with cap (Fisher, cat. # 03-337-4) or, for larger volumes, Corning media/solution bottle (Fisher, cat. # 06-423-3A).
6. Syringe, plastic, 3 mL.
7. Ultrapure water.
8. Methanol (Fisher, cat. # A412-1).
9. Stir plate (Fisher, cat. # 11-510-49SHQ).
10. Fume hood (chemical hood).
11. Glass graduated cylinder (Fisher, cat. # 08-555B).

2.5 Silk Electrospun Mats

1. Aqueous silk solution, concentration 8% (w/v) (see Note 3).
2. Poly(ethylene oxide) or PEO, with average M_v ~ 900,000 (Sigma-Aldrich, cat. # 189456).
3. Conical tube, polypropylene, 50 mL.
4. Stir bar and stir plate (Fisher, cat. # 11-510-49SHQ) or bench platform rocker (Fisher, cat. # 09-047-112Q).
5. Ultrapure water.
6. Syringes, 3, 5, and 10 mL.
7. Electrospinning setup that consists of four components: a high-voltage supplier, a capillary needle, a grounded collector, and a syringe pump.
8. Aluminum foil.
9. Methanol (Fisher, cat. # A412-1).
10. Tweezers.
11. Fume hood.

2.6 Silk Nano- and Microspheres (from Silk/PVA Blend Films)

1. Aqueous silk solution, concentration 5% (w/v) (see Notes 3 and 6).
2. Polyvinyl alcohol or PVA, with average M_v 30,000–70,000, 87–90% hydrolyzed (Sigma, cat. # P8136).
3. Glass beakers.
4. Filter paper.
5. Conical tube, polypropylene, 15 mL (Fisher, cat. # 352096).
6. Branson 450 Digital Sonifier (Branson Ultrasonics Co., Danbury, CT): Model 450 power supply, converter (Part No. 101-135-022), externally threaded disruptor horn (Part No.

101-147-037), and 1/8″ (3.175 mm) diameter tapered microtip (Part No. 101-148-062).

7. 100 mm diameter plastic Petri dish, polystyrene (Fisher, cat. # 08-757-13).
8. Fume hood.
9. Centrifuge tubes, Nalgene, polypropylene, with screw caps (Fisher, cat. # 05-529-1D).
10. Tweezers.
11. Ultrapure water.
12. High-speed centrifuge, refrigerated.
13. Lyophilizer.
14. Microcentrifuge tubes, 1.5 mL (Fisher, cat. # 02-682-550).

2.7 Silk Microspheres Lipid-Template Method

1. Aqueous silk solution, concentration 8% (w/v) (see Note 3).
2. 1,2-Dioleoyl-*sn*-glycero-3-phosphocholine (DOPC, Avanti Polar Lipids, cat. # 850375P).
3. Chloroform (Fisher, cat. # C298-1).
4. Methanol (Fisher, cat. # A412-1).
5. Nitrogen gas (Airgas or other gas vendor).
6. Ultrapure water.
7. Liquid nitrogen.
8. Water bath (set at 37°C).
9. FisherBrand disposable culture tubes (glass tubes) (Fisher, cat. # 14-961-29).
10. Microcentrifuge tubes, 1.5 mL (Fisher, cat. # 02-682-550).
11. Glass beakers (100 mL or similar).
12. Magnetic stir bar.
13. Conical tube, polypropylene, 15 mL (Fisher, cat. # 352096).
14. Conical tube, polypropylene, 50 mL (Fisher, cat. # 14-432-22).
15. Polycarbonate centrifuge tubes (Fisher, cat. # 05-529C).
16. Pipette tips, 1,000 μM.
17. Branson 450 Digital Sonifier (Branson Ultrasonics Co., Danbury, CT): Model 450 power supply, converter (Part No. 101-135-022), externally threaded disruptor horn (Part No. 101-147-037), and 1/8″ (3.175 mm) diameter tapered microtip (Part No. 101-148-062).
18. Fume hood.
19. High-speed centrifuge, refrigerated.
20. Lyophilizer.

2.8 Silk Films: Water-Based, Non-patterned

1. Aqueous silk solution, concentration 7–8% (w/v) (see Note 3).
2. 100 mm diameter plastic Petri dish, non-tissue culture treated (Fisher, cat. # 08-757-12).
3. Tweezers.
4. Vacuum desiccator (Wheaton Science Products, Fisher, cat. # 08-634C).
5. 70% ethanol (Fisher, cat. # 2546701).
6. Laminar flow hood.

2.9 Silk Films: Patterned

1. Aqueous silk solution, concentration 8% (w/v) (see Note 3).
2. Polydimethylsiloxane (PDMS) (Sylgard 184, Ellsworth Adhesives, cat. # 184 Sil Elast Kit 0.5 kg).
3. Diffraction grating: size, number of grooves, and depth of grooves can vary upon requirements, for example, 600 grooves/mm, 1,000 nm ruled diffraction grating, 50 × 50 mm (Edmund Optics cat. # NT43-208).
4. Vacuum oven (Fisher, cat. # 13-262-280A).
5. Tweezers.
6. Paper cup.
7. Aluminum foil.
8. 70% ethanol (Fisher, cat. # 2546701).
9. Needle (size not important, for example, BD Medical 18 G, Fisher cat. # 305195).
10. Serological pipette, 5 mL (Fisher cat. # 357543).
11. Pressurized canned air (Ultrajet Duster, Fisher, cat. # 19-003-233).
12. 100 mm diameter plastic Petri dish, polystyrene (Fisher, cat. # 08-757-13).
13. Hole punch (for example, 14 mm diameter punch, McMaster-Carr, cat. # 3418A14).
14. Vacuum desiccator (Wheaton Science Products, Fisher, cat. # 08-634C).

3 Methods

Silk Purification. The purification process comprises five procedures/techniques: (1) extracting fibroin (degumming of the silk cocoons or removal of sericin, the glue-like protein that holds the fibroin fibers together), (2) dissolving/solubilizing raw fibroin fibers in lithium bromide, (3) dialyzing the lithium bromide/silk fibroin solution to remove the salts or chemicals, (4) centrifuging

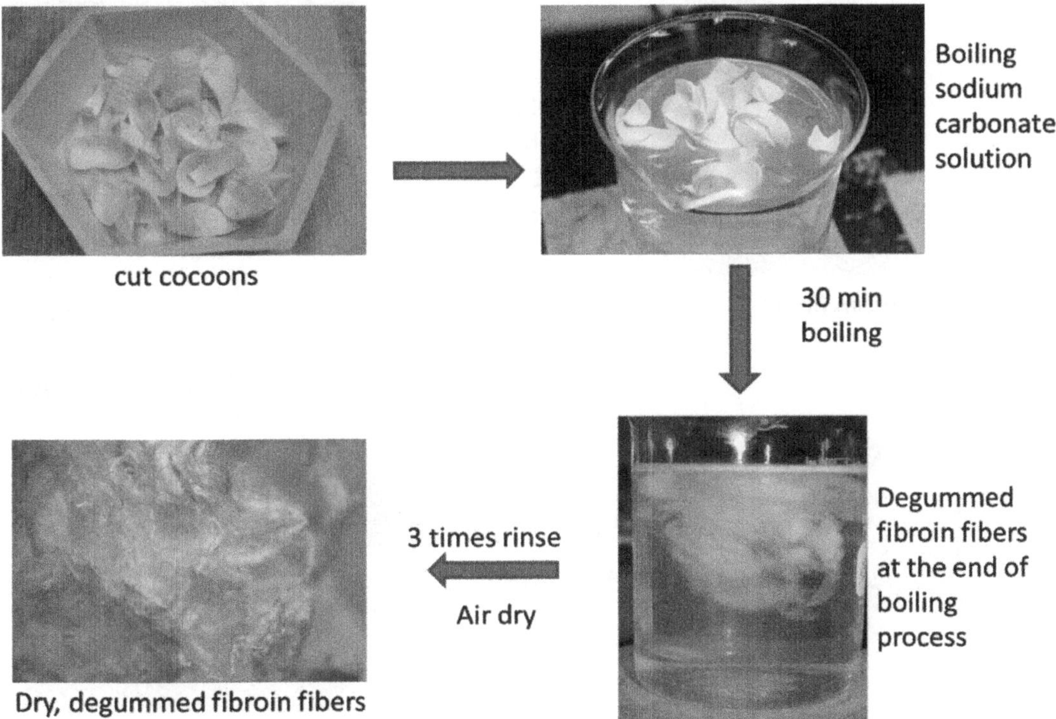

Fig. 1 Silk purification process, the first steps. Obtaining degummed fibroin fibers from silkworm cocoons

the aqueous silk solution to remove the impurities and silk aggregates formed during dialysis to clarify the solution, and (5) calculating the concentration of silk protein in the aqueous solution (Figs. 1 and 2) (see Note 7).

Silk Hydrogels. This method describes the procedure used to induce silk gelation through ultrasonication (31). Sonication is used to transform the random coil structure of the silk fibroin in aqueous solution into a beta-sheet (crystalline, physical cross linked) structure, inducing insolubility. Depending on the sonication parameters, including power output and time, along with silk fibroin concentration, gelation can be controlled from minutes to hours, allowing the post-sonication addition of cells or other bioactive molecules prior to the final setting of the gel. Silk hydrogels can also be obtained using other methods: vortexing, lowering the pH of the silk solution, and application of a direct electrical current (32–34).

Porous Silk Sponges (Scaffolds). Regenerated silk fibroin solutions, both aqueous and solvent, have been utilized in the preparation of porous sponges. Solvent-based sponges can be prepared using sodium chloride particles as the porogen. Solvents such as 1,1,1,3,3,3-hexafluoro-2-propanol (HFIP) do not solubilize the sodium chloride; therefore, pore sizes in the sponges reflect the size of the porogen used in the process (35). Aqueous-based

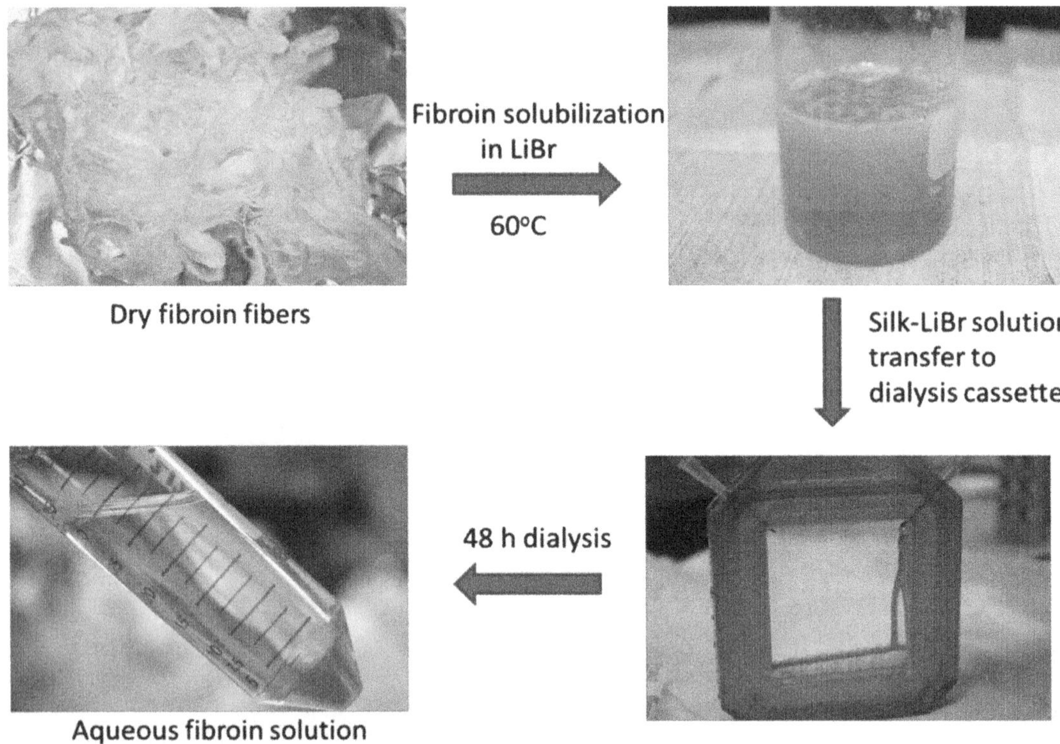

Fig. 2 Silk purification process, the final steps. Solubilizing fibroin fibers, dialysis, and obtaining the aqueous fibroin solution

porous silk sponges can be prepared using variable sized sodium chloride particles as porogen, with control of pore sizes from ~500 to ~1,000 μm, by manipulating the percent silk solution and size of salt crystals (36). Pore sizes are 80–90% of the size of salt crystals due to the limited solubilization of the surface of the crystals during supersaturation of the silk aqueous solution prior to solidification. Aqueous-based sponges have rougher surface morphology than solvent-based sponges due to this partial solubilization, but also provide better interconnects between the pores for the same reason. Also, aqueous-based scaffolds degrade faster than HFIP-based scaffolds (6, 37).

Silk Electrospun Mats. Electrospinning enables the development of nanofiber-based scaffolds by introducing the charged polymer (silk) solution into an electric field. Electrospun scaffolds/mats can be formed from aqueous silk solutions mixed with poly (ethylene oxide) (PEO), or prepared from solvents such as HFIP. Fiber diameter sizes range from 100 nm to 1 mm depending on the solvent used and the spinning conditions (8, 16, 38, 39).

Silk Micro- and Nanospheres (from Silk/PVA Blend Films). Silk fibroin-based micro- and nanospheres provide options for drug delivery due to their tunable drug loading and release properties (40). The preparation method using polyvinyl alcohol (PVA) is based on phase separation between silk fibroin and polyvinyl alcohol at a weight ratio of 1/1 and 1/4. Water-insoluble silk spheres are obtained from the blend in a three-step process: (1) air-drying the blend solution into a film, (2) film dissolution in water, and (3) removal of residual PVA by subsequent centrifugation. Other options for particle preparation include lipid-templating techniques or the use of ultrasonication in concert with the above PVA system to generate nano- vs. micro-sized particles.

Silk Microspheres: Lipid-Templating Method. Preparation of these silk microspheres involves the use of lipid vesicles such as 1,2-dioleoyl-*sn*-glycero-3-phosphocholine (DOPC), a phosphatidylcholine (class of phospholipids). DOPC serves as a template to form the microspheres. Subsequently methanol is used to remove the lipid templates and induce the structural transition of the silk to the silk β-sheet crystals, thus locking in the stability of the microspheres (41).

Silk Films. Films can be made transparent, biocompatible, biodegradable, mechanically strong, and surface-patterned or non-patterned. Silk fibroin films can be cast from aqueous or organic solvent systems, as well as after blending with other polymers such as poly(ethylene oxide) to yield porous films. Silk films can be used to entrap other molecules for improved stabilization (42, 43), or used as a substrate for cell culture (10). Film thickness can be controlled by changing the concentration of the silk solution. Two methods are presented to make non-patterned (44) and patterned silk films from aqueous solutions (10, 45).

3.1 Silk Purification: Obtaining Aqueous Silk Solution from Silkworm Cocoons

3.1.1 Extraction

1. Cut dried cocoons with scissors into dime-sized pieces and weigh 5 g of the cut pieces.
2. Weigh 4.24 g of sodium carbonate.
3. Fill a 2,000 mL glass beaker with 2 L of ultrapure water and heat it up until water starts to boil.
4. Add the sodium carbonate (4.24 g) to the boiling water and let it completely dissolve, obtaining a dilute sodium carbonate solution (concentration 0.02 N).
5. Add the cocoon pieces (5 g) to the boiling sodium carbonate solution and boil for 30 min. Occasionally, poke the fibers with the spatula to promote good dispersion of the fibroin fibers and efficient sericin removal (see Note 8).
6. After 30 min of boiling, discard the sodium carbonate/sericin solution in the sink (use heat-resistant gloves to hold the beaker), and move the degummed silk fibers in a plastic beaker filled with about 1 L of ultrapure water. The fibers will cool down and can be rinsed by hand at this time. Discard the water.

7. Add 1.5 L ultrapure water and keep the fibers in the beaker for 20 min; occasionally rinse the silk fibers by hand.
8. Change the water and repeat the same procedure (step 7) two times.
9. After the third (three times each of 20 min) rinse, take out the silk fibers, squeeze the water out, and then untie fibers by hand.
10. Air dry degummed silk fibers (raw fibroin fibers) on the bench or in a fume hood (faster) for at least 12 h (see Note 9).

3.1.2 Solubilization in Lithium Bromide

1. Prepare a 9.3 M lithium bromide solution (see Note 10).
2. Make a 20% w/v solution of dry silk fibers in 9.3 M lithium bromide solution (e.g., 2.5 g fibers will be dissolved in 10 mL of 9.3 M LiBr solution). Place the calculated amount of silk fibers in a small glass beaker and pack them tight. Pour the calculated volume of lithium bromide on top of the fibers and cover the beaker with aluminum foil.
3. Place the beaker in the oven at 60°C and let the silk dissolve completely for up to 4 h. The silk-LiBr solution obtained is a viscous, yellowish, and honey-like solution.

3.1.3 Dialysis

1. Fill up the plastic 1,000 mL beaker with 1 L water; place a dialysis cassette in water for 2–3 min to wet its membrane (see Note 11).
2. Fill the 20 mL syringe (no needle necessary in this step) with 12 mL of the silk-LiBr solution (obtained in the previous Subheading 3.1.2) (see Note 12).
3. Attach the 18 G needle to the syringe and push the 12 mL silk-LiBr solution into the dialysis cassette; attach a float buoy with the elastic band to the cassette and return the cassette to the beaker with 1 L water to start the dialysis.
4. Change the water after 1, 3, and 5–6 h (three changes first day); the next morning and evening (two changes second day); and the following day in the morning (one change on the third day). Total number of water changes = 6.

3.1.4 Centrifugation

1. Attach a new 18 G needle to a new 20 mL syringe and remove ~20 mL of the silk solution from the dialysis cassette. Remove the needle (to avoid shearing the silk solution) and dispense the silk solution in a conical/centrifuge tube (the tube should be safe to use in the centrifuge—see next step).

   Repeat this step to remove all the silk solution from the cassette; a total of ~26–28 mL silk solution should be collected in the conical tube.
2. Spin down the solution for 20 min at 4–6°C and at 9,000 rpm (8,500–9,500 × *g*). Transfer the silk solution into a new, clean conical tube and repeat this step one more time.

After the second centrifugation, transfer the silk solution into a new clean (or sterile) conical tube for storage. The silk solution obtained should be a clear yellowish liquid (see Note 13).

3. The silk solution is stored at 4°C.

3.1.5 Concentration Calculation

1. Weigh an empty weighing boat (W1).
2. Add 1 mL silk solution (measured accurately with a 1,000 μM micropipetter) and record the weigh (W2).
3. Leave the weighing boat in a 60°C oven overnight.
4. Next day, weigh the weighing boat again (W3).
5. The concentration of the silk solution (w/v) is:

$$\% = (W3 - W1 / W2 - W1) \times 100$$

3.1.6 Lyophilization

This step is done when the silk is desired in a dry form.

1. Cover a 50 mL conical tube (with the aqueous silk solution to be lyophilized) with tissue paper and secure with an elastic band.
2. Place the 50 mL conical tube containing the aqueous silk solution in a −80°C freezer, overnight. The 50 mL conical tube can be filled with up to 40 mL of silk solution.
3. Next day, place the conical tube with the frozen silk solution in the lyophilizer (under vacuum and a temperature around −80°C).
4. Remove the dried silk after 2–3 days.

3.2 Silk Hydrogels

1. If working with sterile silk solution, the sonicator disruptor horn (with the tip attached) has to be placed in the sterile laminar flow hood. If the silk solution is not sterile, the sonication can be done on the bench top.
2. Prepare silk samples in either 1.5 mL Eppendorf tubes, 1 mL silk solution/tube, or in 15 mL conical tubes, 5 mL silk solution/tube. The most common silk concentration used is 2–4% (w/v) (see Note 14).
3. To determine silk gelation under various sonication durations, sonicate (using the Branson Sonifier) several samples of 1 mL silk solution in a 1.5 mL Eppendorf tube. The silk concentration can be varied from 2 to 8% (w/v) and sonication time can be varied from 5 to 30 s at the 20% amplitude setting (see Note 15).
4. Choose one set of gelation parameters based on the results of the screening (for silk gelation parameters) done in the previous step.
5. Repeat sonication with the chosen parameters (silk concentration, power output/amplitude, and sonication time) for a

larger sample, for example, 5 mL silk solution in a 15 mL conical tube (see Note 16).

6. Immediately transfer the sonicated solution to a well plate or a Petri dish (for easier handling/removal of the hydrogel). It is also possible to gel the sonicated silk solution in a syringe (in this case transfer the sonicated silk to a syringe and cap the syringe) (see Notes 17 and 18).
7. Incubate the well plate, Petri dish, or syringe at 37°C (humidified incubator) or leave at room temperature. Visually monitor the sol-gel transition.
8. Once the silk is gelled in the plates, punch out, using a biopsy punch, small plugs with the desired diameter.

3.3 Silk Porous Sponges: Aqueous-Based Silk Scaffolds

1. Prepare silk fibroin solution 8% (for large pore size scaffolds) or 6% (for smaller pore size scaffolds). To obtain the 6% fibroin solution, dilute concentrated fibroin sol. with ultrapure water. For silk fibroin solution preparation (see Note 3).
2. Sieve sodium chloride to obtain desired salt particle size (e.g., for sponges with the pore size of 500–600 μM, sodium chloride should be sieved using the 600 μM opening sieve placed on top of the 500 μM opening sieve).
3. Weigh 4 g of salt in weighing boats.
4. Aliquot silk fibroin solution into the plastic containers, 2 mL/container.
5. Slowly pour salt on top of fibroin solution in the plastic container while rotating the container, to obtain a uniform layer of salt per unit surface area of the container.
6. Tap the container gently on a bench top surface to remove air bubbles.
7. Close the containers with a cap and leave them overnight at room temperature (for fibroin to gel); it should gel in 1–2 days (see Note 19).
8. When the silk has gelled and the scaffolds have formed, immerse the opened containers in ultrapure water in a beaker to remove/extract the salt. Add a stir bar and place the beaker on a stir plate.
9. Change the water 2–3 times per day, for 2–3 days; after 1–2 days, the scaffolds should come out of their plastic containers (salt leaches out and the sponge becomes soft).
10. Store scaffolds in ultrapure water at 4°C until needed for use.
11. When needed for use, cut the scaffolds at the desired dimensions while they are wet (use razor blades, surgical blades, biopsy punches).

12. Silk scaffolds can be sterilized by autoclaving (dry cycle, 20 min, 121°C, and 15 psi) or by soaking them in 70% ethanol (see Notes 20 and 21).

3.4 Silk Porous Sponges: Solvent-Based Silk Scaffolds

1. Weigh lyophilized silk and place it in the glass vial.
2. Prepare 17% silk-HFIP solution (e.g., for a volume of 10 mL final solution: 1.7 g lyophilized silk+~9 mL HFIP): in the fume hood add the calculated volume of HFIP over the silk in the vial (see Note 22).
3. Close vial with the cap, seal with parafilm, and keep it at RT for 1–2 days, until all silk is dissolved (the silk-HFIP solution is a viscous, yellowish solution).
4. Weigh 3.4 g sodium chloride (with desired particle size) and place in each plastic vial (Teflon or Polyethylene) (see Note 23).
5. Tap the container gently on the bench top surface to level the salt in the container.
6. Bring the containers with the sodium chloride, the glass vial with 17% silk-HFIP solution, and the 3 mL syringe into a fume hood.
7. Fill the 3 mL syringe with 2 mL of 17% silk-HFIP solution and close the vial quickly to limit HFIP evaporation.
8. Add very quickly the 17% silk-HFIP solution, 1 mL to each plastic vial, on top of the sodium chloride particles; close the cap on the plastic vial immediately to limit HFIP evaporation.
9. Repeat steps 7 and 8 until all the silk-HFIP solution is used.
10. Keep the capped plastic vials in the hood for 24 h.
11. Check the plastic vials to see if the silk-HFIP solution reached the bottom of the containers and all the salt appears "wet."
12. Open the caps and leave the containers in the hood for at least 24 h (for the HFIP to evaporate).
13. When the salt-silk mix inside the containers appears dry and detached from the container walls, place the open containers in a beaker with methanol. Cover beaker with aluminum foil to prevent evaporation and let the beaker sit in the hood for at least 24 h.
14. Remove the containers from methanol and let them dry in the hood (1 or 2 days).
15. Place the open containers in a beaker with 2 L of ultrapure water (~6 containers per 2 L water), add a stir bar, and place it on a stir plate (see Note 24).
16. Change water two to three times/day, for 2–3 days, to ensure that all the sodium chloride was removed.

17. Store the scaffolds in ultrapure water, in 50 mL tubes, at 4°C until needed for use (can keep 2–6 months), or dry scaffolds and store dry.
18. When needed for use, cut the scaffolds at the desired dimensions while they are wet (use razor blades, surgical blades, biopsy punches).
19. Silk scaffolds can be sterilized by autoclaving (dry cycle, 20 min, 121°C, and 15 psi) or by soaking them in 70% ethanol (see Notes 20 and 21).

3.5 Silk Electrospun Mats

1. Prepare a 5% PEO solution by dissolving 1 g PEO into 20 mL ultrapure water (see Note 25).
2. Using a 3 mL syringe (the solution is very viscous, the use of a micropipetter could be difficult), add 2 mL of the 5% PEO solution to 8 mL of 8% aqueous silk solution (for example in a 50 mL conical tube), to obtain a 7.5% silk/PEO solution.
3. Stir very gently/at a very slow rate, for about 10–15 min, to obtain a homogeneous solution. Alternatively, the mixing can be achieved by placing the conical tube on a platform rocker.
4. Fill a 5 mL syringe (or a 10 mL syringe for thicker electrospun mats) with the silk/PEO solution and connect it to the steel capillary tube.
5. Mount the syringe on the syringe pump and hold the capillary tube on an adjustable, electrically insulated stand.
6. Adjust the solution flow rate (0.01–0.03 mL/min), electric potential (12–14 kV), and the distance between the capillary tip and the collection disk (10–25 cm) to obtain a stable jet.
7. Collect the silk nanofibers on the aluminum-covered collection disk until the desired thickness is achieved (or until all the silk/PEO solution has been used) (see Note 26).
8. Remove the aluminum foil with the silk mat from the collection disk.
9. Immerse the silk mat (still on the aluminum foil) in 100% methanol for 20–30 min to obtain water-insoluble silk mats.
10. Remove the silk mat/aluminum foil from the methanol and let it dry in a fume hood for at least 2–3 h.
11. Wash the methanol-treated silk mats with water for 48 h (two to three water changes/day) to remove PEO.
12. Cut to the desired shape using either scissors or dermal biopsy punches.
13. Detach the silk mat from the aluminum foil using a pair of tweezers. Store in ultrapure water at 4°C.
14. The silk mat can be sterilized by soaking in ethanol (two times, 20 min each time). Let the alcohol evaporate and rinse well with sterile water, buffer, or cell culture media.

3.6 Silk Nano- and Microspheres (from Silk/PVA Blend Films)

1. Prepare a 5% (w/v) silk solution from a more concentrated silk solution by diluting with ultrapure water (see Note 6).
2. Prepare a 5% (w/v) PVA solution: add 5 g PVA to 100 mL water and let it stir overnight at room temperature. If necessary, filter the solution to remove undissolved matter.
3. For a weight ratio of 1/4 between silk fibroin and polyvinyl alcohol (PVA), add slowly 1 mL silk solution over 4 mL PVA solution in a 15 mL conical tube. Mix gently with a pipette (see Note 27).
4. Sonicate the solution with a Branson Sonifier for 30 s at an energy output of 10–25% amplitude. Slowly move the probe up and down in the liquid during the sonication. At lower sonication amplitudes, a mix of nano- and microspheres will be obtained. At higher amplitude (30%) sonication, mostly nanospheres will form.
5. Immediately after sonication, move the solution (5 mL) to a 100 mm diameter plastic Petri dish; make sure that the whole bottom of the dish is covered with the solution. Place the dish (open, without the lid) in a fume hood and leave it overnight (see Note 28).
6. Using tweezers, peel off the dry film and remove it from the Petri dish. Place it in a centrifuge tube. Add a little amount of water in the Petri dish (use a micropipetter and pipette tip) and wash to remove all the film material. Collect this water and add it to the centrifuge tube.
7. Add 30 mL ultrapure water to the centrifuge tube and shake gently for 10 min at room temperature to allow the film to completely dissolve. The solution will turn cloudy due to the suspension of silk microspheres. This step is done to remove the PVA.
8. Centrifuge at 15,000 rpm 27,000 × *g* for 20 min, at 4°C, and carefully discard the supernatant.
9. Wash/resuspend the pellet in 30 mL ultrapure water and centrifuge again (as in step 8). Discard the supernatant.
10. Resuspend the final pellet in 5 mL of ultrapure water and transfer the solution to a 15 mL conical tube (see Note 29).
11. Sonicate at 10% energy output for 15 s to disperse the clustered silk nano- and microspheres and make the suspension homogenous.
12. To store the silk microspheres for a longer time, lyophilize the suspension and keep the dry material at 4°C. If no storage is desired, skip this step and step 13.
13. Weigh a certain amount of dry material in a 1.5 mL Eppendorf tube, suspend in ultrapure water or buffer, and pipette to homogenize. If big aggregates still remain, apply a short probe-ultrasonication (low-energy output, such as 10% amplitude for 5 s). This step is done to resuspend the lyophilized microspheres.

3.7 Silk Microspheres: Lipid-Template Method

1. Weigh 100 mg of 1,2-dioleoyl-*sn*-glycero-3-phosphocholine powder (DOPC) and add to a glass culture tube.
2. In the fume hood, add 1 mL of chloroform and gently shake the tube until the DOPC completely dissolves.
3. Apply a slow flow of nitrogen gas in the tube while rolling the tube continuously, until the solvent completely evaporates leaving a thin film on the tube wall (homogeneous layer of lipid). Leave the tube to dry in the hood for 30 min.
4. Using a 1,000 μL pipette tip, add 0.5 mL of 8% (w/v) silk solution in the tube and hydrate the lipid film with this solution (see Note 30).
5. When the lipid layer is rehydrated, add 1.5 mL ultrapure water to the tube, mix with the lipid suspension, and move the diluted mixture to a 15 mL conical tube. Use the water in two increments of 750 μL to wash the pipette tip and the glass tube well.
6. Place the conical tube in liquid nitrogen for freezing, keep it for 10 min.
7. Move the conical tube to the 37°C water bath for approximately 5–10 min. Gently shake the tube during thawing and check that the suspension is completely thawed.
8. Repeat the freeze-thaw cycle (steps 6 and 7) twice. The result is a milky white suspension.
9. Fill a 100 mL glass beaker with 25 mL ultrapure water, add a stir bar, and place the beaker on a stir plate. While stirring at high speed, slowly pipette the thawed silk-lipid solution into the beaker. Continue stirring for 10 min, at room temperature.
10. Transfer the solution to a 50 mL conical tube and freeze it at −80°C overnight.
11. Lyophilize the material for 2–3 days (see Note 31).
12. Move the dried material into a centrifuge tube (organic solvent tolerant).
13. Add 20 mL methanol to the centrifuge tube and shake for 30 min at room temperature. Allow the methanol to dissolve the material (see Note 32).
14. Remove the insoluble fibers carefully (see Note 33).
15. Centrifuge at 10,000 rpm 12,857 × *g* for 20 min, at 4°C. Discard the supernatant.
16. Dry the pellet in the tube, overnight, in a fume hood.
17. Store the dry pellet of pure silk microspheres at 4°C.
18. To resuspend the microspheres, weigh the desired amount of dry material in a microcentrifuge tube and add ultrapure water or buffer. Mix using a micropipette, to homogenize.

19. If big aggregates still remain after resuspending the microspheres, they can be dispersed with a Branson Sonifier (low-energy output, e.g., 10% amplitude, for 5–10 s).

3.8 Silk Films: Water Based

1. Prepare silk fibroin solution at 8% (w/v). For silk fibroin solution preparation, (see Note 3).
2. Add 4 mL of 8% (w/v) silk solution into a 100 mm polystyrene Petri dish; this volume will generate a film with a thickness of ~50 μm.
3. Air dry overnight or until the silk film is detaching from the Petri dish (slow drying at room temperature, with the lid partly off the dish will result in a uniform film).
4. Fill the bottom of the vacuum desiccator with ultrapure water.
5. Place the opened Petri dish (with the film) in the desiccator.
6. Apply a vacuum for 10 min and then close the vacuum line.
7. Keep the film in the desiccator for 24 h, allowing the film to water anneal.
8. Gently remove the film from the Petri dish (see Note 34).
9. Cut to desired dimensions using scissors or a biopsy punch.
10. To sterilize the cut films: place the silk films in a new Petri dish and soak three times with 70% ethanol, 20–30 min each time. This step is done in a laminar flow hood.
11. Wash with sterile PBS (or sterile water) to remove traces of ethanol.
12. If films are not used immediately and need to be stored, let them dry in the laminar flow hood and then store in a sterile container at room temperature.

3.9 Silk Films: Patterned

1. Make sure that the diffraction grating is free of debris. Spray with ethanol, wipe, and blow dry with pressurized canned air.
2. Place a paper cup on a balance and weigh out the desired amount of base and then curing agent from the PDMS Sylgard kit (the ratio is 10:1). For one mold (5 cm × 5 cm) use 4.5 g base and 0.5 g of curing agent.
3. Mix with the pipette for 1 min.
4. Place the paper cup with the PDMS mixture (base and curing agent) in the vacuum oven for 30–60 min to remove air bubbles.
5. Place the diffraction grating in the Petri dish and slowly pour the PDMS over the diffraction grating; avoid introducing bubbles (see Note 35).
6. To cure the PDMS, place the Petri dish in a 60°C oven for 4–6 h or leave at room temperature for over 24 h.

7. Gently remove the PDMS from the diffraction grating (care should be taken not to rip the PDMS mold) (see Note 36).
8. Using the hole punch, punch out 14 mm disks from the PDMS mold. Different size disks can be cut, depending on the need.
9. Place the 14 mm PDMS disks into a Petri dish and add 100 μL of 8% (w/v) silk solution onto each disk.
10. Let the films dry overnight.
11. Fill the bottom of the vacuum desiccator with ultrapure water.
12. Place the opened Petri dish (with the film) in the desiccator.
13. Apply a vacuum for 10 min and then close the vacuum line.
14. Keep the dry films in the desiccator for 12–24 h, allowing the films to water anneal.
15. Remove films from the high humidity atmosphere and allow them to dry for 20 min. Films can now either be removed from the PDMS molds, step 16, or be kept wet for immediate use, step 17.
16. Gently peel films from the PDMS mold using tweezers. The films can be stored dry, at room temperature.
17. If desired, pour water on top of the films on PDMS molds, in the dish, to keep films wet and flat.

4 Notes

1. All solutions should be prepared in ultrapure water—water with a resistivity of 18.2 MΩ-cm. This water is referred to as "water" throughout in this text.
2. Subheading 3.1 describes the procedure used to purify the silk from one "batch" of cocoons = 5 g cocoons. 5 g cocoons will yield ~3.6–3.8 g of degummed fibroin fibers. Solubilizing 2.5 g of the degummed fibroin fibers in lithium bromide, followed by dialysis to remove the lithium bromide, will produce about 26–28 mL of aqueous silk solution with a concentration of 7–8% (w/v). All quantities listed in Subheading 2.1, are necessary for one batch of cocoons.
3. Aqueous silk fibroin solution preparation is described in Subheading 3.1.
4. Silk solution can be sterilized by autoclaving it in a steam autoclave at 121°C and 15 psi, for 20–30 min. It is normal for the silk solution to change appearance after being autoclaved; it will become turbid and whitish in color.
5. For pore sizes of 100–700 μM, the silk fibroin solution used should be ~6% (w/v), and for pore sizes >700 μM, silk fibroin solution used should be ~8% (w/v).

6. Silk solution at 5% concentration (w/v) is prepared by diluting a 8% (w/v) silk solution using ultrapure water.
7. "Silk solution," throughout the text, means aqueous fibroin solution.
8. Loose fibroin fibers will be observed after the first few minutes of boiling the cocoon pieces; sericin is water soluble and as it dissolves in the sodium carbonate solution, fibroin fibers will become free from the cocoon pieces and the solution will become yellowish.
9. Dry silk fibers can be stored in a clean ziplock bag for a long time if necessary; one can choose to extract silk fibroin fibers and not continue with the solubilization step immediately.
10. Solubilization of lithium bromide powder in water is an exothermic reaction, the beaker will become hot! Start by solubilizing the calculated/necessary amount of lithium bromide powder in a volume of water that represents about 65% of the final volume of LiBr solution. Let it stir in a glass beaker until the solution becomes clear and then pour this solution in a graduated cylinder and add water to obtain the desired final volume of LiBr solution. For example: To make 50 mL of LiBr solution, 9.3 M concentration, the calculated amount of LiBr powder (40.4 g) should be dissolved in ~32–33 mL water.
11. One cassette needs to be dialyzed against 1 L water; for more than one cassette, the amount of water should be increased proportionally.
12. Using a needle in this step is not necessary; physical shear forces can induce β-form (silk II) formation and silk gelation. Also, it is normal to feel a lot of resistance when pushing the 20% LiBr-silk solution from the syringe into the cassette.
13. Centrifugation will remove the particulates that are suspended in the silk solution (these particulates will be visible on the centrifuge tube's wall).
14. For each batch of silk solution it is useful to generate a set of data for gelation times: different silk concentrations and volumes can be tested to determine the gelation time when using different energy outputs and times for the sonication.
15. If desired, different values for the amplitude setting can be used, such as 30, 40 or 50%. The higher the silk concentration or the amplitude setting, the faster the gelation, therefore shorter sonication time should be used. Depending on the batch of silk, sometimes the solution needs to be sonicated again under the same conditions (for example gelation can be achieved by sonicating for 4 × 15 s bursts—total of 60 s—at 15% amplitude).
16. During the sonication process, the solution temperature can increase from room temperature to ~40–70°C for a short period of time.

17. Pouring the 5 mL sonicated silk in a well plate or Petri dish will result in a thicker or thinner gel layer (this is a function of the surface area of the well or Petri dish).
18. If a bioactive molecule or cell suspension needs to be mixed with the silk gel: after sonication (step 5) allow the sonicated silk solution to cool down for ~5 min and then add the bioactive molecule solution or the cell suspension and mix well. The mixture is then quickly poured into a well plate (step 6) and allowed to gel.
19. After 1 day, open the containers and check with a pipette tip if the top layer of the silk scaffold feels solid (gently poke with the pipette tip on the side of the container). If the pipette tip can go easily through the silk/salt mix, then place the containers for 1–3 h in a 60°C oven; temperature speeds up the silk gelation process.
20. For sterilizing the scaffolds by autoclave: loosely wrap the dry or wet scaffolds in aluminum foil and place them in an autoclave pouch. Alternatively, scaffolds can be autoclaved in a glass Petri dish. Sterile scaffolds need to be conditioned/hydrated in sterile water, culture medium, or PBS overnight (or at least 5–8 h), in an incubator at 37°C, prior to using them. Dry, sterile scaffolds will not absorb liquid immediately; they need time to get hydrated. Handle the dry—non-sterile or sterile—scaffolds with care, because they are brittle.
21. For sterilizing scaffolds by ethanol: use this method for small size scaffolds (e.g., 5 mm diameter × 2–3 mm thick) and work within a sterile biohood. Remove scaffolds from the ultrapure water and aspirate the water to empty the pores. Soak the scaffolds in 70% ethanol for 20 min (in a sterile conical tube or Petri dish), then aspirate the ethanol, and add a fresh volume of ethanol. Leave scaffolds in the ethanol for another 20 min and then aspirate the alcohol and wash the scaffolds extensively with sterile water, PBS, or medium to remove traces of alcohol.
22. HFIP is a volatile solvent. Keep the bottle in the fridge for ~1 h prior to use, to cool it down and reduce its volatility. Use a glass graduated cylinder to measure the volume needed.
23. The ratio of silk protein/sodium chloride needed is 1/20 (w/w); therefore for 3.4 g sodium chloride, the protein amount is 0.17 g (or 1 mL of the prepared 17% silk-HFIP solution). For smaller molds/vials, proportionally smaller amounts of silk-HFIP and sodium chloride can be used. When using the 7 or 10 mL plastic vials with 3.4 g sodium chloride and 1 mL silk-HFIP, the scaffolds obtained are ~15 mm in diameter and 10–12 mm thick.

24. This step is necessary to remove the sodium chloride particles from the silk scaffolds. Water will easily dissolve the sodium chloride particles, thus leaving empty scaffold pores.
25. The 5% PEO solution is very viscous; when preparing it, use a stir bar and mix the solution at very low rpm. Using warm water helps the dissolving process.
26. Cover the collection disk with aluminum foil for an easy handling of the electrospun silk mat. Make sure that the surface of the aluminum-covered collection disk is flat and smooth.
27. For bigger volumes (e.g., 5 mL silk + 20 mL PVA), use a glass beaker and stir bar to mix. Stir at 150 rpm for 2 h, at room temperature.
28. A volume of 5 mL (silk + PVA) in a 100 mm Petri dish will form a 50–100 μM thick film.
29. The pellet can be suspended in a smaller volume (1–5 mL) and used immediately after sonication (low energy, 10% and 10–15 s) to make the suspension homogenous.
30. Cut the pipette tip at the end, so that mixing the silk solution will not generate many air bubbles (silk solution can foam easily when mixed). Scratch the tube wall with the pipette tip to help the lipid film detach from the tube and dissolve in the silk solution. The solution will become turbid as the lipid film is hydrated and the glass tube wall should remain clear at the end of this step (all lipid film being dissolved in the silk solution).
31. After lyophilization, the amount of dry material should be ~200–300 mg.
32. It is not necessary to dissolve the whole amount of dry material in methanol. For example, 10–20 mg of lyophilized silk-lipid microspheres can be dissolved in 1.5 mL methanol, in a 2 mL microcentrifuge tube.
33. To remove the insoluble fibers, use tweezers or a 50–70 μm cell strainer.
34. After the water annealing, if the silk film is not cut immediately and it is stored at RT, it will dry and become hard. To cut smaller dimension films, soak the large film (Petri dish size) in PBS or water until it is soft, 1–2 h.
35. Once the grating is covered, remove any bubbles with the needle.
36. Wipe the PDMS mold clean with ethanol. PDMS molds can be stored at room temperature and used many times. Discard the molds when there is damage present on the casting face.

References

1. Altman GH, Diaz F, Jakuba C, Calabro T, Horan RL, Chen J, Lu H, Richmond J, Kaplan DL (2003) Silk-based biomaterials. Biomaterials 24:401–416
2. Omenetto FG, Kaplan DL (2010) New opportunities for an ancient material. Science 329: 528–531
3. Vepari C, Kaplan DL (2007) Silk as a biomaterial. Prog Polym Sci 32:991–1007
4. Leal-Egana A, Scheibel T (2010) Silk-based materials for biomedical applications. Biotechnol Appl Biochem 55:155–167
5. Horan RL, Antle K, Collette AL, Wang Y, Huang J, Moreau JE, Volloch V, Kaplan DL, Altman GH (2005) In vitro degradation of silk fibroin. Biomaterials 26:3385–3393
6. Park SH, Gil ES, Kim HJ, Lee K, Kaplan DL (2010) Relationships between degradability of silk scaffolds and osteogenesis. Biomaterials 31:6162–6172
7. Wang X, Zhang X, Sun L, Subramanian B, Maffini MV, Soto A, Sonnenschein C, Kaplan DL (2009) Preadipocytes stimulate ductal morphogenesis and functional differentiation of human mammary epithelial cells on 3D silk scaffolds. Tissue Eng Part A 15:3087–3098
8. Zhang X, Baughman CB, Kaplan DL (2008) In vitro evaluation of electrospun silk fibroin scaffolds for vascular cell growth. Biomaterials 29:2217–2227
9. Unger RE, Wolf M, Peters K, Motta A, Migliaresi C, James Kirkpatrick C (2004) Growth of human cells on a non-woven silk fibroin net: a potential for use in tissue engineering. Biomaterials 25:1069–1075
10. Gil ES, Park SH, Marchant J, Omenetto F, Kaplan DL (2010) Response of human corneal fibroblasts on silk film surface patterns. Macromol Biosci 10:664–673
11. Hofmann S, Knecht S, Langer R, Kaplan DL, Vunjak-Novakovic G, Merkle HP, Meinel L (2006) Cartilage-like tissue engineering using silk scaffolds and mesenchymal stem cells. Tissue Eng 12:2729–2738
12. Kim HJ, Kim UJ, Vunjak-Novakovic G, Min BH, Kaplan DL (2005) Influence of macroporous protein scaffolds on bone tissue engineering from bone marrow stem cells. Biomaterials 26:4442–4452
13. Mauney JR, Nguyen T, Gillen K, Kirker-Head C, Gimble JM, Kaplan DL (2007) Engineering adipose-like tissue in vitro and in vivo utilizing human bone marrow and adipose-derived mesenchymal stem cells with silk fibroin 3D scaffolds. Biomaterials 28:5280–5290
14. Wang Y, Kim UJ, Blasioli DJ, Kim HJ, Kaplan DL (2005) In vitro cartilage tissue engineering with 3D porous aqueous-derived silk scaffolds and mesenchymal stem cells. Biomaterials 26: 7082–7094
15. Schneider A, Wang XY, Kaplan DL, Garlick JA, Egles C (2009) Biofunctionalized electrospun silk mats as a topical bioactive dressing for accelerated wound healing. Acta Biomater 5:2570–2578
16. Wharram SE, Zhang X, Kaplan DL, McCarthy SP (2010) Electrospun silk material systems for wound healing. Macromol Biosci 10:246–257
17. Hofmann S, Hagenmuller H, Koch AM, Muller R, Vunjak-Novakovic G, Kaplan DL, Merkle HP, Meinel L (2007) Control of in vitro tissue-engineered bone-like structures using human mesenchymal stem cells and porous silk scaffolds. Biomaterials 28:1152–1162
18. Kim HJ, Kim UJ, Kim HS, Li C, Wada M, Leisk GG, Kaplan DL (2008) Bone tissue engineering with premineralized silk scaffolds. Bone 42:1226–1234
19. Kim HJ, Kim UJ, Leisk GG, Bayan C, Georgakoudi I, Kaplan DL (2007) Bone regeneration on macroporous aqueous-derived silk 3-D scaffolds. Macromol Biosci 7:643–655
20. Rockwood DN, Gil ES, Park SH, Kluge JA, Grayson W, Bhumiratana S, Rajkhowa R, Wang X, Kim SJ, Vunjak-Novakovic G, Kaplan DL (2011) Ingrowth of human mesenchymal stem cells into porous silk particle reinforced silk composite scaffolds: An in vitro study. Acta Biomater 7:144–151
21. Wang Y, Blasioli DJ, Kim HJ, Kim HS, Kaplan DL (2006) Cartilage tissue engineering with silk scaffolds and human articular chondrocytes. Biomaterials 27:4434–4442
22. Kardestuncer T, McCarthy MB, Karageorgiou V, Kaplan D, Gronowicz G (2006) RGD-tethered silk substrate stimulates the differentiation of human tendon cells. Clin Orthop Relat Res 448:234–239
23. Altman GH, Horan RL, Lu HH, Moreau J, Martin I, Richmond JC, Kaplan DL (2002) Silk matrix for tissue engineered anterior cruciate ligaments. Biomaterials 23:4131–4141
24. Moreau JE, Bramono DS, Horan RL, Kaplan DL, Altman GH (2008) Sequential biochemical and mechanical stimulation in the development of tissue-engineered ligaments. Tissue Eng Part A 14:1161–1172
25. Lovett M, Cannizzaro C, Daheron L, Messmer B, Vunjak-Novakovic G, Kaplan DL (2007) Silk fibroin microtubes for blood vessel engineering. Biomaterials 28:5271–5279
26. Soffer L, Wang X, Zhang X, Kluge J, Dorfmann L, Kaplan DL, Leisk G (2008) Silk-based electrospun tubular scaffolds for

tissue-engineered vascular grafts. J Biomater Sci Polym Ed 19:653–664

27. Zhang X, Wang X, Keshav V, Johanas JT, Leisk GG, Kaplan DL (2009) Dynamic culture conditions to generate silk-based tissue-engineered vascular grafts. Biomaterials 30: 3213–3223
28. Subramanian B, Rudym D, Cannizzaro C, Perrone R, Zhou J, Kaplan DL (2010) Tissue-engineered three-dimensional in vitro models for normal and diseased kidney. Tissue Eng Part A 16:2821–2831
29. Wang X, Sun L, Maffini MV, Soto A, Sonnenschein C, Kaplan DL (2010) A complex 3D human tissue culture system based on mammary stromal cells and silk scaffolds for modeling breast morphogenesis and function. Biomaterials 31:3920–3929
30. House M, Sanchez CC, Rice WL, Socrate S, Kaplan DL (2010) Cervical tissue engineering using silk scaffolds and human cervical cells. Tissue Eng Part A 16:2101–2112
31. Wang X, Kluge JA, Leisk GG, Kaplan DL (2008) Sonication-induced gelation of silk fibroin for cell encapsulation. Biomaterials 29:1054–1064
32. Kim UJ, Park J, Li C, Jin HJ, Valluzzi R, Kaplan DL (2004) Structure and properties of silk hydrogels. Biomacromolecules 5:786–792
33. Leisk GG, Lo TJ, Yucel T, Lu Q, Kaplan DL (2010) Electrogelation for protein adhesives. Adv Mater 22:711–715
34. Yucel T, Cebe P, Kaplan DL (2009) Vortex-induced injectable silk fibroin hydrogels. Biophys J 97:2044–2050
35. Nazarov R, Jin HJ, Kaplan DL (2004) Porous 3-D scaffolds from regenerated silk fibroin. Biomacromolecules 5:718–726
36. Kim UJ, Park J, Kim HJ, Wada M, Kaplan DL (2005) Three-dimensional aqueous-derived biomaterial scaffolds from silk fibroin. Biomaterials 26:2775–2785
37. Wang Y, Rudym DD, Walsh A, Abrahamsen L, Kim HJ, Kim HS, Kirker-Head C, Kaplan DL (2008) In vivo degradation of three-dimensional silk fibroin scaffolds. Biomaterials 29: 3415–3428
38. Jin HJ, Fridrikh SV, Rutledge GC, Kaplan DL (2002) Electrospinning Bombyx mori silk with poly(ethylene oxide). Biomacromolecules 3:1233–1239
39. Zhang X, Reagan MR, Kaplan DL (2009) Electrospun silk biomaterial scaffolds for regenerative medicine. Adv Drug Deliv Rev 61: 988–1006
40. Wang X, Yucel T, Lu Q, Hu X, Kaplan DL (2010) Silk nanospheres and microspheres from silk/pva blend films for drug delivery. Biomaterials 31:1025–1035
41. Wang X, Wenk E, Matsumoto A, Meinel L, Li C, Kaplan DL (2007) Silk microspheres for encapsulation and controlled release. J Control Release 117:360–370
42. Hofmann S, Foo CT, Rossetti F, Textor M, Vunjak-Novakovic G, Kaplan DL, Merkle HP, Meinel L (2006) Silk fibroin as an organic polymer for controlled drug delivery. J Control Release 111:219–227
43. Lu S, Wang X, Lu Q, Hu X, Uppal N, Omenetto FG, Kaplan DL (2009) Stabilization of enzymes in silk films. Biomacromolecules 10: 1032–1042
44. Jin HJ, Park J, Karageorgiou V, Kim UJ, Valluzzi R, Kaplan DL (2005) Water-stable silk films with reduced beta-sheet content. Adv Funct Mater 15:1241–1247
45. Lawrence BD, Marchant JK, Pindrus MA, Omenetto FG, Kaplan DL (2009) Silk film biomaterials for cornea tissue engineering. Biomaterials 30:1299–1308

Chapter 3

Aqueous-Based Spinning of Fibers from Self-Assembling Structural Proteins

Steven Arcidiacono, Elizabeth A. Welsh, and Jason W. Soares

Abstract

There has been long-standing interest in generating fibers from structural proteins and a great deal of work has been done in attempting to mimic dragline spider silk. Dragline silk balances stiffness, strength, extensibility, and high energy to break. Mimicking these properties through aqueous-based spinning of recombinant silk protein is a significant challenge; however, an approach has been developed that facilitates the formation of fibers approaching the mechanical properties seen with natural dragline silk. Due to the multitude of solution, spinning and post-spinning variables one has to consider, the method entails a multivariate approach to protein solution processing and fiber spinning. Optimization to maximize mechanical integrity of the fibers is performed by correlating the solution and spinning variables to mechanical properties and using this information for subsequent fiber spinning studies. Here, the method is described in detail and emphasizes the lessons learned during the iterative variable analysis process, which can be used as a basis for aqueous-based fiber spinning of other structural proteins.

Key words Spider silk, Recombinant proteins, Fiber, Spinning, Mechanical properties

1 Introduction

Spiders have captured the interest of scientists for many years because spider silks are among the toughest of materials, having properties that surpass some man-made synthetic materials. Examination of natural silk in the gland has gained an understanding of how the spider keeps a large supply of protein with propensity for self-assembly soluble and ready for fiber formation. Changes occurring at the molecular level as the protein moves through the gland to become a fiber have been studied extensively and summarized by Vollrath et al. (1). The study of regenerated natural spider silk and silkworm silk has demonstrated numerous variables to consider that influence mechanical properties of the formed fibers. These variables include draw rate and temperature (2, 3), presence of water in the protein solvent (4), spinneret diameter (5), coagulation bath composition (6, 7), and draw ratio (6–8).

Juliet A. Gerrard (ed.), *Protein Nanotechnology: Protocols, Instrumentation, and Applications*, Methods in Molecular Biology, vol. 996, DOI 10.1007/978-1-62703-354-1_3, © Springer Science+Business Media New York 2013

Biophysical characterization also revealed that secondary structure, which influences self-assembly of the silk proteins, is dependent on solution composition and time (7). While natural spider silks possess desirable mechanical properties, acquisition of sufficient quantities is a constant challenge. Unlike silkworms, farming spiders to harvest silk is not feasible due to their territorial behavior. Therefore, recombinant silk proteins provide an opportunity to generate sufficient material to spin synthetic silk fibers that can be manipulated to have mechanical properties approaching those of natural spider silks.

Over the last two decades, we have recombinantly produced several spider silk proteins based on dragline silk, purified under denaturing conditions and spun fibers with mechanical properties similar to those seen with natural spider silks (9, 10). The silk solutions were at low protein concentration and processed into concentrated spin solutions in dilute aqueous denaturing buffers. Unlike a method to process recombinant spider silk by protein dissolution (11), the proteins were kept soluble throughout the preparation. The goal was to mimic the method of spinning employed by the spider, in which the protein remains in solution until fiber formation. Similar to natural silk, a series of variables are present during spin solution preparation, fiber spinning and post-spinning manipulation that can potentially influence the fiber's mechanical properties, both individually and in combination. Individual variable effects can be determined empirically, although investigating the interaction and dependence between variables is more difficult.

Presented here is the method derived from the aqueous-based spinning of recombinant silk proteins. Specifically, purified ADF-3 (9), a truncated dragline protein from *Araneus diadematus* produced in tissue culture, will be used as an example to describe the method. However, this method has also been employed to spin fibers from other purified recombinant silk proteins including NcDS (10), a truncated *Nephila clavipes* dragline protein spidroin I, produced in bacteria, and $[(SPI)_4/(SPII)_1]_4$ (10), a synthetic hybrid protein from consensus sequences of *N. clavipes* dragline proteins spidroin I and II, produced in bacteria. An iterative approach is employed that enables optimization of variables for aqueous spinning of fibers with mechanical integrity. There are three major aspects to the approach: (1) spin solution preparation, (2) fiber spinning, and (3) fiber characterization (microscopy and mechanical testing). Within spin solution preparation, recombinant silk solutions are dialyzed into dilute aqueous denaturing buffers supplemented with glycine to enhance protein stability and concentrated through ultrafiltration. During fiber spinning, a concentrated protein solution is spun through an aqueous-based spinning technique. The technique encompasses extrusion of the protein solution through a spinneret into a coagulation bath. Fibers are collected after undergoing a series of post-spinning manipulations. After collection, the fibers are characterized through microscopy under polarizing and common white light to investigate molecular alignment. Scanning electron microscopy (SEM) is employed to examine the surface of the fiber and the inner core under high magnification. Lastly, select fibers that exhibit minimum defects and a consistent diameter are subjected to mechanical testing for elucidation of tensile properties. For each major procedural aspect, variables for consideration are identified. A systematic, multivariate approach to correlate the solution, spinning and post-spinning variables to the generation of fibers with mechanical integrity is the basis of the described method.

2 Materials

2.1 Spin Solution Preparation

2.1.1 Spin Dope Processing

1. Purified recombinant silk ADF-3 protein solution (concentrations of 0.027–0.16% (w/v)): Purity was at least 70% or greater based on Reverse Phase-High Performance Liquid Chromatography (RP-HPLC) using peak area integration (see Note 1).
2. Spin solution buffer (i.e., dialysis buffer): 160 mM urea containing 10 mM NaH_2PO_4, 1 mM Tris and 20 mM NaCl, pH 5 supplemented with 10 or 100 mM glycine (see Note 2).
3. Dialysis tubing: Spectra/Por regenerated cellulose 12,000–14,000 Molecular Weight Cut-off (MWCO) tubing (Spectrum Laboratories, Inc., Rancho Dominguez, CA).
4. Ultrafiltration (UF) devices: Amicon stirred cell, Centriprep10, Centricon10, Microcon10 (Millipore, Bedford, MA) with YM10 membranes having 10,000 MWCO (see Note 3).

2.1.2 SDS-Polyacrylamide Gel Electrophoresis

1. Running buffer (10×): 1 M Tris base, 1 M Tricine, 1% (w/v) sodium dodecyl sulfate (SDS).
2. Sample buffer (5×): 250 mM Tris, 10% (w/v) SDS, 0.5% (w/v) bromophenol blue, 50% (w/v) glycerol; store solution at room temperature.
3. Novex Xcell II system (Invitrogen Corporation, Carlsbad, CA) for precast gels.
4. Gel: Novex pre-cast 10–20% Tricine (Invitrogen Corp.).
5. Molecular weight markers: Novex Mark12 markers (Invitrogen Corp.).
6. Conventional gel staining/destaining solutions (see Note 4): Stain—40% (v/v) methanol, 10% (v/v) acetic acid, 0.6% (w/v) Coomassie blue R-250; destain—10% (v/v) methanol, 7.5% (v/v) acetic acid.

2.2 Fiber Spinning

1. Coagulation vials: Screw top glass vials with height of approximately 3 in. (7.62 cm).
2. Hamilton syringe (100 μl) (Hamilton Co., Reno, NV).
3. Harvard Apparatus Infusion/Withdrawal pump (Harvard Apparatus, Natick, MA).
4. A specialized microspinner (cavity volume 0.5 ml, 5 mm I.D.) equipped with a removable 6 cm (0.125 mm I.D.) piece of PEEK HPLC tubing (Sigma-Aldrich, St. Louis, MO) (see Note 5).
5. Coagulation bath was made of Lexan with a painted black bottom (see Note 6); dimensions 24 in. (61.0 cm) long × 8 in. (20.3 cm) wide × 5 in. (12.7 cm) high.
6. Coagulation bath solution was composed of various concentrations of methanol water (concentration varied but was typically 70–80% (v/v) methanol). Approximately 2 L is required (see Note 7).
7. Tweezers – Style 2A or 2B (Electron Microscope Sciences, Hatfield, PA) (see Note 8).
8. Humidity controlled storage box (Fisher Scientific Corp, Pittsburgh, PA) with magnesium nitrate (Sigma) to maintain 50–55% humidity.

2.3 Fiber Characterization

2.3.1 Light Microscopy

1. Optiphot2-pol polarizing light microscope (Nikon Corp., Garden City, NY) equipped using a standard scale bar accessory calibrated with an optical micrometer (Swift Inc., San Antonio, TX) to approximate fiber diameter; diameter was converted to linear density (denier) using standard spreadsheet software (see Note 9).
2. Fiber images were acquired with a Nikon D1 digital camera (Nikon Corp.) attached to the polarizing microscope and processed using Nikon Capture software through a FireWire connection.
3. Standard glass microscopy slides.
4. Tweezers—Style 2A or 2B.

2.3.2 Scanning Electron Microscopy

1. CSM 950 SEM (Carl Zeiss, Jena, Germany).
2. Conductive carbon tape.
3. SEM sample stub.
4. Gold/palladium to fix samples to stub for imaging.
5. Tweezers—Style 2A or 2B.

2.3.3 Mechanical Testing

1. Mechanical testing was performed using the Instron Model 55R4201 (Instron Corp., Canton, MA) equipped with accessories as follows:

   2525-815 Drop Thru Static Load Cell (2.5N).

2712-013 Pneumatic Side Action Grips (general purpose static tensile grips); capacity = 0.5 kgf (Instron Corp.) (see Note 10).

2. 3H-6B2-D4 Hercules Air Val pneumatic foot pedal control attachment (Linemaster Switch Corp., WoodStock, CT) for grip closure connected to a gas cylinder containing compressed nitrogen equipped with a standard regulator to maintain constant pressure (50 psig) (see Note 11).
3. Single fiber as a control specimen (see Note 12).
4. Mechanical data were processed using Instron software Series IX Automated Materials Testing System.
5. MI-150 Fiber-Lite High Intensity Illuminator (Dolan-Jenner Industries, Lawrence, MA) to ease loading of single fibers into grips.
6. Tweezers—Style 2A or 2B.
7. Low adhesive, clear Scotch tape (3M Corp., St. Paul, MN).

3 Methods

3.1 Spin Solution Preparation

The preparation of the spin solution influences both fiber formation and the ability to impart post-spinning manipulations to the fiber. Protein solutions were processed under denaturing conditions with the goal of maintaining solubility throughout the process. Volume of the dilute protein solution was reduced until a suitable protein concentration was achieved.

Solution variables for consideration: protein molecular weight, purity and concentration, buffer composition, temperature, and molecular interaction (see Note 13).

3.1.1 Spin Dope Processing

1. Purified ADF-3 protein solutions were added to dialysis tubing and dialyzed into 50 volumes (vol) of spin solution buffer (see Notes 14 and 15).
2. After 1 h, exchange buffer with new 50 vol of spin solution buffer.
3. After 1 h, change into a third 50 vol and dialyze overnight.
4. Remove sample from tubing and clarify by centrifugation (16,000 × g, 5 min, 25 °C).
5. Transfer supernatant (see Note 16).
6. Determine protein concentration spectrophotometrically at A_{280} using the protein extinction coefficient (ε_m) (see Note 17):

   Concentration (mol/l) = A_{280}/protein ε_m.

   Protein $\varepsilon_m = x(\varepsilon_m$ tyrosine$) + y(\varepsilon_m$ tryptophan$) + z(\varepsilon_m$ cysteine$)$, where x = # tyrosine, y = # tryptophan, z = # cysteine.

7. Add soluble fraction to UF device with 10,000 MWCO membrane and concentrate to 10–28% (v/v) (see Notes 18–20). UF device used depends on sample volume.
8. At each UF stage, clarify any gelled material by centrifugation as above (see Notes 21 and 22).
9. Dilute samples in spin solution buffer to determine protein concentration as described above (see Note 17).
10. Analyze soluble and insoluble fractions of samples from dialysis and each stage of UF by SDS-PAGE.
11. Determine purity by RP-HPLC (see Note 1).
12. UF retentates (i.e., spin dopes) were stored at 4 °C to minimize propensity for gelation.

3.2 SDS-Polyacrylamide Gel Electrophoresis

1. Prepare samples by adding 2–5 μg protein/well of sample (see Note 23), water to 32 μl, 8 μl 5× sample loading buffer. As a control, run 10 μl Mark12 marker.
2. Heat samples for 5 min in a boiling water bath. Do not heat Mark12 protein markers.
3. Run gel approximately 90 min at 125 V (constant) until bromophenol blue dye front reaches end of gel.
4. Disconnect power supply and disassemble gel.
5. Stain gel for 30 min to overnight with Coomassie blue on rocking platform. Destain background to desired level (see Notes 24 and 25).
6. View Coomassie stained gels on light box; capture image with digital camera.

3.3 Fiber Spinning

Fibers ideally were spun such that they are capable of undergoing draw. Fibers without adequate structure due to insufficient coagulation or molecular entanglement will break during post-spinning manipulation. Coagulation at too high a rate should be avoided; protein molecules locked into a random orientation would preclude the draw required for improved mechanical properties, resulting in a brittle fiber.

Spinning and post-spinning variables for consideration: spinneret dimensions, coagulation bath composition, spinning rate, draw ratio and speed, draw solvent composition, and temperature (see Note 13).

1. Centrifuge samples as above to remove any insoluble material (see Note 22).
2. Test small aliquot (approx. 10 μl) of spin solution in various concentrations of methanol in water to determine the optimal methanol/water composition to be used for the coagulation bath (see Note 26). Spin solution was applied to 7 ml methanol/water solution using a Hamilton syringe (submerge only the tip).

3. Load 25–50 μl of spin solution into the microspinner, taking care not to introduce air.
4. Screw PEEK tubing into place; place syringe into position.
5. Begin pump prior to placing PEEK tubing into coagulation bath to avoid clogging the tube by spin solution precipitation.
6. Spin solutions were extruded into a methanol coagulation bath (typically 70–80%) at a rate of 2–10 μl/min (see Notes 27 and 28).
7. Fibers were hand-drawn in the methanol bath with tweezers (single draw). Double-drawn fibers were drawn in methanol and then again in water. Fibers were held under constant tension while transferring to the water bath for further draw (see Note 29).
8. Fibers were removed from the bath under constant tension and allowed to dry.
9. Single fibers were placed in petri dishes and stored at ambient temperature with 50–55% relative humidity.
10. To reuse coagulation bath solution, filter through Whatman #4 paper to remove protein (see Note 30).

3.4 Fiber Characterization

3.4.1 Light Microscopy

1. Single fibers stored in petri dishes as indicated above are removed, transferred to standard glass slides, and secured with clear tape (see Notes 31–34).
2. Analyze fiber under 10× or 20× magnification with common white light along the entire length of fiber to visualize surface defects (i.e., severe thinning, bends, breaks), core defects (e.g., crack along fiber length or circumference), and diameter (see Note 35).
3. Using the digital camera attached to the microscope, take representative common white light images of the fiber. As an example, ADF-3 recombinant spider silk as-spun fibers are shown in Fig. 1a.
4. Molecular alignment analysis is performed by investigating the birefringent nature of the single fibers under polarizing light with a first-order red plate at 530 nm. Using the attached digital camera, take representative images of the fiber under polarizing conditions to visualize birefringent or non-birefringent nature. Examples of fiber exhibiting birefringence are shown in Fig. 1b (undrawn) and Fig. 1c (drawn) (see Note 36).
5. Estimate an average fiber diameter using a scale bar that is based on converting the number of divisions the fiber occupies to diameter (mm) as shown in Fig. 1a (see Notes 37–39).

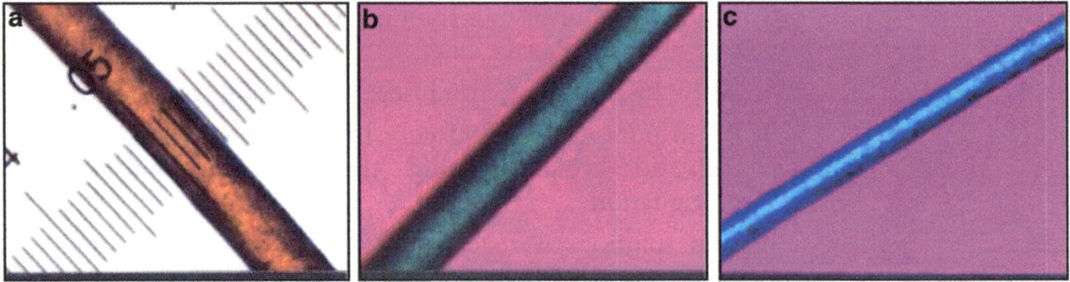

Fig. 1 Microscopy of recombinant silk fibers (20×). Fibers have a smooth surface free of defects and show birefringence under polarizing light, indicative of molecular orientation. (**a**) ADF-3 undrawn under common white light with a standard scale bar; (**b**) ADF-3 under polarizing light using a red plate (530 nm), as spun; (**c**) ADF-3 under polarizing light using a red plate (530 nm), double drawn. Scale bar shown in (**a**) is used to approximate fiber diameter by measuring the divisions (div) occupied by the fiber (10× = 0.0097 mm/division; 20× = 0.0067 mm/div). In this image, the fiber occupies 4.5 divisions corresponding to a 30 μm diameter

6. Convert fiber diameter (cm) to linear density (denier = *g*/9,000 m):

$$\rho_{\text{linear}} = A \times \rho_{\text{specific}} \times \text{constant}$$

where: ρ_{linear} = linear density (denier; unitless); A = area (πr^2; cm^2); r = radius (cm); ρ_{specific} = specific gravity (g/cm^3); constant = 9×10^5 (cm) (see Note 40).

7. Place fiber back in humidity box, keeping the fiber mounted on the glass slide within the petri dish.

3.4.2 Scanning Electron Microscopy

1. Place conductive carbon tape on the cross-sectional SEM stub (see Note 41).
2. Using tweezers, remove fiber from petri dish.
3. Fiber is trimmed with a razor blade and mounted lengthwise onto conductive carbon tape.
4. Sample is coated with gold/palladium (35 nm coating).
5. SEM is conducted at 20 keV between 1,000 and 5,000× magnifications. Typical SEM images of the fiber surface can be seen in Fig. 2a.
6. For a cross-sectional view, fiber is cut with the razor blade and prepared as indicated above. Typical solid core of the fiber is seen in Fig. 2b.

3.4.3 Mechanical Testing

1. Samples that are a minimum of 1.5 in (3.75 cm) in length and that have few defects, as determined through microscopy above, are selected for mechanical testing.

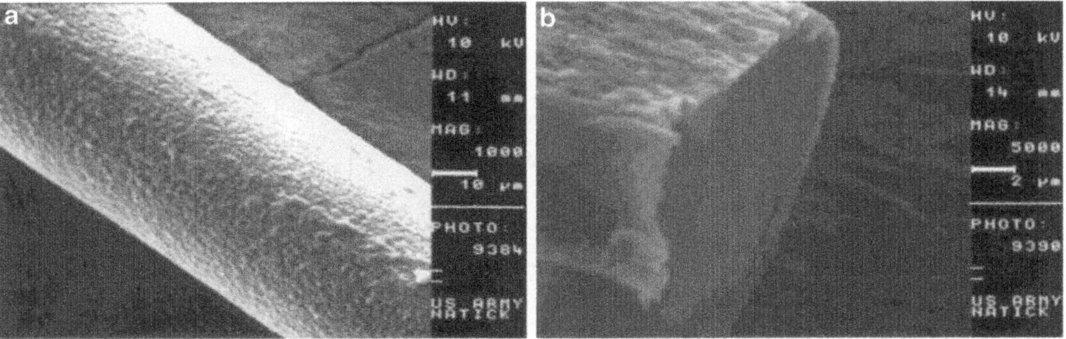

Fig. 2 Scanning electron microscopy of recombinant silk fiber, ADF-3. (**a**) Fiber surface (1,000×) and (**b**) interior at break point (5,000×). Fibers have a relatively smooth surface and a solid core

2. Mechanical testing parameters using a 2.5 N (0.2549 kgf) Load Cell are as follows:
   (a) Specimen length = 0.5 in. (12.70 mm) (see Note 42).
   (b) Crosshead speed = 12.70 mm/min.
   (c) Sampling rate = 100 pts/s.
   (d) Prompt for linear density prior to test.
   (e) High load level = 0.2447 kgf.
   (f) High extension = 1,016 mm.
   (g) Load threshold = 0.2243 kgf.
   (h) No slack correction.
3. Initially, perform test on control fiber to ensure instrument operation is acceptable and to generate comparative data.
4. Remove single fiber from glass slide by cutting the fiber ends at the intersection of fiber and tape.
5. Use tweezers to transfer the fiber to the Instron grips (see Note 33).
6. Using compressed nitrogen (50 psig), secure the fiber in the top grip by slowly pressing the foot pedal. Upon engaging the foot pedal, nitrogen will be regulated through the feed lines that go directly to the pneumatic grips. One click shuts/locks the top grip.
7. With the tweezers, remove slack from the fiber then further depress the foot pedal until the second click, which shuts/locks the bottom grip.
8. Input appropriate linear density (denier) and proceed with test for collection of mechanical properties (see Note 43). An example of typical stress-strain curves can be seen in Fig. 3.
9. Upon completion of test, the crosshead will return to original gauge length position.

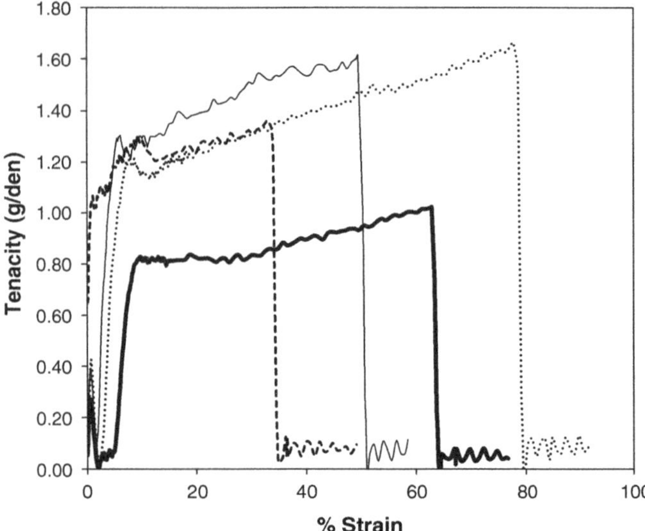

Fig. 3 Stress–strain curves from a fourfold drawn, ADF-3 fiber (four individual tests from a single fiber). All specimens demonstrated extensive elongation, which illustrates susceptibility to further draw. Average mechanical properties: Tenacity = 0.958 g/den, % Strain = 56.732%, Modulus = 93.680 g/den, Toughness = 0.662 g/den

10. Typical mechanical properties that are generated include: Maximum Load (*g*); Tenacity at Maximum Load (*g*/den); % Strain at Maximum Load (%); Displacement at break (mm); Tenacity at break (*g*/den); % Strain at break (%); Automatic Young's Modulus (*g*/den); Toughness (*g*/den).
11. If fiber remains in the grips after testing, depress the foot pedal located further inside the "shoe" of the housing of the pedal to release/open both grips. Collect fiber with tweezers and place in petri dish, store in humidity box for post-testing analyses. Otherwise, just use pedal to open grips for next sample.

3.5 Iterative Variable Analysis

1. Mechanical properties can be correlated to previously described solution, spinning and post-spinning variables. These correlations are the basis of subsequent spin trials for the optimization of fibers to maximize mechanical performance. An example of the impact of draw on mechanical properties can be seen in Fig. 4 (see Note 44).
2. Fibers collected after testing, as described above, can be transferred to glass slides and analyzed under both common and polarizing light, as previously described, and correlated to post-spinning variables. An example of birefringent analysis of a post-tested recombinant ADF-3 fiber can be seen in Fig. 5 (see Note 45).

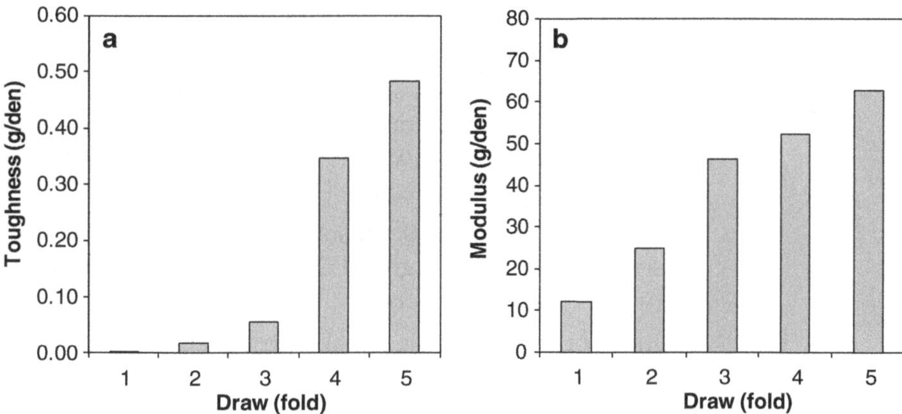

Fig. 4 Influence of post-spinning draw on (**a**) Toughness and (**b**) Young's Modulus of ADF-3 fibers. Draw is a critical post-spinning variable, as evidenced by the increasing mechanical integrity of the fibers as draw ratio increases. The correlation of mechanical properties to solution and spinning and post-spinning dependent variables is an iterative process and essential for optimizing fiber formation and mechanical properties (average of ~100 single fiber tests)

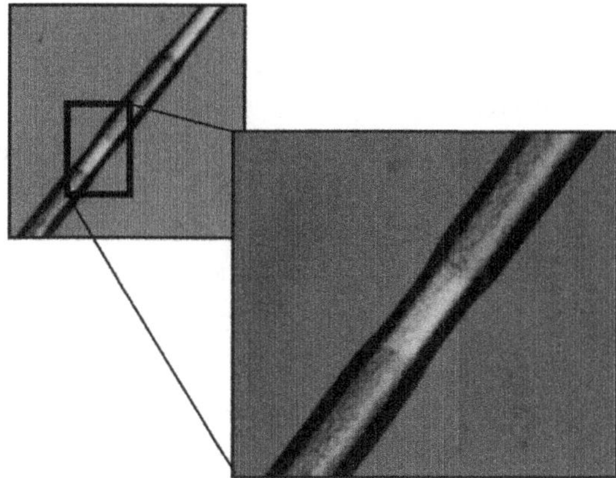

Fig. 5 Polarizing light microscopy of ADF-3 fiber after mechanical testing. Drawing of the fiber was not optimal since areas of elongation were evident that were not associated with the break point. These results demonstrate that the fiber is susceptible to additional draw during post-spinning manipulation, which is important information for subsequent optimization studies

4 Notes

1. Purity can also be determined by densitometry from SDS-PAGE (using TotalLab image analysis software, Madison, WI) or quantitative amino acid composition analysis.
2. Additional denaturing buffers have also been used to prepare spin dopes from other recombinant silks, e.g., 1 M urea for NcDS and 160 mM urea without glycine for $[(SPI)_4/(SPII)_1]_4$.

3. Volumes for UF devices: Stirred cell 50—500 ml, Centri-prep 10—15 ml, Centricon 10—2 ml, Microcon 10—0.5 ml.
4. For enhanced gel staining/destaining, use SYPRO Red (Invitrogen Corp.) protein stain in 7.5% (v/v) acetic acid, destain with 7.5% (v/v) acetic acid, and image with Storm 860 optical scanner (Molecular Devices, Sunnyvale, CA).
5. The spinneret must be of sufficient length for fiber formation; solutions spun through a 0.5 cm long spinneret were incapable of forming intact fibers.
6. Black bottom made it possible to visualize and manipulate the fibers.
7. Adjust solution depth within the coagulation bath to approximately the same depth used in the coagulation vials.
8. Flat-end tweezers allow for easier manipulation and removal of fibers from the bath. Sharp-end tweezers result in fiber damage or breakage.
9. As an alternative to microscopy, denier determination could be acquired directly using a VIBROMAT M (TEXTECHNO Herbert Stein GmbH Co., Monchengladbach, Germany) using pre-tension metal, tong-shaped weights applied to the fibers.
10. More brittle fibers, in particular the first-generation fibers that are not optimized, tend to break due to the excessive force during pneumatic grip closure; in this situation it is recommended to use manual side action grips.
11. Compressed air can also be utilized.
12. Typically, Kevlar 29 was used as a control fiber. However, the control specimen could be any single fiber including other forms of Kevlar, Nylon 6,6, or polyethylene.
13. Each variable can affect a fiber's mechanical properties, both individually and collectively. The challenge is to identify those variables most important for maximizing fiber properties.
14. 50 vol refers to 50× the sample volume. For example, for a 10 ml sample, dialyze into 500 ml of buffer.
15. Presence of glycine stabilized spin solutions and mitigated gelation.
16. If desired, retain precipitate for analysis via SDS-PAGE. Characterization of insoluble material may provide information regarding protein susceptibility to precipitation under the processing conditions.
17. Samples diluted in spin solution buffer such that A_{280} is between 0.1 and 1 (typically 1:10 for dialysates, 1:100–1:500 for UF retentates).

18. The use of 10,000 MWCO membrane was to minimize loss of silk protein while removing small molecular weight contaminates to further increase purity of spin dope. The choice of MWCO membrane is dependent upon the size of the protein of interest and if further purification is desired. For example, if only volume reduction is needed, then a low MWCO membrane should be selected to minimize protein loss, but if further purification is desired then a MWCO membrane at a maximum of half the size of the target protein is suggested. Judicious selection of the MWCO membrane should be considered because loss of target protein can easily occur. For instance, a 55 kDa recombinant silk protein was found in the filtrate after ultrafiltration using a 10,000 MWCO membrane.
19. Increasing protein concentration by UF resulted in visible streaks in the solution of different density and refractive index, a phenomenon termed schlieren. A portion of this population of nonhomogeneous protein molecules had some protein on the verge of gelation, while the remainder was soluble.
20. Initial estimates of final protein concentration can be achieved by using the pre-UF concentration and estimating the desired concentration based on reduction in volume. The actual achievable final concentration will be dependent upon volume restrictions and susceptibility of the protein to gelation. For example, NcDS solution could be concentrated to 20–25% while $[(SPI)_4/(SPII)_1]_4$ could only reach 6.5–18%. Any additional processing of $[(SPI)_4/(SPII)_1]_4$ resulted in protein aggregation and irreversible gelation. Gelation could have been a result of the nature of the protein or volume limitations, which induced aggregation (<50 μl).
21. If desired, resuspend gel fraction in buffer at a volume equal to the sample volume used for the UF device for analysis on SDS-PAGE. Characterization of the gelled insoluble material may provide information regarding loss of protein during processing.
22. To the extent possible, protein was always kept soluble. This was not always feasible because silk solutions are inherently unstable, forming gels as a function of concentration, time, and temperature. Centrifugation is critical to removing gelled material that can nucleate further gelation/self-assembly.
23. For soluble sample, use sample diluted for spectrophotometry above. Choose appropriate volume to load 2–5 μg protein.
24. To reduce time for staining and destaining, gels can be gently heated in microwave (15 s for 50 ml solution). Do not boil as toxic fumes may result. In addition, placing a Kimwipe in the destain solution will further reduce the time needed.

25. Not all silk proteins stain well with Coomassie blue. SYPRO Red has enhanced sensitivity compared to Coomassie and results in less protein-to-protein variation in staining. For SYPRO Red, stain gel for 30 min at room temperature on rocking platform in a covered container. Destain for 10–60 min until desired background is achieved. Image gel on Storm 860 scanner using the red fluorescence/chemiluminescence mode at 800 V.
26. Coagulation test may not always accurately predict the optimal coagulation bath composition, but will provide a basis for coagulation bath selection.
27. The extrusion rate of 2–10 μl/min was the general range used. Actual rate used allowed the spin solution to form a fiber just prior to reaching the bottom of the coagulation bath. Too slow a rate resulted in tube clogging; too high a rate resulted in the spin solution not forming a fiber but instead an insoluble mass at the bottom of the bath.
28. Once the fibers have formed, they are initially tacky and will accumulate at the bottom of the bath causing difficulty in removing single fibers. To avoid this, use the side of the tweezers to pull the fibers as they are coming out of the spinneret toward the rear of the bath. This not only avoids agglomeration but creates long fibers.
29. When hand drawing the as-spun fibers with tweezers, ensure all handling of the fiber is at each end to minimize introduction of defects along the fiber. Drawing is most easily accomplished by holding one end still and applying a slow, steady pace when pulling the fiber at a predetermined length by the other end. In instances where significant draw (greater than fourfold) is possible, exchange the ends that are held still to facilitate even stretching.
30. After several uses, the methanol concentration will decrease due to high volatility. Weigh an aliquot and determine the concentration by specific gravity. Add additional methanol to bring the solution to the desired concentration.
31. When using tweezers, ensure all handling of the fiber is at the ends to minimize introduction of defects along the fiber. Also, recommend keeping fiber relaxed during transfer.
32. Single fibers are very difficult to visualize especially under normal room lighting. To facilitate mounting of the fibers to glass slides, recommend placing glass slide in a large petri dish, which is placed on top of black construction paper. This will also assist in preventing loss of fibers.
33. Minimize room air turbulence by shutting windows, turning off heating/cooling units and preventing movement by others

in the vicinity of the work area. Single fibers are difficult to handle and the slightest air flow can cause fiber breakage and subsequent loss of fiber. Also, minimize distance to glass slide by having a centralized location with all supplies in close proximity to each other. By handling the fiber at both ends, loss can be minimized.

34. Especially for first-generation fibers when brittleness is anticipated, use tape for fiber transfer as opposed to tweezers. In the original petri dish, place small pieces of tape on both ends of the fiber and transfer to glass slide securing fiber with tape already on fiber ends. In cases in which tape is difficult to remove from dish or slide, release tape ends using tweezers.

35. Mark on glass slides the location of defects and inconsistent diameter (indicative of uneven draw). If possible, avoid placing these sections in the mechanical testing area since these portions of the fiber will be more susceptible to breakage.

36. Using the red plate, birefringence will be more easily visualized; however, a red plate is not necessary especially for highly birefringent fibers. Birefringent behavior is typically represented by a blue color with instances of yellow color either in unevenly stretched areas or along the edges of the fiber.

37. The number of readings will be based on the length of the fiber and should be sufficient to estimate the average diameter of the entire fiber. For example, a one inch (2.54 cm) fiber should have a minimum of five diameter readings for averaging.

38. Average diameter reading should be recorded that will be representative of the mechanical testing area. Therefore, if areas of the fiber will not be in the testing area due to defects or inconsistent stretching, then do not include these areas in the diameter approximations. For a representative example of an inconsistently stretched region that you would not include in diameter approximations or mechanical testing area, see Fig. 6.

39. Alternatively, VIBROMAT M could be used for a direct linear density determination. Measurement is based on vibration using the technology of frequency interval method with acoustic excitation of the fiber. The fiber is placed in the specimen holder and pre-tension weights are applied. VIBROMAT range = 15–200 denier.

40. Scale bar will provide diameter in mm, so conversion to cm will be required. For all single silk fibers, specific gravity was approximated as 1.35.

41. One piece of the conductive carbon tape is placed on the side wall and another piece on the top flat part of the stub, in order to image the sample in both planar and cross-sectional views using the same stub.

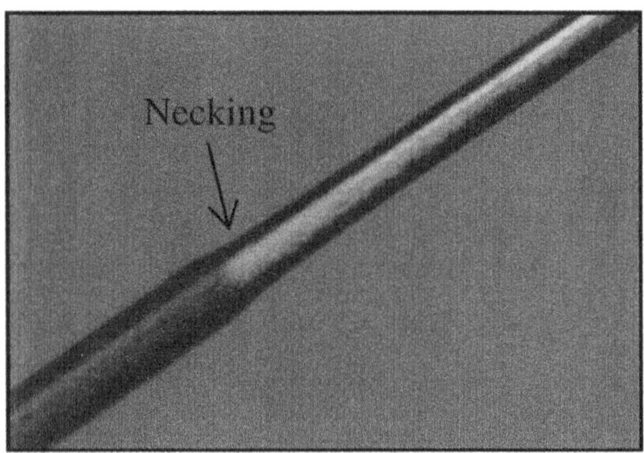

Fig. 6 Polarizing light microscopy image of ADF-3 silk fiber. Fiber defects, such as inconsistent stretching that occurred during drawing, must be identified and marked accordingly to avoid inclusion in both diameter approximations and the mechanical testing area. The necking region of this fiber would be more susceptible to breakage during mechanical testing

42. Sample gauge length is constant at 0.5 in. (12.7 mm) regardless of the length of the fiber. This is the portion of sample that will undergo the mechanical load required for testing. Typical fibers are 6–12 in. (15.2–30.5 cm) in length; therefore, up to 5–6 specimens from a single fiber can be analyzed for averaging of mechanical properties.
43. As previously discussed, the testing area should consist of minimal defects and a known linear density.
44. Draw has been seen as the predominant variable in influencing mechanical properties (9); however, the ability to draw a fiber is dependent on the solution and spinning variables.
45. This should only be done on select fibers that exhibit extensive elongation during mechanical testing prior to breaking. Microscopy of fibers after testing may reveal regions of the fiber that are susceptible to further draw. This information is important for subsequent optimization since increasing draw has demonstrated increasing mechanical integrity. For example, a twofold drawn fiber that exhibits 50–60% elongation during mechanical testing may be susceptible to further draw if the elongation is not associated with the break point. Additional draw would decrease elongation and increase tensile strength.

References

1. Vollrath F, Knight DP (2001) Liquid crystalline spinning of spider silk. Nature 410:541–548
2. Vollrath F, Madsen B, Shao ZZ (2001) The effect of spinning conditions on the mechanics of a spider's dragline silk. Proc R Soc Lond B Biol Sci 268:2339–2346
3. Shao ZZ, Vollrath F (2002) Materials: Surprising strength of silkworm silk. Nature 418:741–741
4. Jin HJ, Kaplan DL (2003) Mechanism of silk processing in insects and spiders. Nature 424:1057–1061

5. Seidel A, Liivak O, Jelinski LW (1998) Artificial spinning of spider silk. Macromolecules 31:6733–6736
6. Seidel A, Liivak O, Calve S, Adaska J, Ji GD, Yang ZT, Grubb D, Zax DB, Jelinski LW (2000) Regenerated spider silk: Processing, properties, and structure. Macromolecules 33:775–780
7. Trabbic KA, Yager P (1998) Comparative structural characterization of naturally- and synthetically-spun fibers of *Bombyx mori* fibroin. Macromolecules 31:462–471
8. Liivak O, Blye A, Shah N, Jelinski LW (1998) A microfabricated wet-spinning apparatus to spin fibers of silk proteins. Structure-property correlations. Macromolecules 31:2947–2951
9. Lazaris A, Arcidiacono S, Huang Y, Zhou JF, Duguay F, Chretien N, Welsh EA, Soares JW, Karatzas CN (2002) Spider silk fibers spun from soluble recombinant silk produced in mammalian cells. Science 295:472–476
10. Arcidiacono S, Mello CM, Butler M, Welsh E, Soares JW, Allen A, Ziegler D, Laue T, Chase S (2002) Aqueous processing and fiber spinning of recombinant spider silks. Macromolecules 35:1262–1266
11. Fahnestock S (2001) Recombinantly produced spider silk , 31 Jul 2001. US patent 6,628,169

Chapter 4

Fibrous Protein Nanofibers

Jeffrey E. Plowman, Santanu Deb-Choudhury, and Jolon M. Dyer

Abstract

One of the promising new techniques in the production of biomaterials is the electrospinning process, whereby fibers of uniform thickness down to the nanoscale can be produced from solutions of polymeric material in a high electric field. At the same time there has been increasing interest in the manufacture of biodegradable nanomaterials from nonfood sources and this has led to investigations into the use of proteins such as collagen, keratin, and fibroin. Explorations into the use of these proteins in the generation of mats suitable for filtration purposes or scaffolds with applications for tissue engineering form the subject of this review.

Key words Collagen, Keratin, Fibroin

1 Introduction

With the increasing interest in the development of biopolymers from renewable sources, attention is turning to the use of protein materials from nonfood sources. Included among these are keratin proteins from materials such as wool, feathers and horns, and fibroin fibers from the pupae of the silk moth, *Bombyx mori*. Additionally, while there are ethical issues relating to the conversion of food materials into biopolymers, investigations have been made into the use of collagen for this purpose, because its primary source, the connective tissue of mammals, is a by-product of the meat processing industry. Recently, attention has focused on utilizing these materials to generate nanofibers, for which there is a wide range of applications including the fabrication of mats for filtration and scaffolds for tissue engineering. Nanofibers are useful for mat fabrication and scaffolds because of their high surface area, which results from their small diameter. These properties can result in highly porous nanofiber mats with excellent interconnection between the pores. Nanofibers are particularly suited for tissue scaffolds in tissue engineering due to their high surface area and porosity. A number of approaches have been made to manufacture

Juliet A. Gerrard (ed.), *Protein Nanotechnology: Protocols, Instrumentation, and Applications*, Methods in Molecular Biology, vol. 996, DOI 10.1007/978-1-62703-354-1_4,

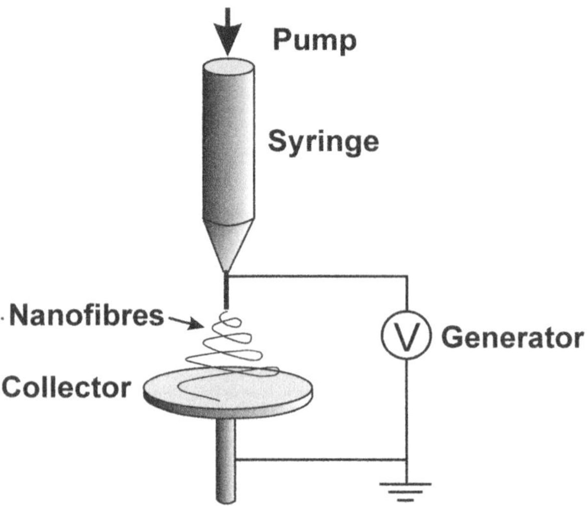

Fig. 1 A basic electrospinning apparatus where the polymer solution or melt is charged to a high potential and then directed towards a grounded target, in this case a spinning disk. As the polymer solution is driven across the gap the carrier solvent evaporates resulting in the production of fine fibers that are deposited on the target

nanofibers, including self-assembly, phase separation, and melt-blown techniques but the most promising technique for producing them on a large scale is electrospinning (1) as it is a relatively easy and fast process, and the fiber diameter can be adjusted from micro- down to nanometers.

Electrospinning was developed in the first half of the twentieth century (2, 3), a schematic representation being shown in Fig. 1. It relies on the fact that when a sufficiently high electric voltage (typically of the order of 10–20 kV) is applied to a liquid droplet (from a polymer solution drawn into a spinneret) it becomes charged. As it does so, the electrostatic repulsion works against the surface tension until, at a critical point known as the Taylor cone, the electrostatic repulsion of the polymer solution exceeds the surface tension at the spinneret tip, forming an electrically charged fine jet of entangled polymer chains. This stream will remain intact so long as the molecular cohesion in the liquid stream is high enough, and as the solvent evaporates in flight, the charge migrates to the surface of the fiber. Electrostatic repulsion develops at small bends in the fiber, leading to elongation and thinning via a whipping process until it is deposited on a collector plate (usually a rotating disk 15–20 cm away). This results in fibers of a uniform thickness ranging from 50 nm to 10 μm in diameter depending on the polymer solution and the solvent used.

In this chapter we will review the application of electrospinning to the production of nanofibers utilizing collagen, keratin, and fibroin. Included in this survey will be details of protein extraction from the base materials, production of nanofibers both with and without the inclusion of other polymeric material, and issues of biodegradability and potential applications.

2 Fibrous Proteins

2.1 Collagen

Collagen is the most abundant protein and is the primary structural component of the extracellular matrix (ECM) of vertebrates. The ECM is a complex structural component found within mammalian tissues that plays a vital role in the structural integrity of various tissues and organs. It not only provides support but also acts as a scaffold for arranging cells within connective tissues and is a dynamic entity that defines tissue function and cellular behaviour (4). Collagen plays a vital role in the repair and restoration of the structure and function of tissue following injury. However, a balance is necessary between excessive collagen deposition at the site of injury, resulting in fibrosis and loss in function, to insufficient deposition, making the repair process weak and suboptimal (5).

The collagen family constitutes over 44 different gene products, giving rise to at least 26 genetically distinct types of collagen. Most of these collagens associate either with one another or with other proteins in the ECM. Collagens exist in various morphological forms ranging from complex interwoven structures to long fibrillar forms and is characterized by its unique amino acid composition. It consists of three polypeptide chains forming a triple helical structure with a repeating motif at 67 nm. The 67 nm interval imparts a characteristic banding pattern to the molecule. The primary sequence of amino acids is characterized by Gly-Xaa-Yaa repeats, where every third amino acid is a glycine. Proline typically occupies the Xaa position and approximately every seventh position of Yaa is occupied by 4-hydroxyproline, formed as a result of post-translational modification of the peptide-bound prolyl residue.

Collagen molecules require the correct alignment of three polypeptides to form the characteristic triple helical structure of the protein (6). The three polypeptide chains are called α-chains and are either identical (homotrimeric collagens) or genetically distinct (heterotypic collagens). Fibrillar collagen types I, II, and III are the most abundant and in association with small quantities of collagen V and nanosized hydroxyapatite (HA) crystals, form the framework for bone tissue. Collagen type I is also predominantly present in tendons, blood vessels, cornea, dentin, skin, and muscle. Collagen types III and V are present in the skin and are primarily responsible for its tensile strength. Collagen type III is a homotrimer consisting of three α1-chains, and is also present in ligaments and blood vessels. This type of collagen is expressed during the initial wound healing process. Non-fibrillar collagens such as type IV associate with fibrillar collagens, forming microfibrils and network structures, and are a major component of the basal lamina and basement membranes.

2.2 Keratin

Keratin proteins are ubiquitous in nature, two types generally being known, the soft keratins and the hard keratins. The former are found in the *stratum corneum* of the skin while the latter form the major component of wool, hair, feathers, nails, and horns of mammals, birds, and reptiles. The cysteine content of the former group is low, generally less than 3% (7), while the latter group of keratins is noted for its high cysteine content, ranging from 5 to 35%. Four different classes of keratins from these fibers have been identified, with the major component being the low sulfur or intermediate filament proteins, having a cysteine content of 4–6% and which have molecular weights ranging from 45 to 60 kDa (8). In addition to these there are two classes of high sulfur proteins, based on whether their cysteine content is greater or less than 30 moles %, with molecular weights ranging from 10 to 35 kDa. There is also a class of proteins high in glycine and tyrosine, with molecular weights ranging from 7 to 9 kDa (9). The intermediate filament proteins are highly structured proteins, the central region characterized by a heptad repeat moiety, and are noted for their high degree of α-helicity, while the keratin-associated proteins are noted for their lack of structure (10). The keratins further assemble into coiled-coil structures, stabilized by hydrophobic interactions and salt bridges (10). In contrast, the keratin-associated proteins are thought to have a largely amorphous structure, though structural studies have suggested that some high sulfur proteins form strings of disulfide bond-stabilized pentapeptide loops (11). Hard keratins owe their high stability to the preponderance of these cysteines, which through intra- and intermolecular bonding gives rise to a compact three-dimensional structure (10).

2.3 Fibroin

Silk fibroin belongs to the fibrillar or fibrous protein class (12). It is comprised structurally of macrofibrils with a width up to 65 μm. These macrofibrils are constructed from helically packed nanofibrils 90–170 nm in diameter. The silk fibers produced by the widely cultivated *Bombyx mori* mulberry silkworm are made up mainly from two proteins, fibroin and sericin. Two fibroin threads are adhered together with sericin gum, with the resultant single thread about 10–25 μm in diameter (13). The fibroin primary structure is dominated by four amino acid residues—glycine, alanine, serine, and tyrosine—that combine to make up approximately 90 mol% (14). Consequently, fibroin consists of approximately 75% hydrophobic amino acid residues.

Fibroin secondary structure is characterized by three major structural types: α- helical (silk I) and ß-folded (silk II) structures within the crystalline areas, and disordered globules within the amorphous areas (15). This secondary structure is stabilized by various kinds of interactions, notably hydrogen bonding.

Fibroin proteins produced by the silkworm are a natural polymer that is utilized worldwide as a textile fiber (16). Materials constructed from silk fibers are characterized by high durability and comfort (15). During the processing of silk fibers made by cultivated *Bombyx mori* mulberry silkworms, a large amount of waste material is generally produced, including from fibers that are difficult to extract from the cocoon. This material represents a potentially valuable stream for the production of protein biomaterials.

3 Collagen Nanofibers

Favorable properties of collagen such as low antigenicity, water affinity, and the ability to promote tissue regeneration, dissipate energy, and transmit forces contribute towards the structural integrity and biological function of the ECM. These characteristics make collagen an ideal biopolymer for tissue engineering applications. Apart from being bio-resorbable, collagen fibers can also prevent premature mechanical failure and provide signals to adjacent cells for functional responses (17).

The ideal goal of tissue engineering is the development of three-dimensional scaffolds with appropriate biomechanical and biological properties for restoring both the structural integrity and the remodelling process of native ECM, vital for promoting tissue regeneration. Scaffold properties require careful optimization for porosity and pore connectivity and must be mechanically stable to protect cells from factors such as mechanical overload. Apart from being bio-resorbable and biodegradable, scaffolds should also have the ability to provide sufficient nutrient supply and have appropriate strength and viscoelasticity. Collagen scaffolds exhibit some of these ideal properties in addition to being noninflammatory and providing a large surface area for cell attachment, supporting the vascularization process. It has been reported that collagen type I does provide extracellular support and assist in organizing cells in a three-dimensional lattice in vitro (18).

The electrospinning process used in tissue engineering applications can potentially produce collagen fibers closely mimicking the biological and structural properties of the natural collagen ECM. While there is much potential for the electrospinning of biopolymers, a reproducible process creating ideal fiber diameter and pore size still needs to be realized. Lack of reproducibility in fiber diameter and pore size can adversely affect the reconstituted ECM in its essential temporal and spatial complexity. The biocompatibility of these tissue scaffolding materials is dependent on their surface chemistry, which in turn is influenced by their material properties. To achieve optimal scaffold properties, modifications of nanofibers could be a prerequisite in tissue engineering such as polymer blending of collagens with synthetic polymers, cross-linking to increase mechanical integrity and coating or grafting of proteins (19–22).

Collagens have been used in the development of bioengineered tissues, such as vascular grafts, heart valves, and ligaments. Nanofibers formed from collagen types I and III dissolved in 1,1,1,3,3,3 hexafluoroisopropanol (HFP) have been shown to exhibit a linear correlation between the polymer concentration and the fiber diameter. Scaffolds composed of fibers ranging from 100 nm to 5 μm in diameter have been produced from collagen type I solutions varying in concentration from 0.03 to 0.1 mg/ml of HFP. Electrospinning of type IV has also been attempted and fibers obtained with diameters ranging from 100 nm to 2 μm (23).

It has been suggested that collagen dissolved in HFP, a strong organic solvent, may result in structures composed merely of gelatin fibers (24). However, studies also revealed that only minor differences were observed between the structures of electrospun collagen dissolved in HFP, in PBS, or in PBS/ethanol binary mixtures, indicating that the structural properties of collagens are retained even with the use of this strong solvent. Furthermore, distinct morphological differences were observed between scaffolds formed from electrospun collagen and electrospun gelatin. Electrospun gelatin fibrils exhibit an amorphous structure, whereas the 67 nm banding pattern typical of collagen fibrils is only seen on electrospun collagen. A difference was also observed in the average pore size in collagen, ranging from 1,500 to 4,000 nm^2, compared to 2,000–6,000 nm^2 observed in gelatin. The in vivo behavior of these two structures was also different—when a collagen type I construct was implanted in the vastus lateralis of rat and recovered after 7 days, a fully integrated construct was revealed, with the surrounding tissue infiltrated with interstitial cells and free from fibrotic encapsulation. A gelatin implant, on the other hand, delaminated from the host tissue and developed fibrotic capsules with lymphocyte infiltration (25).

While electrospun collagen serves as a favorable biopolymer, like other protein-based nanofibers, its weakness is its lack of mechanical integrity upon hydration. Cross-linkers such as glutaraldehyde have been previously used in successfully increasing its strength, but the risk of cytotoxicity and calcification in vivo is of major concern with this type of compound.

To increase the mechanical stability while maintaining a high level of collagen bioactivity, biodegradable synthetic polymers have also been incorporated in the electrospun collagen scaffolds. Polylactic acid (PLA) has been used to create scaffolds for vascular engineering in combination with collagen. Polycaprolactone (PCL)/collagen type I composite scaffolds with high tensile strength (4.0 ± 0.4 MPa) and elasticity (2.7 ± 1.2 MPa) have also been used in tissue engineering to resist high degrees of pressurized flow over extended periods of time (26, 27). Scaffolds generated from PCL/collagen composites have been successfully seeded with smooth muscle cells, which show regularity in their alignment, improving the speed with which grafts can be incorporated into the vasculature. Electrospun PCL and collagen type I and II blends have also been used to form nanofibers with diameters in the range of 210–225 nm and have demonstrated a tensile modulus of 18 MPa with a tensile strength of 7.79 MPa that is ideal for blood vessel conduits (22). Collagen-blended polymers have also been used to enhance the attachment and spreading of human coronary artery endothelial cells and preserve their phenotype and viability. A three-layered mesh composed of type I collagen, styrenated gelatin, and segmented polyurethane, using multilayering electrospinning, has been used to provide compliance matching of native arteries and vascular grafts (21, 28).

Collagen plays an important role in these polymer blends by increasing the biocompatibility of the scaffolds. Though coating and grafting of proteins have been trialled on nanofibrous structures, these demonstrate a slow mass transfer of proteins into the porous structure of the scaffolds. Electrospun-blended collagen mixtures avoid the slow mass transfer process, and the presence of collagen on the surface and within the scaffold provides continuous cell recognition signals crucial for cell function and proliferation (29).

Collagen and elastin, being highly abundant proteins, have been blended to form scaffolds for vascular applications. Though these scaffolds mimic native ECM, they lack mechanical viability due to the absence of any synthetic polymer in the blend. Combining synthetic polymers with collagen and elastin has introduced desirable characteristics such as sustained mechanical integrity and bioactivity. Nanofiber scaffolds obtained using 40% collagen type I, 15% elastin, and 45% polylactide-*co*-glycolide (PLGA) demonstrated a uniform distribution of collagen and elastin throughout the scaffold. These scaffolds demonstrated biocompatibility in vivo when implanted subcutaneously in mice, no systemic or neurological toxicity, and minimal inflammatory response. The incorporation of biodegradable synthetic polymers such as PLGA, PCL, and poly(d,l-lactide-*co*-ε-caprolactone) (PLCL) with collagen/elastin blends demonstrated improved mechanical properties and biocompatibility. The addition of synthetic polymers was shown to be necessary for sustained stability of these blends, as without these the collagen/elastin blends completely dissolved after 28 days in culture medium at 37°C (27).

Collagen plays a vital role in maintaining the biological and structural integrity of the ECM. The morphological and functional versatility demonstrated by collagen makes it an ideal candidate for creating electrospun biopolymers. The electrospinning technique permits mimicking of the architecture of the native ECM to a great extent, though ideal scaffolds are yet to be created. Challenges yet to be met include creating reproducible scaffolds with ideal fiber and pore sizes. More knowledge is required to understand the behavior of these biopolymers in vivo, before they can become fully applicable in medical science. Though novel synthetic biomaterials have been developed resembling the bioactivities and structure especially of collagen type I, they remain oversimplified mimics. They do, however, represent advances in material engineering and provide valuable information into the biomaterial-cellular interactions. Further refinements are required to optimize the fabrication and application of these biomaterials for satisfactory interaction and imitation of biological functions. Knowledge of polymers has led scientists to create electrospun structures but a thorough knowledge of the remodelling of the biopolymers in vivo is required. A balance thus needs to be struck between the remodelling and integration of these electrospun structures at an accelerated pace without any adverse effects and premature degradation in vivo.

4 Keratin Nanofibers

One of the drivers for the use of keratin in the manufacture of nanomaterials has been an increasing interest in the use of natural polymers because of policies promoting the use of renewable resources, particularly those that are biodegradable. Keratin is a renewable raw material that may be sourced as by-products of the textile industry or as poor-quality wools and feathers from slaughterhouses. Estimates have put the total keratin obtained worldwide from the poultry industry alone at more than four million tons per year (30). Disposal by burning for fuel is not an option because it is inefficient and the high sulfur content could lead to a substantial pollution hazard. There are also obvious advantages for its use as a biomaterial such as its biocompatibility and biodegradability.

Two potential uses of keratin-based nanofibers exist, that of filters for heavy metals or volatile organic compounds, and of scaffolds for tissue engineering. A disadvantage of keratin for biomaterial production is the relatively low molecular weight range of its component proteins, meaning that they have poor mechanical properties and need to be blended with other polymeric material using a common solvent system (31). To overcome this, investigations have been conducted into combinations of keratin with a variety of other biocompatible polymeric materials. Possible contenders investigated were polyamide 6 (PA6), for active filtration of air and water, and polyethylene oxide (PEO), fibroin (32) and polyhydroxybutyrate-*co*-hydroxyvalerate (PHBV) (33), and PLA (1) for biomedical applications, both PHBV and PLA having the advantage that they are also biodegradable.

To determine the best mixture of polymer to protein in the preparation of nanofiber mats, aqueous keratin solutions obtained via the process of oxidative sulfitolysis (34) were mixed with PEO (molecular weight, 40,000 Da) to obtain the following keratin:PEO blends: 10:90, 30:70, 50:50, 70:30, and 90:10. Pure keratin solutions were not used as it was discovered that a 7 weight% aqueous solution turned into a gel within a matter of a few hours (31). Scanning electron microscopy (SEM) analysis showed that the 90:10 keratin:PEO blend produced the finest fibers of around 95 ± 20 nm, but droplets and bead-like effects were evident among them. This was thought to be due to the low viscosity of the solution giving rise to a break in the jet at the capillary tip. This effect disappeared once the proportion of PEO had increased above 10 weight% and was thought to be because there were sufficient molecular chain entanglements to prevent the breakage in the electrically driven jet. Nevertheless, it was apparent that nanofibers produced from solutions rich in keratin were observed to be more homogeneous that those rich in PEO. Moreover, the composition of the blend appears to have a direct effect on the mean diameter of the nanofibers, with this decreasing from 268 to 95 nm as the keratin composition increases from 10 to 70%. This is the result of two things, the lower viscosity of the solution, and the higher charge density carried by the jet that promoted a stronger whipping motion and enhanced filament stretching.

A similar set of blends for keratin:PA6, dissolved in concentrated formic acid, was also evaluated. Again, the PA6 and keratin:PA6 blends were found to be defect free, while the pure keratin produced many bead defects (35). There was no significant difference in the mean diameter of the nanofibers produced with this polymer, though those rich in PA6 appeared to be more homogeneous than those higher in keratin composition. The nanofiber structure in the mats high in keratin was almost totally destroyed after soaking for 24 h in water, those composed of 30:70 keratin:PA6 blend were found to start to swell and become flat, and the pure PA6 and keratin:PA6 10:90 blends were unaffected.

The problem of bead defects in pure keratin nanofibers was also observed when keratin extracted with a solution of urea, sodium dodecyl sulfate, and mercaptoethanol and alkylated with iodoacetic acid (IAA) was blended with PHBV to produce mixes of either 0:10, 3:7, 7:3, or 10:0 keratin:PHBV and electrospun from 6 weight% solutions in HFIP (34). Once again there was a decrease in the mean diameter of the nanofibers with increasing keratin content, ranging from 487 ± 161 nm, through to 720 ± 124 and 815 ± 98 for the keratin:PHBV 7:3, 3:7, and 0:10 mixes, respectively. As the mats readily dissolved in water investigations were conducted into introducing cross-links by treating them in a glutaraldehyde-vapor-saturated chamber, with the unreacted aldehyde groups blocked in aqueous glycine solution. The loss of weight of the mats in water was shown to decrease steadily with the increase in time of glutaraldehyde treatment, finally levelling out after the cross-linking had been carried out for 7 h. In vitro enzyme degradation studies were carried out using trypsin or *Pseudomonas stutzeri* BM190 depolymerase to determine the propensity of the mats to biodegrade after use. After 12 h of digestion with depolymerase, the mats had become rather brittle, with only large chunks of degraded material remaining. In contrast, the mats were almost unaffected by treatment with trypsin over the same time period, and after 24 h only some fibers showed evidence of having broken down.

IAA-alkylated keratin was also blended with PLA, and nanofibers were electrospun from different blends of these two components dissolved in HFIP to a final concentration of 15 weight%. Only one ratio of keratin:PLA of 2:8 was actually reported as this mixture produced fine fibers of around 500 nm (1). It was found that when the keratin content was high, the nanofibers were rough with a few defects. The mats had poor mechanical properties and the nanofibers lost their morphology, becoming swollen and partly dissolved on coming in contact with water. Biodegradability was also assessed by tryptic digestion of the mats: 12 h was found to be sufficient to erode some to the point where they started to break up, while others completely disappeared. Thus it would appear that keratin/PLA mats can be readily biodegraded with enzymes like trypsin.

The ability of the keratin-based nanofibers to reduce heavy metal contamination was also assessed. Keratin:PA6 nanofiber mats electrospun onto polyester plates were found to readily absorb Cr^{3+} (the concentration of it on the mats increasing with increasing keratin concentration), while the concentration of chromium on the mats was found to be generally higher than that on films produced from the same mixture of keratin and PA6 (35). When spun onto the surface of a polypropylene filter, it was found that the keratin:PA6 nanofibers improved the effectiveness of the nanofiber by reducing the formaldehyde concentration by up to 70%, whereas the polypropylene sheet on its own only reduced it by 30% and the PA6 nanofibers by 40% (36).

To assess the utility of keratin nanofiber mats as scaffolds for tissue engineering keratin:PHBV nanofiber mats were seeded with penicillium G-streptomycin cells and monitored by SEM (34). It was evident that the ability of the cells to adhere to the mats was directly related to the proportion of keratin in the mixture, cell proliferation being significantly greater on the pure keratin fibers than on the pure PHBV or 3:7 keratin:PHBV fibers. Furthermore, it was apparent that the fibers provided a guide to the development of cell growth. Cells initially grew along the direction of the fiber and then formed a three-dimensional, multicellular network based around the architecture of the nanofiber mat. Over the period of a whole day, the cells continued to grow among the fibers and eventually covered the whole surface of the mat. Likewise, with keratin:PLA composites it was apparent after 4 h that the cells attached to the keratin:PLA fibers and spread largely along the fibers, something not so evident on the PLA mats (1). After 20 h, the proliferation of cells over the keratin:PLA mats was considerably higher than on the PLA mats, while the cell viability on the keratin:PLA mats after 3 days was significantly enhanced over that of the PLA-only mats.

Thus, it would appear that the poor mechanical properties of keratin can be overcome by the incorporation of other polymeric materials to produce robust nanofiber mats or scaffolds. Furthermore by selecting biodegradable polymers environmentally friendly materials can be produced. The advantage of incorporating keratin into these biomaterials is that they are more effective in the removal of volatile organic compounds or heavy metals than the pure polymer alone, while there is much potential in biomedical applications such as wound dressings and in scaffolds for tissue engineering.

5 Fibroin Nanofibers

Silk fibroins have been extensively studied with respect to potential biomedical utilization, due to their good biocompatibility and biodegradability (36). For instance, silk is currently used for surgical sutures (16). Other applications that have been, or are being, investigated include use as a substrate for the immobilization of enzymes (37, 38), as an anti-thrombogenic material (39, 40), in wound dressings (41, 42), and in bone regeneration scaffolds (43).

The first step in preparing silk fibroin nanostructures is fibroin extraction. The formation of nanofibers and films with tailored structural organization is highly dependent on the solvent systems and the conditions of extraction (15). For example, *N*-methylmorpholine *N*-oxide has been identified as a particularly effective solvent in the development of fibroin nanofiber technologies as it allows the complicating salt removal step usually required subsequent to extraction to be eliminated.

After extraction, aqueous fibroin solutions are usually utilized for nanofiber production. A range of processes for producing silk fibroin nanostructures have been reported, with many involving freezing and thawing fibroin aqueous solutions with small quantities of organic solvent (16, 44). Through processing adjustments such as varying fibroin concentration and thawing time, the physico-mechanical properties of the resultant fibroin scaffolds can be controlled, including porosity and pore size.

Since silk fibroins show such excellent promise as materials for biomedical applications, the development of effective protocols for the construction of fibroin nanofibers with tailored properties is a key goal of research and development. Dry fibroin films are usually brittle and lacking mechanical strength (45). There has been significant effort directed at improving the spinnability of silk fibroins, with various blends utilized to achieve this. For instance, PEO has been used with fibroins (46). Porous silk fibroin scaffolds were generated by removing PEO from nanofibrous blend matrices, after crystallization of the fibroin. Chitin/silk fibroin blends also show excellent potential, with a blend containing 75% chitin and 25% fibroin producing promising biomimetic 3D structures.

Silk fibroin/polyvinyl alcohol gels have been investigated, prepared both with air- and freeze drying (45). The air-dried gels displayed higher crystallinity and strength, suitable for mechanical applications, while the freeze-dried gels had a highly porous structure, potentially suitable as a cell culture substrate. The porosity could be varied by systematically changing the freezing temperature.

Proteinaceous biomaterials, such as silk fibroin, are susceptible to physical and chemical degradation during storage; however, this is considerably slowed in the dry state (47). For this reason, lyophilization or freeze drying of fibroins is considered important prior to transport or storage.

6 Keratin: Fibroin Nanofibres

A range of keratin:fibroin mixtures of 15 weight% in formic acid were investigated in the preparation of nanofibers, ranging from pure keratin to pure fibroin as well as the following blends: 10:90, 30:70, 50:50, 70:30, and 90:10 (48). The mean fiber diameter decreased as the proportion of keratin increased, starting with diameters of 945 ± 483 nm for pure fibroin, while very thin and homogeneous fibers were produced from the 90:10 keratin:fibroin and pure keratin fibers (233 ± 57 and 169 ± 49 nm, respectively), though the latter suffered from the development of beads as observed in related studies (31). The finest nanofibers were actually observed with the 50:50 keratin:fibroin blend, where the mean diameter was determined to be 207 ± 66 nm. This was thought to be due to the higher viscosity of this solution. It was also noted that the nanofibers produced from pure fibroin and those of the 10:90 and 30:70 keratin:fibroin blends tended to have a flat cross section (48).

Pure fibroin and 50:50 keratin:fibroin nanofibers, the latter prepared from keratin extracted with performic acid, were found to have average diameters of 270 ± 40 and 280 ± 60 nm, respectively, while the porosity was determined to be between 60 and 80% (49). Some of these nanofiber mats were then soaked in methanol for an hour and dried to prevent shrinkage and then stored in a chamber saturated with formaldehyde vapor for 24 h to allow the formation of cross-links. Some swelling was observed in the presence of methanol but the cross-linking was not found to affect the mean diameter of the fibers. These keratin:fibroin nanofiber mats were found to be far more effective than fibroin alone in removing copper from the solution, a steady state being reached after only ten minutes. This was thought to be due to the fact that keratin has far more hydrophilic residues than fibroin. It also proved to be easy to desorb the copper from the mats by pumping 1 mM ethylenediaminetetraacetic acid through them, a period of 30 min proving to be sufficient. Furthermore the cycling efficiency was found to be as high as 97% over seven adsorption desorption cycles.

As an alternative, three-dimensional porous keratin:fibroin scaffolds were fabricated by collecting the nanofiber dispersions in a methanol bath (50). After this, the methanol was replaced by 1,4-dioxane in which 300–440 μm NaCl particles were stirred before pouring into a cylindrical glass vessel. The solution was lyophilized and cross-linked with glutaraldehyde in a sealed chamber over a 24-h period. Subsequent to this the scaffolds were treated with a 2.45 GHz waveguide-based, microwave-induced argon plasma system at atmospheric pressure. The keratin:fibroin scaffolds were found to have a uniform distribution of pores and a porosity of 91%, the pore sizes being of the order 390–500 μm (50). The plasma treatment was found to increase the surface roughness and pore depth of the scaffolds without adversely affecting the morphology of the nanofibers themselves. The main difference observed was that inverted cone-shaped pores of the scaffolds became more cylindrical as a result of plasma treatment.

Neonatal human knee articular chondrocytes (nHAC-kn) were seeded onto the keratin:fibroin scaffolds both before and after treatment with plasma. Following incubation for 4 h, attachment to the scaffolds was found to be higher on the plasma-treated samples, though the result was not statistically significant (50). After incubation for 1, 3, and 7 days the proliferation rates were not significantly increased by the plasma treatment; however, the rate of cell growth on the plasma-treated scaffolds was observed to increase markedly after 3 and 7 days, something not observed in the non-plasma-treated group. This was reflected in SEM images that showed only a partial coverage of the non-plasma-treated scaffolds by the cells; furthermore these cells were forced to alter their shape to fit on the scaffold. In contrast, the nanofibrils on the plasma-treated scaffolds formed a single layer of highly organized patterns and shapes while maintaining the original morphology of the cells. This suggests that the enlargement of the pores provided a larger space for the cells to grow and also made cell migration easier. The plasma treatment was also thought to aid cell adhesion to the scaffolds by increasing the hydrophilicity of the scaffolds.

Both keratin and fibroin have poor mechanical properties; yet when mixed together they can be electrospun with relative ease and a more mechanically stable nanomaterial produced by applying cross-links with formaldehyde or glutaraldehyde. Thus stabilized, the resulting keratin:fibroin mats have proved to have potential in the areas of filtration for the removal of environmental contaminants such as heavy metals from water, while the scaffolds, after plasma treatment, appear to have good prospects in areas of tissue engineering, though what is still needed are in vivo studies and clinical trials in human patients to test the relevance of these studies.

7 Conclusions

With the increasing interest in the development of biodegradable nanofibers for environmental and biomedical applications, the utilization of fibrous proteins in these materials shows a lot of promise. Fibrous proteins such as collagen, fibroin, and keratin exhibit poor mechanical properties when used alone, being brittle or, as in the case of keratin, resulting in nanofibers with numerous defects. However, the application of cross-linking agents like glutaraldehyde and formaldehyde or mixing with biodegradable synthetic polymers has resulted in the production of nanofibers that have proved robust enough in their proposed application areas. Furthermore, by keeping the proportion of fibrous protein high it has been possible to produce finer nanofibers than would be possible with the pure polymer they were mixed with. Naturally enough, the incorporation of protein material into these fibers is important when it comes to the issue of biodegradability, but it also has other benefits. Not only are these nanofibers better at absorbing heavy metals or volatile organic chemicals such as formaldehyde, but they also form better scaffolds for the adhesion of cells, considerably enhancing cellular proliferation over the surface of the scaffold. They also appear to integrate better into living tissues. Thus the incorporation of fibrous proteins into nanofibers has great potential in the development of filters for removing environmental contaminants or tissue and bone implants.

Acknowledgments

We wish to thank Anita Grosvenor for her assistance with proofreading this chapter and Carol Thomas for carrying out the literature search.

References

1. Yuan J, Shen J, Kang I-K (2008) Fabrication of protein-doped PLA composite nanofibrous scaffolds for tissue engineering. Polym Int 57:1188–1193
2. Formhals A (1932) Improvements in or relating to processes and apparatus for the production of artificial filaments. United Kingdom Patent 364,780
3. Taylor G (1969) Electrically driven jets. Proc R Soc Lond A Math 313:453–475
4. Aszodi A, Legate KR, Nakchbandi I et al (2006) What mouse mutants teach us about extracellular matrix function. Ann Rev Cell Dev Biol 22:591–621
5. Beckman MJ, Shields KJ, Diegelmann RF (2004) Collagen. In: Encyclopedia of biomaterials

and biomedical engineering. Marcel Dekker, New York, pp 324–334

6. Hulmes DJS (2002) Building collagen molecules, fibrils, and suprafibrillar structures. J Struct Biol 137:2–10
7. Fraser RDB, MacRae TP, Rogers GE (1972) Keratins: their composition, structure and biosynthesis. Charles C Thomas, Springfield, IL
8. Powell BC, Rogers GE, Jollès P et al (1997) The role of keratin proteins and their genes in the growth, structure and properties of hair. Formation and structure of human hair. Birkhäuser, Basel, pp 59–148
9. Fratini A, Powell BC, Rogers GE (1993) Sequence, expression, and evolutionary conservation of a gene encoding a glycine/tyrosine-rich keratin-associated protein of hair. J Biol Chem 268:4511–4518
10. Parry DAD, Steinert PM (1995) Intermediate filament structure. Springer, Heidelberg
11. Parry DAD, Smith TA, Rogers MA et al (2006) Human hair keratin-associated proteins: Sequence regularities and structural implications. J Struct Biol 155:361–369
12. Finkel'shtein AV, Ptitsyn OB (2002) Fizika belka (protein physics). Universitet Moscow, Moscow
13. Lin F, Li Y, Jin J et al (2008) Deposition behavior and properties of silk fibroin scaffolds soaked in simulated body fluid. Mater Chem Phys 111:92–97
14. Kaplan D, Adams WW, Farmer B et al (1993) Silk: biology, structure, properties, and genetics. In: Kaplan D, Adams WW, Farmer B, Viney C (Eds) Silk polymers, vol 544. American Chemical Society, pp 2–16
15. Sahina ES, Bochek AM, Novoselov NP et al (2006) Structure and solubility of natural silk fibroin. Russ J Appl Chem 79:869–876
16. Tamada Y (2005) New process to form a silk fibroin porous 3-D structure. Biomacromolecules 6:3100–3106
17. Kolacna L, Bakesova J, Varga F et al (2007) Biochemical and biophysical aspects of collagen nanostructure in the extracellular matrix. Physiol Res 56(Suppl 1:S):51–60
18. Kemp PD (2009) Tissue engineering and cell-populated collagen matrices. In: Ram SEAVV (Ed) Methods in molecular biology. PSI Publications, Washington, DC, pp 363–370
19. Stitzel J, Liu J, Lee SJ et al (2006) Controlled fabrication of a biological vascular substitute. Biomaterials 27:1088–1094
20. Barnes CP, Pemble CW, Brand DD et al (2007) Cross-linking electrospun type II collagen tissue engineering scaffolds with carbodiimide in ethanol. Tissue Eng 13:1593–1605
21. Barnes CP, Sell SA, Boland ED et al (2007) Nanofiber technology: designing the next generation of tissue engineering scaffolds. Adv Drug Deliv Rev 59:1413–1433
22. Venugopal J, Low S, Choon AT et al (2005) Fabrication of modified and functionalized polycaprolactone nanofibre scaffolds for vascular tissue engineering. Nanotechnology 16:2138
23. Matthews JA, Wnek GE, Simpson DG et al (2002) Electrospinning of collagen nanofibers. Biomacromolecules 3:232–238
24. Zeugolis DI, Khew ST, Yew ESY et al (2008) Electro-spinning of pure collagen nano-fibres—just an expensive way to make gelatin? Biomaterials 29:2293–2305
25. Telemeco TA, Ayres C, Bowlin GL et al (2005) Regulation of cellular infiltration into tissue engineering scaffolds composed of submicron diameter fibrils produced by electrospinning. Acta Biomater 1:377–385
26. Lee SJ, Liu J, Oh SH et al (2008) Development of a composite vascular scaffolding system that withstands physiological vascular conditions. Biomaterials 29:2891–2898
27. Lee SJ, Yoo JJ, Lim GJ et al (2007) In vitro evaluation of electrospun nanofiber scaffolds for vascular graft application. J Biomed Mater Res A 83A:999–1008
28. Kidoaki S, Kwon IK, Matsuda T (2005) Mesoscopic spatial designs of nano- and microfiber meshes for tissue-engineering matrix and scaffold based on newly devised multilayering and mixing electrospinning techniques. Biomaterials 26:37–46
29. He W, Yong T, Teo WE et al (2005) Fabrication and endothelialization of collagen-blended biodegradable polymer nanofibers: potential vascular graft for blood vessel tissue engineering. Tissue Eng 11:1574–1588
30. Yin J, Rastogi S, Terry AE et al (2007) Self-organization of oligopeptides obtained on dissolution of feather keratins in superheated water. Biomacromolecules 8:800–806
31. Aluigi A, Vineis C, Varesano A et al (2008) Structure and properties of keratin/PEO blend nanofibres. Eu Polym J 44:2465–2475
32. Tonin C, Aluigi A, Varesano A et al (2010) Keratin-based nanofibres. In: Kumar A (ed) Nanofibers. INTECH, Rijeka, pp 139–158
33. Yuan J, Xing Z-C, Park S-W et al (2009) Fabrication of PHBV/Keratin composite nanofibrous mats for biomedical applications. Macromol Res 17:850–855
34. Thomas H, Conrads A, Phan KH et al (1986) In vitro reconstitution of wool intermediate filaments. Int J Biol Macromol 8:258–264
35. Aluigi A, Vineis C, Tonin C et al (2009) Wool keratin-based nanofibres for active filtration of air and water. J Biobased Mater Bioenergy 3:311–319

36. Park KE, Kang HK, Lee SJ et al (2006) Biomimetic nanofibrous scaffolds: preparation and characterization of PGA/chitin blend nanofibers. Biomacromolecules 7:635–643
37. Miyairi S, Sugiura M, Fukui S (1978) Immobilization of ß-glucosidase in fibroin membrane. Agric Biol Chem 42:1661–1667
38. Demura M, Asakura T (1989) Immobilization of glucose oxidase with Bombyx mori silk fibroin by only stretching treatment and its application to glucose sensor. Biotechnol Bioeng 33:598–603
39. Furuzono T, Ishihara K, Nakabayashi N et al (1999) Chemical modification of silk fibroin with 2-methacryloyloxyethyl phophorylcholine I. Graft-polymerization onto fabric using ammonium persulfate and interaction between fabric and platelets. J Appl Polym Sci 73:2541–2544
40. Furuzono T, Ishihara K, Nakabayashi N et al (2000) Chemical modification of silk fibroin with 2-methacryloyloxyethyl phophorylcholine II. Graft-polymerization onto fabric through 2-methacryloyloxyethyl isocyanate and interaction between fabric and platelets. Biomaterials 21:327–333
41. Sugihara A, Sugiura K, Morita H et al (2000) Promotive effects of a silk film on epidermal recovery from full-thickness skin wounds. Proc Soc Exp Biol Med 225:58–64
42. Minoura N, Tsukada M, Nagura M (1990) Physicochemical properties of silk fibroin membrane as a biomaterial. Biomaterials 11:430–434
43. Kim HJ, Kim U-J, Leisk GG et al (2007) Bone regeneration on macroporous aqueous-derived silk 3-D scaffolds. Macromol Biosci 7: 643–655
44. Lv Q, Feng Q (2006) Preparation of 3-D regenerated fibroin scaffolds with freeze drying method and freeze drying/foaming technique. J Mater Sci 17:1349–1356
45. Li M, Lu S, Wu Z et al (2002) Structure and properties of silk fibroin-poly(vinyl alcohol) gel. Int J Biol Macromol 30:89–94
46. Jin H-J, Fridrikh SV, Rutledge GC et al (2002) Electrospinning Bombyx mori silk with poly(ethylene oxide). Biomacromolecules 3:1233–1239
47. Weska RF, Vieira WCJ, Nogueira GM et al (2009) Effect of freezing methods on the properties of lyophilized porous silk fibroin membranes. Mater Res 12:233–237
48. Zoccola M, Aluigi A, Vineis C et al (2008) Study on cast membranes and electrospun nanofibres made from keratin/fibroin blends. Biomacromolecules 9:2819–2825
49. Ki CS, Gang EH, Um IC et al (2007) Nanofibrous membrane of wool keratose/silk fibroin blend for heavy metal ion adsorption. J Membr Sci 302:20–26
50. Cheon YW, Lee WJ, Baek HS et al (2010) Enhanced chondrogenic responses of human articular chondrocytes onto silk fibroin/wool keratose scaffolds treated with microwave-induced argon plasma. Artif Organs 34: 384–392

Chapter 5

Self-Assembling Nanomaterials: Monitoring the Formation of Amyloid Fibrils, with a Focus on Small-Angle X-Ray Scattering

Elizabeth B. Sawyer and Sally L. Gras

Abstract

Amyloid fibrils are attractive targets for applications in biotechnology. These thin, nanoscale protein fibers are highly ordered structures that self-assemble from their component proteins or peptides. This chapter describes the use of several biophysical techniques to monitor the formation of amyloid fibrils including a common dye-binding assay, turbidity assay, and small-angle X-ray scattering. These techniques provide information about the assembly mechanism, the rate and reproducibility of assembly, as well as the size of species along the assembly pathway.

Key words Light scattering, Turbidity, Aggregation, SAXS, ThT, Peptide, Protein, Cross-β, β-Sheet, Fiber

1 Introduction

The term "amyloid fibril" describes a fiber-like protein assembly in which proteins adopt a β-sheet secondary structure, with stacks of β-sheets arranged perpendicular to the fibril axis forming a cross-β core. These fibrils represent low-energy conformations that are thought to be accessible to all proteins regardless of sequence (1–3).

Amyloid fibrils have a number of features that make them attractive targets for biotechnology: they self-assemble from their component peptides or proteins; they are incredibly strong and robust; and the physicochemical properties of the fibril core and surface can be manipulated with relative ease by changing the sequence of the component protein/peptide. Unlike many globular proteins, they do not denature upon interaction with surfaces and they can adopt a variety of topographies—from straight or twisted fibers to linear lateral assemblies, rings, or spherulites. Further discussion of these properties is outside the scope of this

Juliet A. Gerrard (ed.), *Protein Nanotechnology: Protocols, Instrumentation, and Applications*, Methods in Molecular Biology, vol. 996, DOI 10.1007/978-1-62703-354-1_5, © Springer Science+Business Media New York 2013

chapter, but the reader is directed to a recent review in the Australian Journal of Chemistry (4) for a comprehensive discussion.

Although the cross-β structure of amyloid fibrils is expected to be accessible to all proteins, certain sequences are known to have a particularly high propensity to adopt this conformation. In some cases these sequences have been identified by comparing the protein sequences of healthy individuals with those of individuals suffering from protein misfolding diseases. Differences in a single mutation such as the "arctic" mutation E693G in Alzheimer's Disease (5) or the number of repeated residues, as occurs for glutamine in the poly-Q tract of the Huntington protein (6), greatly affect the age of onset and rate of progression of the disease.

More recently amyloid fibrils have been identified in organisms ranging from bacteria to fungi, for example the curli proteins from *Escherichia coli* (7) and the reproductive prion protein Het-s from the filamentous fungus *Podospora anserina* (8, 9). Amyloidogenic proteins have also been found in barnacle cement (10) and in human cells, where Pmel17 is thought to play a role in melanin biosynthesis (11, 12). The identification of these functional amyloid fibrils suggests that such structures may be amenable to exploitation in biotechnology. Furthermore, examination of the relationship between the sequences and properties of these fibrils informs the design of new sequences for self-assembly.

A number of algorithms have been developed to predict the propensity of certain amino acid sequences to form β-aggregates. These algorithms employ different strategies and a combination of experimental data and modelling to evaluate the likelihood of a given amino acid sequence to form β-aggregates. They consider parameters such as hydrophobicity, charge, secondary structure propensities, aromatic residues, the accessible surface area of side chains, and the energetic cost of fully burying residues in the fibril core (desolvation). Zyggregator, TANGO, and AGGRESCAN are three examples of such algorithms; the reader is directed to two reviews for an evaluation of these algorithms and other predictive tools (13, 14).

Several options exist for the generation of amyloidogenic proteins and peptides. In many cases these proteins can be obtained by purification from their natural source—such as lysozyme from hen egg white or insulin from cattle. Recombinant expression in bacterial or yeast cells has also been employed to produce large quantities of protein, although in some cases purification is limited by the formation of insoluble inclusion bodies. Smaller peptides may be expressed as recombinant fusion proteins and yields of 4 mg/l of culture have been obtained for the $TTR_{105-115}$ peptide following cleavage and purification by high-performance liquid chromatography (15). The authors of this study suggest the yield of peptide could be further improved by optimization of the conditions used for fermentation.

Alternatively, peptides may be obtained by using FMOC or BOC solid-phase synthesis. This method is used for many *de novo* designed peptides, although highly hydrophobic sequences can make synthesis and purification challenging. Examples of amyloidogenic peptides obtained from synthetic sources include Aβ peptides (16), the $TTR_{105-115}$ (17) and variant peptides TTR1-RGD and TTR1-RAD (18), and *de novo* designed peptide STVIIE and variants (19). Recent interest in the potential of amyloid fibrils as materials for biotechnology has led to the development of methods for generating large quantities of amyloid fibrils in a cost-effective manner. In practical terms this means forming fibrils from crude mixtures of readily available proteins without extensive purification, such as from mixtures from crude crystallins (20).

The self-assembly of amyloid fibrils is one of the key features recommending these structures for use in nanotechnology. Self-assembly provides a bottom-up approach to manufacturing that exploits the chemical properties of each component, allows control of self-organization, and results in a useful structure. There are several advantages to self-assembling materials, including the small size of the components and structures, potentially cheaper production costs, the ability to produce multiple devices in parallel, and the ability to disassemble devices into their individual components so that they can be recycled or degraded easily.

The formation of amyloid fibrils proceeds via a nucleation-dependent polymerization pathway. In the lag phase the rate of fibril formation is limited by the slow generation of nuclei. These are small amyloidogenic oligomers, from which fibrils then grow rapidly with the addition of monomers to the oligomer ends (the exponential growth phase) (21). Self-assembly is driven by non-covalent interactions and can be controlled by adjusting solution conditions including temperature, pH, ionic strength, and the presence of reducing agents. Vital information can be obtained about the mechanism of assembly and the structures of any intermediates formed on the assembly pathway using techniques that monitor the assembly process. Furthermore, for fibrils intended for commercial use or large-scale production, an understanding of the process of assembly enables the reliability of assembly to be assessed and the end point of fibril formation to be determined.

This chapter describes techniques that can be used to monitor amyloid fibril formation including thioflavin T(ThT) assays, turbidity measurements, and small-angle X-ray scattering (SAXS). The use of seeding to increase the consistency of the nucleated self-assembly pathways is discussed, along with other useful methods in the notes (for seeding, refer to Note 1). The systematic preliminary experiments typically required to find the best conditions for fibril formation, such as buffer, pH, temperature, or salt concentration, are beyond the scope of this chapter. The reader is directed to

Note 2 and the recent literature (22) for information that will assist in the preparation of samples prior to the measurement of self-assembly.

1.1 Thioflavin T Fluorescence Assay

Numerous dyes have been described that act as chemical probes of fibril formation. Assays utilizing these dyes rely on the different optical properties of the dye free in solution compared with the dye bound to amyloid fibrils. The most common dye-binding assay is the thioflavin T (ThT) fluorescence assay. Although the precise mechanism of ThT binding to amyloid fibrils is not completely understood, it is known that the binding of ThT to β-sheet structure causes a red shift in the excitation and emission maxima and a dramatic increase in intensity in the fluorescence emission spectrum of the dye (23, 24). The ThT molecules are suggested to bind to channels along the length of the fibril, intercalating between and orthogonal to the β-strands (25–28). Recent insights into the mechanism of ThT binding to fibrils are reviewed by Biancalana and Koide (29).

The ThT assay is not strictly quantitative. Differences in ThT binding affinity have been observed for fibrils formed from different proteins or peptides resulting in a difference in the concentration of ThT required to give a satisfactory signal above that arising from the unbound dye. The increase in fluorescence intensity with fibril concentration may also be nonlinear, and a dose-response curve should be constructed for more quantitative applications. Despite these limitations this technique is a benchmark assay for monitoring fibril formation *in vitro*.

Whilst this assay is fairly robust, there are a few factors necessary of consideration. Firstly, if the fibrils further aggregate to form clumps the surface area accessible for ThT binding will be reduced leading to a decrease in ThT binding. This behavior may give rise to a false-negative result (22). Some fibrils also fail to bind the ThT dye due to amino acid-dependent interactions (29), which affects the quantum yield of the bound ThT (28). Conversely, ThT may bind to amorphous aggregates in a sample, leading to false-positive results. In protein systems where there is a high proportion of β-sheet structure in the native globular structure, ThT may also bind to the native protein before any fibrils have formed, leading to a false-positive result. Nevertheless, the ThT assay is an excellent starting point for monitoring fibril formation in vitro and for establishing the conditions and timescales for subsequent experiments, such as SAXS. As each of the methods presented in this chapter measure different aspects of fibril formation with different sensitivities, they also provide complementary data on the assembly process.

1.2 Turbidity

The interaction of light with a turbid solution of protein and fibrils can be used to monitor fibril formation. When light is passed through a sample it interacts with protein molecules causing some of the light to deviate from its straight path. This phenomenon is known as scattering. There are three types of scattering, defined according to the size of the scattering particle relative to the wavelength of the scattered radiation. These are:

- Rayleigh scattering (particle < wavelength of light).
- Mie scattering (particle ≈ wavelength of light).
- Geometric scattering (particle > wavelength of light).

The focus of this discussion is geometric scattering.

Solutions containing large particles, including amyloid fibrils, appear turbid because extensive scattering leads to a decrease in the amount of light transmitted through the sample. Measuring the change in absorbance of light with a particular wavelength is therefore indicative of the amount of aggregated material present. Turbidity measurements are nondestructive, accurate for a range of solution conditions, suitable for continuous monitoring of fibril assembly and technically simple, and typically requiring only standard laboratory apparatus.

In this chapter, we focus on static turbidity measurements but the reader should be aware that more sophisticated techniques for measuring scattering also exist. The most common of these is quasi-elastic or dynamic light scattering (QLS/DLS), which measures the Brownian motion of molecules and therefore gives an estimate of their hydrodynamic volume.

1.3 Small-Angle X-Ray Scattering

Small-angle X-ray scattering (SAXS) is a type of turbidity method in which the scattering of X-rays (rather than visible light) through a sample provides information about the size and shape of the scattering particle. For an excellent review of this method and the theory of X-ray scattering the reader is directed to reviews by Svergun and Koch (30) and Putnam and coworkers (31).

SAXS is an accurate, nondestructive, and label-free method for monitoring self-assembly in solution. It does, however, require access to an X-ray source—this may be a laboratory source but often a higher flux of X-rays at a defined wavelength is required, for which a synchrotron source is necessary.

Data from a SAXS experiment can be used to determine a range of parameters for the species on the assembly pathway such as the radius of gyration, R_g, defined as the root mean square distance of each part of the molecule from its center of mass; the zero angle intensity, $I(0)$; the molecular mass; the maximum dimension of the protein D_{max}; and the Porod volume, defined as the volume of the hydrated protein. Like ThT assays and turbidity measurements, SAXS also provides a method to follow the kinetics of self-assembly.

2 Materials

2.1 Thioflavin T Fluorescence Assay

1. Monomeric protein or peptide solution at a concentration of 2.5–10 mg/ml (see Notes 3 and 4).
2. A solution of 10 mM potassium phosphate buffer, 150 mM NaCl, pH 7.0 described as phosphate buffer below.
3. ThT solution:
   (a) To make the stock solution: dissolve the ThT dye in the phosphate buffer to a final concentration of 1–3 mg/ml and filter through a 0.2 μm syringe filter. This stock solution is light sensitive and should be stored in the dark. It is stable for several weeks, though you should refilter it before each use.
   (b) On the day of the experiment: dilute 1 ml of ThT stock solution into 49 ml of phosphate buffer. The concentration of ThT can be determined by absorbance at 416 nm as described in Note 5.
4. Black 96-well plates or fluorescence cuvettes (see Note 6).
5. Fluorescence plate reader or fluorimeter (see Note 7).
6. A cover for the cuvette or multi-well plate that does not promote condensation, such as Thinseal polyester adhesive films (available from Excel Scientific) or mineral oil. These steps will prevent dust entering the sample and slow evaporation.
7. A clean room with little dust, this could be a clean room facility although most experiments are performed in a Physical Containment Level 2/biosafety level 2 (BSL2) or laboratory of equivalent standard.

2.2 Turbidity

1. Protein or peptide preparation without any existing pre-aggregates (see Note 4).
2. Buffer solution for background measurements (see Note 8).
3. Quartz cuvette or suitable multi-well plate (see Note 9).
4. Spectrophotometer (see Note 10).
5. A cover for the cuvette or multi-well plate as described for the ThT assay above.
6. A clean room or Physical Containment Level 2 room with little dust as described in the ThT assay above.

2.3 Small-Angle X-Ray Scattering

Small-angle X-ray scattering experiments require an X-ray source. This can be a laboratory instrument or a synchrotron facility. Basic requirements include:

1. The capability to resolve features on a sub-Angstrom to tens of Angstrom scale using single or multiple camera lengths. For example, on the X33 beamline at EMBL Hamburg a

resolution of 0.6–60 Å can be achieved using a detector distance of 2.7 m. At the Australian Synchrotron a resolution of 0.0012–1.1 Å can be achieved, but this involves use of several camera lengths and the merging of data sets. An estimate of the maximum dimension of your protein or starting and end dimensions for a kinetics experiment should be used to predict the full length of scattering (defined as the q range) required and experimental setup (see Note 11).

2. A 2D ("area") detector for recording the X-ray scattering pattern (see Note 12).
3. The capability to position capillaries containing samples within the beam.
4. A magnifying video camera to visualize the position of the capillary and the position of the sample within the capillary and X-ray beam is ideal.
5. Syringe and syringe pump to move the sample from a container into the X-ray beam for exposure and to return the sample to the container. The control of fluid flow should ideally be remote, allowing changes from outside the beam hutch for rapid sample exposure and data collection.

You will also need to determine the training and safety requirements from the manager of the X-ray facility.

Additional materials you will need include:

6. Thin-walled glass capillary tubes with a diameter of 2 mm and wall thickness of 0.01 mm. These are available from various companies, including Hilgenberg in Germany, Hampton Research in the USA, or Capillary Tube Supplies Ltd. in the UK.
7. Microfuge tubes.
8. Parafilm (or suitable alternative) to prevent evaporation of the sample from microfuge tubes during data collection, especially for prolonged experiments in which the sample is heated.
9. A suitable buffer for suspension of the protein or peptide sample (filtered through a filter with pores of 0.45 μm diameter or smaller) (see Note 13).
10. Any equipment required to induce fibril formation, e.g., heating block, stirring apparatus, specialist buffers.
11. Glycerol for addition to samples that are sensitive to X-ray damage (see Note 14).
12. Solutions of proteins such as lysozyme for use as standards (see Note 15).
13. Protein or peptide preparation without preexisting aggregates (see Note 4). This protein should be of high purity and in monomeric form as SAXS relies on the sample being

monodisperse. Kinetics measurements also require a monodisperse solution to follow assembly, including the association of early aggregates. Note that the sample requirements for a kinetics experiment are much greater than SAXS analysis of a protein at a single time point. Typically, SAXS experiments for a 100 kDa protein at a single timepoint require 15 μl of protein in the concentration range of 1–10 mg/ml, as described by Putnam (31). In contrast, kinetics experiments require approximately 100 μl of protein at the same concentration to allow periodic withdrawal of the sample from a larger stock placed within a microfuge tube. Exact concentrations are protein dependent; the concentration required for peptides is often much higher than for large proteins.

14. An external hard drive. A typical series of SAXS experiments generates several gigabytes of data. To transfer or store data you will need to use either an external hard drive or a file transfer protocol (ftp). It is a good idea to discuss this with the beamline scientists before beginning experiments.
15. Data processing software. SAXS data are collected as images, which must be converted to intensity plots before analysis—we recommend the SAXS15ID program for this conversion. The SAXS15ID program with built-in virtual machine can be downloaded from the Australian synchrotron Web site http://www.synchrotron.org.au/index.php/aussyncbeamlines/saxs-waxs/saxs-data-a-processing. Note that the version of SAXS15ID program used for analysis must be the same version used for later analysis. The instruction manual and software updates are available from http://cars9.uchicago.edu/chemmat/pages/swsoftware.html. Data processing in SAXS15ID generates .dat files, which can then be analyzed using the suite of ATSAS programs available at http://www.embl-hamburg.de/biosaxs/software.html (32). The manuals available on the website are excellent resources for these programs; for more specific questions refer to the online forum at http://www.saxier.org/forum.

3 Methods

The techniques described in this section are primarily used for monitoring the formation of amyloid fibrils. They may also be adapted for the analysis of fibrils that have already aggregated and for the comparison of the size and shape of the monomer peptide or protein prior to fibrillation with mature fibrils.

3.1 Thioflavin T Fluorescence Assay

The method described below is for the periodic measurement of fibril formation in a 96-well plate using a plate reader operating in cycle mode. This method can easily be adapted for use in a cuvette (see Note 6). The method may also be adapted for larger sample volumes by removing aliquots from the sample and measuring the ThT fluorescence at each individual timepoint in a new well or rinsed cuvette.

Set the plate reader or fluorimeter to excite the sample at 440 nm (slit width 5 nm) and collect the emission intensity at 480 nm (slit width 10 nm) (24). Many spectrometers such as Varian or BMG Labtech are equipped with a kinetics software package that can aid the collection of kinetics data. Parameters entered into the software prior to a kinetics experiment typically include the frequency of absorbance measurements, the length of the experiment (see Note 16), the temperature of the measurement, the mode of agitation of the sample (e.g., orbital or linear vibration; see Note 17), the frequency of vibration, and the length of vibration. An example of settings is given below, although different samples will require optimization:

1. Temperature 37°C.
2. Cycle time 300 s.
3. Number of cycles 250.
4. Flashes per well 50.
5. Shaking time 180 s.
6. Shaking width 1 mm.
7. Shaking mode orbital.

Next, the sample must be prepared:

1. Dilute samples to the appropriate concentration in ThT solution. If necessary, make up the remaining volume with buffer. Each well should contain a total volume of 200 μl of solution, of which 125 μl must be the ThT analysis solution. The final protein concentration should be in the range 2.5–10 mg/ml (see Note 3). Each set of experimental conditions should be measured with at least duplicate samples. Avoid using the wells on the edges of the plate as these are more prone to evaporation and other edge effects that make measurements from these wells less reliable.
2. Prepare a blank sample containing the ThT solution and phosphate buffer only.
3. Measure the fluorescence intensity of the blank sample by excitation at 440 nm (slit width 5 nm) and emission 480 nm (slit width 10 nm) and select the averaging time to maximize signal:noise (usually 60 s works well).
4. Seal the plate carefully and ensure that the plastic seal adheres to all sections of the plate to avoid bubbles in the plastic forming during the experiment.

5. Place the plate in the plate reader and initiate the kinetics experiment by applying the conditions necessary to induce fibril formation, e.g., heat.
6. Inspect data to ensure that the kinetics of aggregation have been recorded as indicated by a plateau in the ThT fluorescence intensity. If fibril formation is very slow (i.e., several days) it may be necessary to increase the time between sampling, e.g., to every 4 h (see Note 16).
7. Some samples display a decrease in fluorescence intensity towards the end of the experiment rather than a plateau in ThT binding. See the introduction above for a discussion of factors that may influence ThT binding.
8. A repeat of the experiment outlined above, in addition to replicate data performed during the experiment, will illustrate variability in the rate of fibril assembly. It is common to present data from a series of experiments to give a good representation of any variability in assembly.

3.2 Interpretation of ThT Assay Data

The data collected in a ThT fluorescence assay are best represented in a plot of fluorescence intensity versus time. Fibril growth has three distinct phases: the lag phase, the exponential growth phase, and a reduced growth phase where fibrils mature. In the lag phase the rate of increase in fluorescence intensity is slow as fibril formation is limited by the slow generation of nuclei. Exponential growth follows and is rapid as fibrils assemble and elongate. Once all the monomers have self-assembled or an equilibrium between monomers and aggregated protein has been reached, the fluorescence intensity will plateau as the fibrils mature. Therefore, a typical ThT fluorescence curve is sigmoidal in shape; however, if fibril formation proceeds very rapidly the lag phase may not be detected, resulting in a hyperbolic curve. The data in a ThT assays are often reported normalized to the highest intensity signal obtained, usually the end point of the assay, so that the units of the *y*-axis are % fluorescence intensity. The background signal from blank solution should be subtracted prior to normalization. An example of fibril assembly as measured using the ThT fluorescence assay is given in Fig. 1.

3.3 Turbidity

The method described below is for the periodic measurement of fibril formation in a 96-well plate using a plate reader operating in cycle mode. This method can easily be adapted for use in a cuvette using modifications similar to those described above for the ThT assay (see Note 6). A range of wavelengths may be used to measure turbidity; for example absorbance can be measured at ~340 nm (see Note 18). A range of parameters will need to be entered into the spectrometer software as described above for the ThT assay.

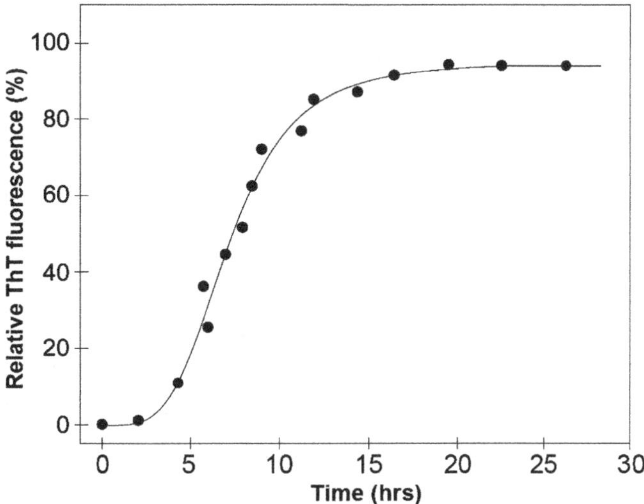

Fig. 1 Fibril formation measured by a ThT fluorescence assay, where the fluorescence of the dye ThT increases as a result of fibril formation by a model protein. The magnitude of ThT fluorescence is expressed as a percentage of the fluorescence recorded at the plateau, i.e., when fibril formation is complete at the end of the experiment. The sigmoidal shape of this growth curve is typical of amyloid fibril formation by most proteins and a sigmodial curve has been fit to the data. Occasionally the lag phase may be very short or absent

To prepare the sample:

1. Dilute the protein/peptide monomers to a concentration of 2.5–10 mg/ml in the buffer for fibril formation and aliquot into the 96-well plate. At least two replicate samples should be measured. As for the ThT assay, users are recommended to avoid use of the wells on the edge of the plate as these are more prone to evaporation and other edge effects that make measurements from these wells less reliable.
2. Add at least two buffer background samples to the plate.
3. Seal the plate carefully and ensure that the plastic seal adheres to all sections of the plate to avoid bubbles in the plastic forming during the experiment.
4. Place the plate in the plate reader and initiate the kinetics experiment.
5. The experiment is complete when there is no change in turbidity with time and the data have reached a plateau.

Repeating the experiment will allow experimental variability to be determined, as described above for the ThT assay data.

3.4 Interpretation of Turbidity Assay Data

The data collected in a turbidity assay are best presented as a plot of intensity at the wavelength measured (e.g., 340 nm) versus time. As for the ThT assay, the growth curve is likely to be sigmoidal,

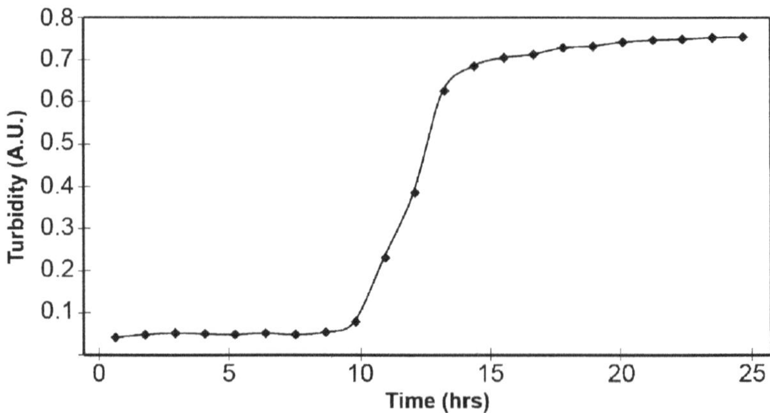

Fig. 2 Fibril formation measured by a turbidity assay, where the turbidity of the sample increases as a result of fibril formation by a model peptide. As for the ThT fluorescence assay, fibril growth is usually observed as a sigmoidal increase in turbidity over time but the lag phase may be short or absent from some fibril-forming systems

though may be hyperbolic if the lag phase is missed. The background signal from the buffer background should be subtracted from the data. An example of a fibril formation monitored by a turbidity assay is given in Fig. 2.

3.5 Small-Angle X-Ray Scattering

This section outlines the steps required to set up a SAXS experiment, how to assess the quality of the setup, and the sample and the methods for data acquisition and analysis. It describes the steps necessary for kinetics experiments but could be easily adapted to monitor the start and end points for a protein aggregation experiment. The beamline scientists at a synchrotron source will usually provide experimentalists with hardware and software training specific to the beamline; users of laboratory X-ray sources should also contact the equipment administrator for training. A detailed description of how to operate the X-ray apparatus is beyond the scope of this chapter. Neither will detailed explanations or instructions for using the software be given—this information can be found in the manuals that accompany these programs (see links provided above).

1. First set up the apparatus to monitor fibril formation within the beam hutch as illustrated in Fig. 3. A microfuge tube is placed in a heating block, which has the dual purpose of holding and heating the tube if required, and a capillary placed into the tube and secured using parafilm (or similar). The capillary is then connected to a 1 ml syringe via a length of silicone tubing using standard Luer adapters. The syringe should be secured in a syringe pump that can be operated from outside the hutch. The pump is used to draw the sample into the capillary during X-ray exposures. It is important to keep the sample moving in the beam to limit any damage from the beam (see Note 20).

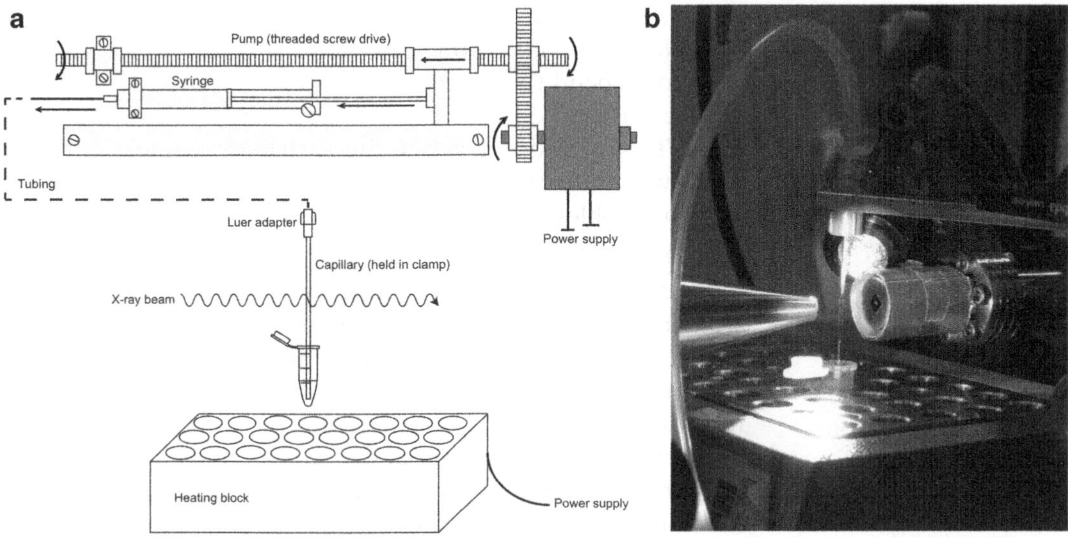

Fig. 3 Experimental setup for SAXS. (**a**) Schematic illustrating the position of the capillary in the X-ray beam and other apparatus. A syringe pump (controllable from outside the hutch) is connected to the capillary via tubing and Luer adapters. The end of the capillary is placed into a microfuge tube that is within a heating block for temperature control. (**b**) Photograph of experimental setup at The Australian Synchrotron. Lights and cameras positioned around the capillary enable the flow of liquid in the capillary to be monitored from outside the hutch

The capillary must be positioned in the X-ray beam with one end right at the bottom of the centrifuge tube to promote mixing of the sample and to ensure an even flow of the sample into and out of the capillary. The length of capillary that is within the X-ray beam should be a section of uniform width and wall thickness, away from the flared end. If any defects are visible in the capillary, ensure that they are well away from the beam. The top of the centrifuge tube can be covered in parafilm or an equivalent material to reduce evaporation if the experiments occur over an extended period of time.

2. SAXS analysis generates a large number of data files very quickly. A good way to keep track of these is to start a spreadsheet to detail the file name, sample, exposure time, and any other conditions or comments to assist analysis.
3. Calibrate the system by recording the scattering patterns of an empty capillary and a water-filled capillary for a series of exposure times, e.g., 1, 2, 5, and 10 s. At this stage the beamline scientist may mask any areas of the detector where there are artifacts. This masking must be performed before calibration with the blank solution to ensure that the same area is subtracted from both buffer and sample files.
4. Once the apparatus is in position, draw some buffer (blank) into the capillary and record the scattering pattern. This will be the blank image you subtract from the data so the exposure time, detector position, buffer temperature, and conditions

must match those used in the experiment as closely as possible (see Note 20). The buffer should also be the exact solution used to suspend the protein or peptide. You should be able to see the buffer rise in the capillary tube and then return to the tube. This provides a visual check that the sample is in the beam when the exposure to X-rays is occurring and that there are no bubbles in the sample which may scatter the beam leading to misleading data.

5. Record the scattering patterns of a dilution series of the protein or peptide to check there is no aggregation within the starting material. Typically, the dilution series should be run with protein or peptide in the concentration range of 1–10 mg/ml, although the extract concentration will depend on protein properties including shape and size. For a sample where no aggregates are present, the scattering curves after buffer subtraction should be parallel across the full q range; aggregates in the solution give rise to an upward curve of the scattering pattern at low q values as shown in Fig. 4. If aggregates are observed in the sample, as shown in Fig. 4, it may be possible to remove these aggregates by centrifugation, filtration, or size exclusion chromatography—although changes in protein concentration after removal of aggregates should be tracked. Another test that can be performed prior to the experiment as a further check for aggregation is to assess the q range over which the data should be linear on a Guinier plot (see Subheading 3.5.2 below). This range will depend on the shape of the starting protein or peptide, with the data ideally linear for $qR_g < 1.3$ for globular proteins and $qR_g < 0.8$ for more elongated proteins (31). The shape of the protein is expected to become more elongated as fibrils form during the course of the experiment. If the range over which the data are linear is substantially less than expected further steps may be required to reduce preexisting aggregates and optimize the sample.
6. Check that the range of scattering data that can be collected with the experimental setup will enable observation of the full range of oligomeric species by recording the scattering pattern of a solution of fully assembled mature fibrils. It can be helpful to test several dilutions to confirm aggregates can be observed even at low concentrations. The q range can be altered by changing the position of the detector relative to the beam centre, the detector distance, or wavelength of the X-ray beam. If data cannot be collected over the ideal length scale, data can be recorded at two different detector distances and combined into a single data set following collection (see Note 11). If any alterations to improve the scattering range are made at this stage, the blank scattering pattern (buffer only) must be rerecorded using the new settings. It is not necessary to repeat the dilution series if the data have been recorded for wider angles (or a larger q range).

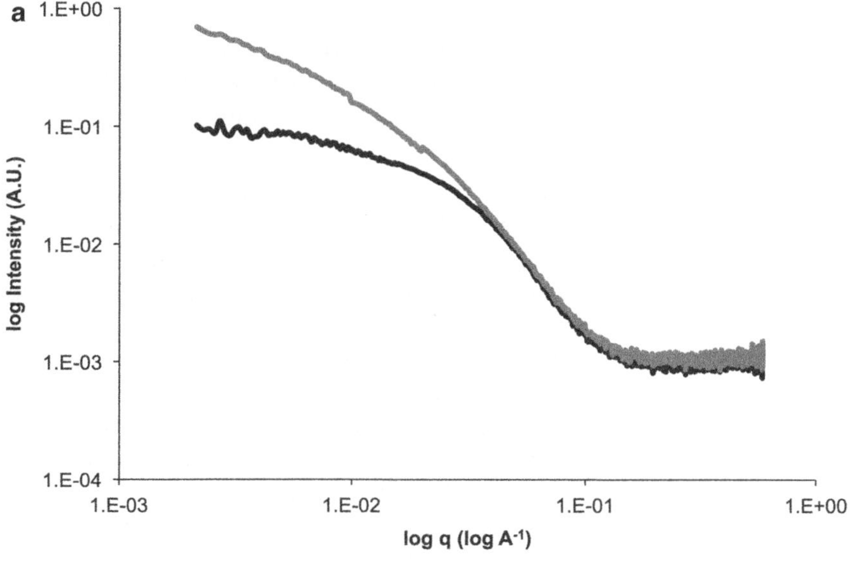

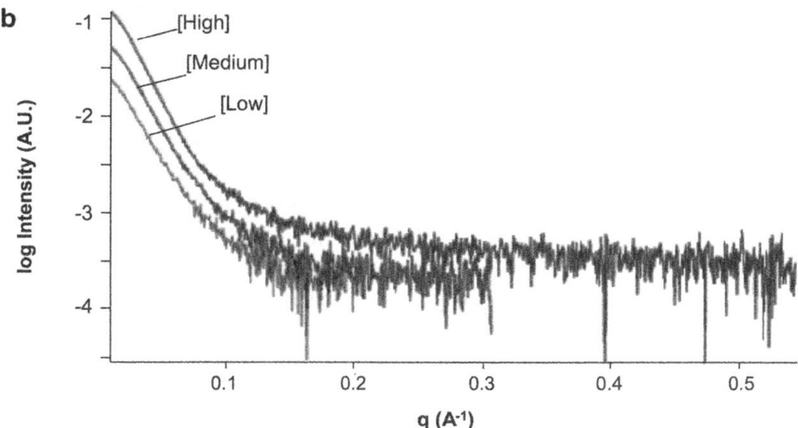

Fig. 4 Preliminary data analysis and evaluation during an SAXS experiment. (**a**) The scattering curve for a solution free from aggregates should be flat at low *q* values (*black line*), whereas an upward turn in the slope at low *q* is indicative of sample aggregation (*grey line*). (**b**) A concentration series obtained by measuring the scattering profile of a protein at several concentrations after serial dilution. Solutions free from aggregates should give rise to scattering curves that are parallel at all *q* values

3.5.1 Data Acquisition

7. Select a protein concentration appropriate for the kinetics experiments using the data from the dilution series experiments. This concentration will need to be sufficient to detect the presence of fibrils at the end of the experiment but low enough to ensure that no aggregates are present at the start of the experiment.
8. Apply the conditions that induce aggregation to initiate the kinetics experiments and collect data that tracks changes in the scattering pattern. The frequency of data collection must be decided in advance—this will depend on the timescale of fibril formation for the system but every 15 min is a good initial

frequency for data collection. As explained in Note 20, we recommend collecting a series of short exposures for each time point and using the average of these data. This approach maximizes the signal:noise whilst limiting radiation damage to the sample.

9. Continue to observe the sample at each time point. Careful monitoring of the sample is necessary to ensure that the liquid is in the beam before each exposure. The capillary should also be scanned to ensure that protein deposits do not form on the capillary surface as a result of beam exposure. If you do observe protein deposition on the capillary, typically apparent as a white deposit, these deposits should be washed off by rinsing the capillary with water between exposures. If the deposits cannot be removed by washing the capillary you will need to pause the experiment and replace the capillary with a new capillary. Any change in capillary requires the scattering patterns of the empty capillary, water, and buffer to be collected again for the new capillary. You may also consider the addition of glycerol to reduce radiation damage to the sample (see Note 14).
10. Aggregation can be assumed to be complete when the scattering curve ceases to change as a function of time.
11. At the end of the experiment record another buffer scattering pattern. If any changes are made to the experimental setup during the course of the experiment (for example, a change of capillary or change in the position of the capillary) a buffer scattering curve should be recorded immediately and data after this point processed by subtraction of the new blank scattering pattern.
12. [Optional extra step] If multiple detector distances are required to capture a wider q range, repeat the experiment (i.e., steps 4–10 above) with the detector set up at a different length. You must ensure that the data collected at the start of the second experiment fits well with the data collected at the start of the first experiment. This fit will be essential for all time points during the kinetics experiment.

3.5.2 Data Processing

The key steps involved in processing SAXS data are summarized in Fig. 5.

13. Use SAXS15ID to convert the image of the scattering pattern to a plot of scattering intensity. The buffer scattering

Fig. 5 (continued) residuals, which are shown on the same plot. The R_g and $I(0)$ determined from the plot are also shown. The Guinier plot can also be generated in AUTORG; compare the parameters obtained using each method to find the optimum fit and best estimate of R_g and $I(0)$. (**d**) Use the R_g value obtained from the Guinier plot to estimate D_{max} (a good first estimate is $D_{max} = 3 \times R_g$) and fit a pair distribution function $P(r)$ using GNOM. (**e**) Generate a Kratky plot (Intensity × s^2 vs. s) to reveal the extent of flexibility and unfolding within the sample

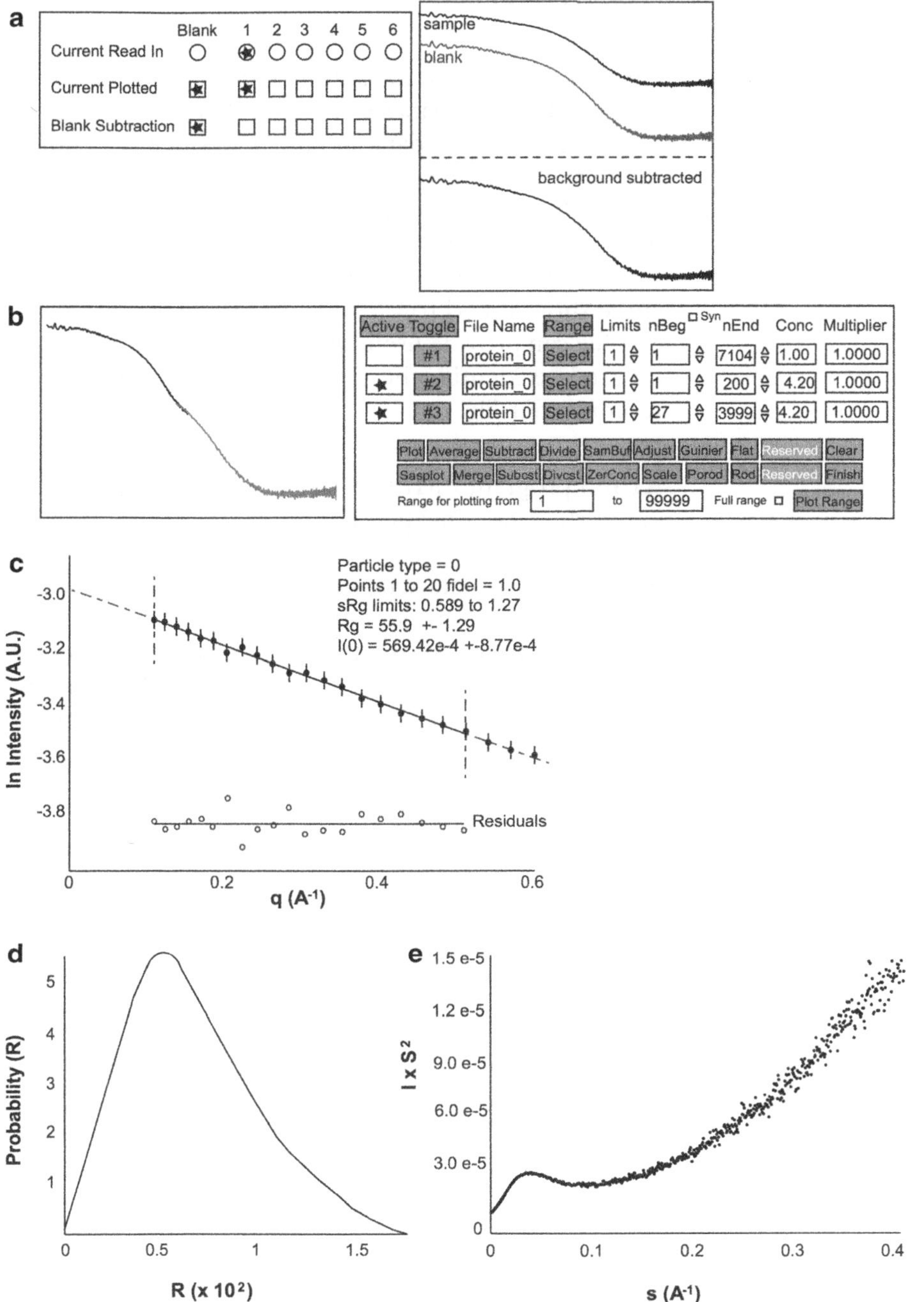

Fig. 5 Processing of SAXS data. (**a**) A schematic representation of the buffer subtraction process in SAXS15ID. Check that the curve resulting from background subtraction does not have negative intensity values, that there are no missing data, and that the curve tends towards zero at lower *q* values. (**b**) A schematic representation of the optional step to merge data sets if multiple detector distances are used to record scattering over a wider *q* range. Open the two .dat files corresponding to the same time point at different detector distances in PRIMUS. Press "plot" to display both scattering curves. The aim is to seamlessly merge the two data sets, so the data sets must overlay to continue. Change the nEnd and nBegin values to remove any data from the extremes of the two sets that do not overlay well. Press "merge." View the merged data set in SASplot to check for any outliers. Create a final merged file for the data and repeat the process above for each time point. (**c**) In PRIMUS open the .dat files and generate a Guinier plot. Exclude the data corresponding to extremely low *q* values as this region is prone to artifacts. Select data that give the optimum fit to the Guinier equation by minimizing the

curve should then be subtracted from the data and each buffer-subtracted data set saved as a separate .dat file (Fig. 5a). The correct buffer data should be chosen for subtraction. If the buffer data recorded before and after the experiment are identical, use an average of both data sets; if they are different use the most appropriate data. A correctly subtracted curve should not have negative intensity or missing data and the curve should tend towards zero at lower q values. The parameter file can provide useful information for subtraction as it can be used to identify the time of data collection and the most recent buffer file.

14. [Optional step if multiple detector distances were used] In the ATSAS program PRIMUS, open the two .dat files corresponding to the same time point at different detector distances. Press "plot" to display both scattering curves. The aim is to seamlessly merge the two data sets, so the data sets must overlay to continue change the nEnd and nBegin values to remove any data from the extremes of the two sets that do not overlay well with each other. Press "merge." View the merged data set in SASplot to check for outliers. Create a final merged file for the data and repeat the process above for each time point (Fig. 5b).
15. Open the [merged] buffer-subtracted data in PRIMUS and generate a Guinier plot (Fig. 5c). The Guinier plot enables the several system parameters to be calculated and evaluated, including:
    (a) The radius of gyration, R_g.
    (b) The zero angle intensity, $I(0)$.
    (c) The molecular mass.
    (d) The Porod volume. This volume, expressed in units of nm^3, is typically about 1.5–2 times the molecular mass in units of kDa, although this approximation is poor for small proteins (31).
16. R_g and $I(0)$ can also be determined using the AUTORG program. See the comment in step 5 above regarding the range over which the data should be linear, depending on protein shape. Note that another approach to determine R_g is via the Debye approximation, where $qR_g < 1\_4$ for elongated macromolecules (31). Check that the values obtained using AUTORG match those obtained from fitting the Guinier plot by hand. The molecular mass can then be estimated from $I(0)$ with reference to the $I(0)$ of a protein of known molecular mass, such as the protein standards described in Note 15. A deviation from the expected molecular weight at $I(0)$ can indicate multimer formation or oligomerization. Use the parameters obtained from the Guinier plot to estimate D_{max} (a good first estimate is $D_{max} = 3 \times R_g$) and fit a pair distribution function $P(r)$ using the

GNOM program available from ATSAS (Fig. 5d). $P(r)$ is a measure of the probability or frequency of finding certain interatomic distances (r) within the particle. The shape of the $P(r)$-function is related to the shape of the particle: if the function is very symmetric, the particle is also highly symmetric. The $P(r)$ function is highly sensitive to the D_{max} and the R_g limits; small changes in D_{max} can dramatically alter the shape of the $P(r)$ curve.

17. Generate a Kratky plot to reveal the extent of flexibility and unfolding within the sample (Fig. 5e).
18. The type and extent of modelling that can be performed on the protein solution structure will depend on what is already known about the fiber-forming system under investigation. If no other structural data are available for the system, the solution structure modelling will be limited to ab initio methods such as DAMMIN and GASBOR. If solution or crystal structures exist for the protein at any stage of aggregation, the SAXS data from the corresponding timepoint can be fitted to this model using rigid body or flexible structure modelling using CRYSOL or SUPCOMB.

4 Notes

1. Protein sequences with aggregation kinetics longer than the timescale over which assembly can easily be monitored can be induced to form fibrils more quickly by the addition of small, preformed fibril “seeds.” Seeds not only accelerate the process of fibril formation but also act as templates, ensuring that the newly made fibrils will be of the same morphology as the seeds. Thus seeding not only promotes rapid fibril formation but also facilitates the production of uniform fibrils. Seeds may be generated by rapid freezing of a solution of fibrils in liquid nitrogen or by sonication (the ultrasound source may be an ultrasonic bath or probe). The frequency of ultrasound can vary considerably between sources and is likely to affect the length of seed generated by breakage of the fibers. Seeds are typically added to a solution of protein at a concentration of about 5% [v/v] seeds.
2. In general the buffer should maximize the stability and solubility of the protein to prevent formation of amorphous aggregates; however, in some systems addition of co-solvents such as trifluoroethanol may help drive fibril assembly. The key factors to consider are:
   - pH—check that the buffer chosen is appropriate for the entire pH range of interest and that the protein is stable

and soluble in this buffer. Fibril formation can often be achieved at physiological pH (around pH 7), although some proteins may require a change in pH to destabilize the native protein structure. Note that different buffers may affect protein solubility/stability differently even at the same pH.

- Reducing agent—if fibril formation is dependent on the formation or reduction of disulfide bonds, addition of a reducing agent will make a large difference to the kinetics if fibril assembly. For example, reduction of insulin and κ-casein accelerates fibril formation whereas fibril formation of the chaplin peptides (natural amyloid fibrils from *Streptomyces coelicolor*) is faster under nonreducing conditions.
- Temperature—usually between 4 and 37°C, although some systems require a higher temperature to destabilize native protein structure. Most systems self-assemble more quickly at higher temperatures. Altering the temperature may allow the experiment to be accelerated/decelerated to a rate appropriate for the experimental timescale.
- Shaking—agitation often accelerates the rate of fibril formation; however, some experiments (e.g., SAXS) cannot easily accommodate the apparatus required to provide shaking/agitation.
- Protein concentration—the rate of fibril formation is usually concentration dependent. Choose a protein concentration high enough to enable accurate detection of fibrils but not so high that aggregation occurs too quickly to monitor. A good starting point is 10 mg/ml using a 100 μl volume of sample for the reaction.

3. As explained in the introduction, differences in ThT binding affinity have been observed for different fibril systems, such that it is hard to estimate the relative concentrations of protein and ThT required to give a satisfactory signal of bound dye compared to the free dye background. The figure of2.5–10 mg/ml is a guideline—try an assay at this concentration of protein and then increase/decrease the protein concentration as required depending on the signal:noise obtained.
4. Proteins or peptides that are prone to aggregation can aggregate during isolation, synthesis, or purification prior to any kinetics assay. Even a small number of aggregates can nucleate fibril formation, so efforts should be made to remove any existing aggregates. Techniques include filtration (although this can lead to a significant decrease in the concentration of hydrophobic proteins that associate with the filter) or centrifugation. A number of techniques have been specifically developed for polyglutamine proteins and the Aβ peptide, including the use of dimethyl sulfoxide (DMSO), trifluoroacetic

acid/1,1,1,3,3,3-hexafluor-2-propanol (TFA/HFIP), or sodium hydroxide (33–35).

5. To determine the concentration of ThT in a stock solution, dilute 100-fold into ethanol and measure the absorbance at 416 nm. The extinction coefficient is ~26,600 $M^{-1}cm^{-1}$. For the first experiments it is sensible to use a ThT concentration higher than the protein concentration (e.g., a fivefold molar excess) such that most of the fibrils formed in solution will be bound by the dye.
6. Be consistent with the type of microplate as the path length can vary between different well geometries. Kinetics measurements made in a cuvette typically require the background sample to be measured first, followed by a single or multiple cuvettes containing the protein or peptide monomer.
7. A fluorimeter or plate reader with a monochromator for continuous wavelength control is ideal but bandpass filters that allow excitation at ~440 nm and emission at ~480 nm with a slit width of 5 and 10 nm are adequate for ThT assay measurements.
8. The buffer used for turbidity measurements should ideally be optically clear without the presence of particles that can scatter light. Filtration of the buffer with a filter containing pores 0.45 μM in diameter or smaller is recommended prior to preparation of the background and the addition of protein or peptide to the buffer.
9. Quartz or fused silica cuvettes are transparent in the ultraviolet and visible regions, whereas some glass and plastic can absorb in these wavelengths. Corning–Costar UV and Greiner UV microplates have good UV transmission properties; check manufacturer details before use. Be consistent with the type of microplate as the path length can vary between different well geometries.
10. A spectrometer or plate reader with a monochromator for continuous wavelength control is ideal but bandpass filters that allow wavelength selection between 320 and 800 nm are adequate for turbidity measurements.
11. The size of the protein or peptide within the sample will determine the experimental setup required for measurements. For kinetics experiments this requires the estimated size of the protein at the start and the end of the experiment to be considered. Putman (31) recommends a theoretical check that $q_{min} \leq \pi / D_{max}$ where q_{min} is the smallest q to be measured and D_{max} is an estimate of maximum dimension of the protein. A second check that $q_{max} \geq 2\pi / D_{max}$ can be used to set the upper limit of the q range, again based on the estimated maximum dimension of the protein. If the minimum and maximum q values estimated for the protein at the start and the end of the

experiment are not possible for the beamline, two different setups will be required to measure the low and the high q values. The actual range required may be wider than that determined theoretically; see Putman (31) for a discussion of the balance that must be made between the collection of high and low q data. Before starting an experiment, it is useful to check the estimates for the q range made on the basis of assumed D_{max} using the starting and end material for your experiment, as described in Subheading 3.3, step 6.

12. The X-ray detector should ideally be free of any artifacts caused by beam damage, although these can sometimes be removed using masking software.
13. The buffer must not scatter X-rays and is best made and filtered immediately before use. Buffers can be screened using turbidity measurements prior to SAXS. Salt should be avoided where possible.
14. Changes in the scattering pattern observed for repeat exposures of a protein can lead to an increase in R_g and $I(0)$ indicating beam damage. The addition of 5–10% (v/v) glycerol to the sample can reduce this effect. Glycerol must also be added at the same concentration to the blank buffer sample.
15. Suitable protein calibrants include lysozyme, bovine serum albumin, and glucose isomerase, which vary in molecular weight between 14 kDa and 172 kDa, or other proteins and peptides of known molecular weight similar to the molecular weight of your protein sample can be used as standards.
16. Some spectrometers will limit the number of data points that can be collected, limiting the frequency of the data collection and the duration of the experiment. You may choose to run an experiment for a long period of time and then run a second experiment where the early time points are taken more frequently than the later time points. Some software allows both frequent and infrequent data collection within the same experiment.
17. Agitation is not necessary for all aggregation experiments but can be used to speed the aggregation process. Vibration can also assist if the sample is observed to sediment on the bottom of the plate between absorbance measurements; a short vibration can assist in suspending any aggregates prior to a measurement increasing the reproducibility of kinetics data. Cuvettes are not typically vibrated but rather a small magnetic flea may be used to speed aggregation and keep any solids suspended.
18. If your spectrometer has a monochromator for continuous wavelength control it is worth measuring the complete absorption spectrum at the start and end of the aggregation experiment. A complete spectrum may also be collected at each time point but will result in a very large data set and may only be

worth collecting if your spectrophotometer has a photodiode array detector that allows rapid data collection.

19. A temperature should be chosen that allows fibril formation to occur within a reasonable timescale for the beamtime you have been allocated. In general fibril formation is accelerated at higher temperatures. This timescale should also be slow enough to allow the sufficient data collection to follow the aggregation process. An ideal SAXS aggregation experiment will last approximately 4 h (i.e., aggregation should just be complete within this time frame) and data should be collected every 15 min. The timescale of aggregation can be optimized for SAXS analysis by systematically examining the effect of temperature and aggregation using the ThT assay or turbidity assays prior to SAXS. You will need to consider the time taken for the heating block (and sample) to reach the desired temperature prior to the start of the SAXS experiment. You will also need to ensure adequate time is allocated to make fine adjustments during the experimental setup. We recommend allocating approximately 3 h to set up time for initial experiments, with a further 1–2 h for minor changes between experiments, such as a change in protein.
20. It is helpful to be able to control the speed of the flow in the capillary with a fast and a slow control. For example, it is ideal to draw up 200 μl of sample quickly and then expel 50 μl of sample slowly at a steady flow rate over a period of about 10 s (depending on the exposure time required for the experiment) whilst acquiring the image. The remaining 150 μl sample can then be expelled quickly allowing the process of sample removal and return to be very quick and facilitating frequent data collection. Radiation damage is minimized during data collection by keeping the sample moving. Multiple frames of short exposures (1–5 s) can also be collected as the sample is slowly expelled. An average of these multiple frames improves the signal:noise ratio. This approach is preferable to a single long exposure and also reduces radiation damage.

References

1. Guijarro J, Sunde M, Jones J, Campbell I, Dobson CM (1998) Amyloid fibril formation by an sh3 domain. Proc Natl Acad Sci U S A 95:4224–4228
2. Dobson CM (1999) Protein misfolding, evolution and disease. Trends Biochem Sci 24: 329–332
3. Fandrich M, Fletcher M, Dobson CM (2001) Amyloid fibrils from muscle myoglobin. Nature 410:165–166
4. Gras SL (2007) Amyloid fibrils: from disease to design. New biomaterial applications for self-assembling cross-β fibrils. Aust J Chem 60:333–342
5. Nilsberth C, Westlind-Danielsson A, Eckman CB, Condron MM, Axelman K, Forsell C, Stenh C, Luthman J, Teplow DB, Younkin SG, Naslund J, Lannfelt L (2001) The 'arctic' APP mutation (E693G) causes Alzheimer's disease by enhanced A-beta protofibril formation. Nat Neurosci 4:887–893
6. Jones L (2002) The cell biology of huntington's disease. Oxford University Press, Oxford OX2 6DP, UK, pp 348–362

7. Olsen A, Jonsson A, Normark S (1989) Fibronectin binding mediated by a novel class of surface organelles on *Escherichia coli*. Nature 338:652–655
8. Saupe SJ (2000) Molecular genetics of heterokaryon incompatibility in filamentous Ascomycetes. Microbiol Mol Biol Rev 64: 489–502
9. Balguerie A, Dos Reis S, Ritter C, Chaignepain S, Coulary-Salin B, Forge V, Bathany K, Lascu I, Scmitter J, Riek R, Saupe SJ (2003) Domain organization and structure-function relationship of the het-s prion protein of *Podospora anserina*. EMBO J 22:2071–2081
10. Barlow D, Dickinson G, Orihuela B, Kulp J III, Rittschof D, Wahl K (2010) Characterization of the adhesive plaque of the barnacle balanus amphitrite: amyloid-like nanofibrils are a major component. Langmuir 26:6549–6556
11. Berson JF, Theos AC, Harper DC, Tenza D, Raposo G, Marks MS (2003) Proprotein convertase cleavage liberates a fibrillogenic fragment of a resident glycoprotein to initiate melanosome biogenesis. J Cell Biol 161:461–462
12. Fowler DM, Koulov AV, Alory-Jost C, Marks MS, Balch WE, Kelly JW (2006) Functional amyloid formation within mammalian tissue. PLoS Biol 4:e6
13. Caflisch A (2006) Computational models for the prediction of polypeptide aggregation propensity. Curr Opin Chem Biol 10:437–444
14. Conchillo-Sole O, de Groot NS, Aviles FX, Vendrell J, Daura X, Ventura S (2007) Aggrescan: a server for the prediction and evaluation of 'hot spots' of aggregation in polypeptides. BMC Bioinformatics 8:65–81
15. Nadaud PS, Sarkar M, Wu B, MacPhee CE, Magliery TJ, Jaroniec CP (2010) Expression and purification of a recombinant amyloidogenic peptide from transthyretin for solid-state NMR spectroscopy. Protein Exp Purif 70:101–108
16. Tickler AK, Clippingdale AB, Wade JD (2004) Amyloid-beta as a 'difficult sequence' in solid phase peptide synthesis. Protein Pept Lett 11:377–384
17. Gustavsson A, Engstrom U, Westermark P (1991) Normal transthyretin and synthetic transthyretin fragments form amyloid-like fibrils *in vitro*. Biochem Biophys Res Commun 175:1159–1164
18. Gras SL, Tickler AK, Squires AM, Devlin GL, Horton MA, Dobson CM, MacPhee CE (2008) Functionalised amyloid fibrils for roles in cell adhesion. Biomaterials 3:22–30
19. de La Paz ML, Goldie K, Zurdo J, Lacroix E, Dobson CM, Hoenger A, Serrano L (2002) *De novo* designed peptide-based amyloid fibrils. Proc Natl Acad Sci U S A 99:16052–16057
20. Garvey M, Gras S, Meehan S, Meade S, Carver J, Gerrard J (2009) Protein nanofibres of defined morphology prepared from mixtures of crude crystallins. Int J Nanotechnol 6: 258–273
21. Wetzel R (2006) Kinetics and thermodynamics of amyloid fibril assembly. Acc Chem Res 39:671–679
22. Nilsson M (2004) Techniques to study amyloid fibril formation *in vitro*. Methods 34:151–160
23. Naiki H, Higuchi K, Hosokawa M, Takeda T (1989) Fluorometric-determination of amyloid fibrils *in vitro* using the fluorescent dye. Thioflavine-T. Anal Biochem 177:244–249
24. LeVine H (1993) Thioflavine-T interaction with synthetic Alzheimer's disease beta-amyloid peptides - detection of amyloid aggregation in solution. Protein Sci 2:404–410
25. Krebs MRH, Bromley EHC, Donald AM (2005) The binding of Thioflavin-T to amyloid fibrils: localisation and implications. J Struct Biol 149:30–37
26. Groenning M, Norrman M, Flink JM, van de Weert M, Bukrinsky JT, Schluckebier G, Frokjaer S (2007) Binding mode of Thioflavin-T in insulin amyloid fibrils. J Struct Biol 159: 483–497
27. Kitts CC, Bout DAV (2009) Near-field scanning optical microscopy measurements of fluorescent molecular probes binding to insulin amyloid fibrils. J Phys Chem B 113: 12090–12095
28. Wolfe LS, Calabrese MF, Nath A, Blaho DV, Miranker AD, Xiong Y (2010) Protein-induced photophysical changes to the amyloid indicator dye Thioflavin-T. Proc Natl Acad Sci U S A 107:16863–16868
29. Biancalana M, Koide S (2010) Molecular mechanism of Thioflavin-T binding to amyloid fibrils. Biochim Biophys Acta 1804: 1405–1412
30. Svergun D, Koch M (2003) Small-angle scattering studies of biological macromolecules. Rep Prog Phys 66:1735–1782
31. Putnam C, Hammel M, Hura G, Tainer J (2007) X-ray solution scattering (SAXS) combined with crystallography and computation: defining accurate macromolecular structures, conformations and assemblies in solution. Q Rev Biophys 40:191–285
32. Konarev P, Petoukhov M, Volkov V, Svergun D (2006) Atsas 2.1, a program package for small-angle scattering data analysis. J Appl Crystallogr 39:277–286
33. Chen S, Wetzel R (2001) Solubilization and disaggregation of polyglutamine peptides. Protein Sci 10:887–891

34. Fezoui Y, Hartley DM, Harper JD, Khurana R, Walsh DM, Condron MM, Selkoe DJ, Lansbury PTJ, Fink AL, Teplow DB (2000) An improved method of preparing the amyloid beta-protein for fibrillogenesis and neurotoxicity experiments. Amyloid 7:166–178

35. Lashuel HA, Hartley DM, Petre BM, Wall JS, Simon MN, Walz T, Lansbury PTJ (2003) Mixtures of wild-type and a pathogenic (E22G) form of $A\beta_{40}$ *in vitro* accumulate protofibrils, including amyloid pores. J Mol Biol 332: 795–808

Chapter 6

Amyloid Fibrils from Readily Available Sources: Milk Casein and Lens Crystallin Proteins

Heath Ecroyd, Megan Garvey, David C. Thorn, Juliet A. Gerrard, and John A. Carver

Abstract

Amyloid fibrils are a highly ordered and robust aggregated form of protein structure in which the protein components are arranged in long fibrillar arrays comprised of β-sheet. Because of these properties, along with their biocompatibility, amyloid fibrils have attracted much research attention as bionanomaterials, for example as template structures (in some cases following modification) that can be used as biosensors, encapsulators, and biomimetic materials. To use amyloid fibrils for such a range of applications will require them to be obtained relatively easily in large quantities. In this chapter, we describe methods for isolating crystallin and casein proteins from readily available sources that contain abundant protein, i.e., the eye lens and milk, respectively, and the subsequent conversion of these proteins into amyloid fibrils.

Key words Crystallins, Small heat-shock proteins, Caseins, Amyloid fibrils, Bionanomaterials, Lens proteins, Milk proteins, Protein aggregation

1 Introduction

Research into the mechanisms of protein misfolding and aggregation has been actively pursued of late because of the association of these processes with many diseases, particularly of aging, e.g., Alzheimer's and Parkinson's and cataract (1, 2). In the majority of these diseases, the aggregates are fibrillar in form (3): the protein self-assembles into long (up to several μm in length), ropelike structures with the polypeptide chain being arranged in a highly ordered cross β-sheet arrangement (Fig. 1). There are many examples of nature overcoming the cytotoxic properties of amyloid fibrils (or, more likely, their soluble prefibrillar species (1, 2)) to utilize this structural motif in so-called functional amyloid. Thus, the strength, stability, insoluble nature, and highly ordered arrangement of amyloid fibrils have been employed in a wide range of organisms for a diversity of tasks such as biofilm formation, modulation of water surface tension, and eggshell and spider silk structure (4).

Juliet A. Gerrard (ed.), *Protein Nanotechnology: Protocols, Instrumentation, and Applications*, Methods in Molecular Biology, vol. 996, DOI 10.1007/978-1-62703-354-1_6, © Springer Science+Business Media New York 2013

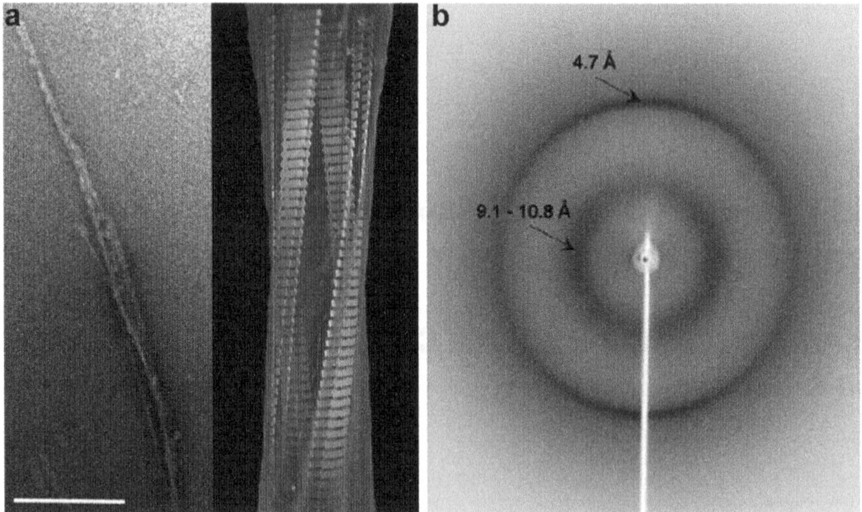

Fig. 1 The generic structure of amyloid fibrils (adapted from ref. 2). Fibrils appear as long, unbranched, ropelike fibers. (**a**) *Left panel*: An α-synuclein fibril viewed by transmission electron microscopy. This magnified view highlights the internal protofilament substructure of the fibrils formed from this protein. Scale bar is 200 nm. *Right panel*: A schematic view of an amyloid fibril formed from insulin (Reproduced from ref. 3) with permission from Proc Natl Acad Sci U S A) which shows the core structure of each filament. The typical cross β-sheet array formed from sheets of β-strands lying perpendicular to the axis of the fibril and the aligning of these β-sheets into individual filaments is shown. (**b**) A typical X-ray fiber diffraction pattern of amyloid fibrils showing the diagnostic meridional and equatorial reflections which form the cross β-sheet array

The presence of functional amyloid, along with the proposal that, under the appropriate solution conditions, any protein can form amyloid fibrils (1), has elicited much interest in the possibility of utilizing in vitro-prepared fibrils as bionanomaterials (5–7). They have potential use in a wide variety of applications such as biosensors, encapsulators, biomimetic materials, and, following functionalization, for tasks such as enhancing cell adhesion. In this chapter, we describe experimental methods that we have developed for the production of amyloid fibrils, in relatively large quantities and in a routine and reproducible manner, from two readily available sources: milk casein and eye lens crystallin proteins.

2 Materials

2.1 Fibril Formation by the Crystallin Proteins

2.1.1 Extraction of Soluble Lens Proteins

1. Fresh bovine eyes from calves less than 2 years old.
2. Solution of 50 mM Tris–HCl, 100 mM NaCl, 0.04% w/v NaN_3 (pH 7.4) buffer.
3. Centrifuge: refrigerated and able to hold 1.5 mL Eppendorf tubes.

2.1.2 Purification of Soluble Lens Proteins by Size-Exclusion Chromatography

1. AKTA FPLC system (GE Amersham, Sweden) equipped with a Sephacryl 300 HiPrep 26/60 column.
2. Dialysis tubing with a molecular mass cut-off of 10 kDa.

2.1.3 SDS-Polyacrylamide Gel Electrophoresis

1. Gel electrophoresis (SDS-PAGE) system and gel (1 mm thick, 12% v/v acrylamide).
2. Resolving gel solution: 4.0 mL of 30% v/v acrylamide/bis solution, 2.5 mL of 1.5 M Tris–HCl (pH 8.8) buffer, 0.1 mL of a 10% w/v sodium dodecyl sulfate solution, 3.4 mL of milliQ-water.
3. Stacking gel solution: 0.65 mL of 30% v/v acrylamide/bis solution, 1.25 mL of 0.5 M Tris–HCl (pH 6.8) buffer, 0.05 mL of a 10% w/v sodium dodecyl sulfate solution, 3.05 mL of milliQ-water.
4. 3× SDS-PAGE loading buffer: 0.5 M Tris–HCl (pH 6.8) 25% v/v glycerol, 2% w/v SDS and 1% w/v bromophenol blue. Add 5% v/v β-mercaptoethanol immediately prior to use.
5. SDS-PAGE running buffer: 25 mM Tris–HCl, 192 mM glycine, 0.1% w/v SDS.
6. Gel Coomassie stain solution: 0.025% w/v Coomassie brilliant blue, 40% v/v methanol, 0.5% v/v acetic acid.
7. Gel destain solution: 40% v/v methanol, 0.5% v/v acetic acid in milliQ-water.

2.1.4 Forming Fibrils of Different Morphologies from Purified Crystallin Proteins

1. Solution of 10% v/v trifluoroethanol (TFE) in water, pH adjusted to 2.0 using concentrated hydrochloric acid (HCl).
2. Solution of 100 mM EDTA stock in milliQ-water.
3. Solution of 0.1 M sodium phosphate buffer (pH 7.4) containing 1 M guanidine hydrochloride.
4. A heating block set to 60°C.

2.2 Fibril Formation by the Casein Proteins

2.2.1 Extracting κ-Casein Protein Fraction from Milk

1. Fresh, unpasteurized bovine milk.
2. 1 M HCl.
3. 1 M NaOH.
4. Ultrafiltration membrane with a molecular mass cut-off of 30 kDa.
5. Solution of 8 M calcium chloride.

2.2.2 Reduction and Carboxymethylation of κ-Casein

1. Disulfide-reducing agents: 60 mg of dithiothreitol (DTT), 19.2 g of urea, 3.6 g of Tris–HCl, 2.1 g of Tris-base (pH 8.0).
2. Disulfide alkylating agent: 150 mg of iodoacetic acid (IAA).

2.2.3 Purification of RCM-κ-Casein by Anion-Exchange Chromatography

1. AKTA FPLC system (GE Amersham, Sweden) equipped with a DEAE HiPrep 16/10 column.
2. DEAE starting buffer: 50 mM Tris–HCl, 6 M urea (pH 8.5).
3. DEAE eluting buffer: 50 mM Tris–HCl, 6 M urea, 0.5 M NaCl (pH 8.5).
4. Dialysis tubing with a molecular mass cut-off of 10 kDa.

2.2.4 Forming Fibrils from RCM-κ-Casein

1. Solution of 50 mM phosphate buffer containing 0.04% NaN_3 (pH 7.0).

2.3 Verifying the Formation of Fibrils by Crystallins and RCMκ-Casein

2.3.1 Examination of Fibrils Using Transmission Electron Microscopy

1. Solution of 2% w/v uranyl acetate (see Note 1) in water.
2. Formvar- and carbon-coated nickel electron microscopy grids.
3. Filter paper.
4. Reverse (self-closing) tweezers.

2.3.2 Monitoring Fibril Formation Using Thioflavin T

1. Solution of 50 mM glycine-NaOH buffer (pH 9.0) with 5 μM thioflavin T (ThT).
2. Fluorimeter.
3. Fluorescence cuvettes.

3 Methods

3.1 Fibril Formation by the Crystallin Proteins

The crystallin proteins are found in very high concentration in the eye lens (around 300–400 mg/mL in the center of the lens). Their well-defined supramolecular arrangement is crucial to proper refraction of light and therefore lens transparency. In mammals, there are three classes of crystallin proteins: α, β, and γ, with each class having a variety of subunits. The β- and γ-crystallins are closely sequence-related, with the subunits of both classes of proteins forming well-ordered and highly structured two-domain all β-sheet structures, with two Greek key motifs per domain (8). The two α-crystallin subunits, A and B, are closely related to each other but not to the β- and γ-crystallins. αA- and αB-crystallin are members of the small heat-shock protein family of molecular chaperone proteins and αB-crystallin is found extensively outside the lens. Structurally, they are comprised of a mixture of β-sheet and disordered regions with little or no α-helix. In the lens, both proteins associate together to form large heterogenous oligomers.

The predominance of β-sheet secondary structure in all crystallin classes may predispose them to amyloid fibril formation. Indeed, it is straightforward to form well-defined amyloid fibrils in vitro under slightly destabilizing conditions from each of the crystallin classes and from the individual crystallin proteins, for example by the addition of guanidinium chloride or trifluoroethanol at elevated temperature or reducing the solution's pH (9, 10). Furthermore, it is possible to make the entire crystallin protein mixture, as isolated from the eye lens, form amyloid fibrils under similar conditions to those for the individual crystallin subunits (5). As with the casein proteins (see below), the ready ability to isolate the crystallin proteins in significant quantities, including the entire crystallin mixture, makes these proteins potentially very useful as bionanomaterials.

3.1.1 Extraction of Soluble Lens Proteins

1. The method described below for the purification of crystallins from whole lens is based on that described previously (11).
2. Fresh bovine (see Note 2) eyes from calves less than 2 years old are obtained from local abattoirs and can be frozen and stored at –80°C or kept on ice and used immediately. Frozen eyes are thawed slowly on ice. Lenses are obtained by decapsulation, i.e., by making a small incision in the lens capsule and extruding the clear lens from the vitreous material.
3. A single lens is homogenized using a Dounce homogenizer in 3 mL of 50 mM Tris–HCl, 100 mM NaCl, and 0.04% w/v NaN_3 (pH 7.4) buffer (see Note 3). This should be done carefully in order to avoid excessive foaming.
4. The homogenate is collected with a pipette (avoiding any large tissue debris that may remain after the homogenization process) and split across two 1.5 mL Eppendorf tubes.
5. The homogenate sample is centrifuged at 18,000×*g* at 4°C for 20 min, and the resulting supernatant is carefully removed from the pellet and transferred to a clean 5–10 mL tube (see Note 4). This sample can be stored at –20°C or directly applied to the size-exclusion column for the purification of the crystallin proteins.
6. Typical yields of soluble proteins from single lenses are 200–250 mg.

3.1.2 Purification of Crystallins by Size-Exclusion Chromatography

1. These directions, for the purification of crystallin proteins by size-exclusion chromatography, assume the use of an AKTA FPLC system (GE Amersham, Sweden) equipped with a Sephacryl 300 HiPrep 26/60 column (dimensions 26 mm×600 mm). However, any fast-protein liquid chromatography system able to perform an isocratic elution is suitable.
2. Prepare approximately 2 L of 50 mM Tris–HCl, 100 mM NaCl, 0.04% w/v NaN_3 (pH 7.4) buffer and filter and de-gas the buffer by passing it through a Millipore Steritop filter connected to vacuum and leave for 10–15 min.
3. Equilibrate the Sephacryl 300 HiPrep 26/60 column with at least two column volumes (640 mL) of this buffer at a flow rate no greater than 1 mL/min.

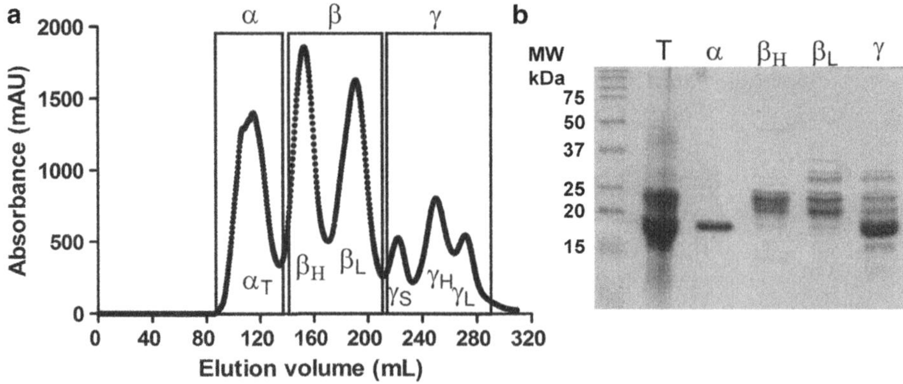

Fig. 2 Purification of crystallin proteins, extracted from bovine lens, using size-exclusion chromatography. (**a**) The three classes of crystallin (α, β, and γ) proteins elute in six major peaks from a Sephacryl 300 HiPrep 26/60 column. Fractions corresponding to peaks within each crystallin class are pooled and fibrils formed as described in Subheading 3.1. (**b**) SDS-PAGE of fractions collected from the α-, βH-, βL-, and pooled γ-crystallin peaks following size-exclusion chromatography. T, total bovine lens homogenate corresponding to that loaded onto the size-exclusion column. The molecular mass markers are given on the *left-hand side* of the gel. The volumes loaded onto the gel correspond to 1 μL of the total lens homogenate and 10 μL of the peak fractions. Note that the two subunits of α-crystallin, αA- and αB-crystallin, cannot be distinguished in this gel due to them having similar molecular masses in their monomeric forms (19.8 and 20.1 kDa, respectively)

4. Inject the sample (up to 3 mL) onto the column (see Note 5) by loading it into a suitably sized sample loop. Typically the elution is performed at a flow rate of 0.5–1.0 mL/min with 10 mL fractions collected.
5. The first peak to elute contains α-crystallin (see Note 6), followed by high molecular mass β-crystallins (β_H-crystallin), low molecular mass β-crystallins (β_L-crystallin), and finally the γ-crystallins (γ_S, γ_H, and γ_L) (Fig. 2). At this point, the composition and purity of the fractions can be assessed by sodium dodecyl-sulfate-polyacrylamide gel electrophoresis (SDS-PAGE, see below).
6. Fractions corresponding to peaks of α-, β- (containing both β_H- and β_L-crystallin), and γ-crystallins are pooled. Using these conditions typically 40 mg α-crystallin, 75 mg β-crystallin, and 60 mg γ-crystallin can be obtained from a single lens.
7. The pooled samples are transferred to dialysis tubing with a molecular mass cut-off of 10 kDa and dialysed extensively against milliQ-water (see Note 7).
8. The samples are frozen, then freeze-dried, and stored at −20°C.

3.1.3 SDS-PAGE of Fractions Collected from the Purification of Crystallin Proteins

1. These instructions assume the use of a BioRad Mini-Protean® gel electrophoresis system and a 1.0 mm thick, 12% v/v gel. However, these methods are easily adaptable to other formats.
2. A sample (20 μL) from peak fractions eluted from the size-exclusion column, as well as from the loaded sample (see

Subheading 2.1.3, item 4), is mixed with 10 μL of 3× loading buffer and heated for 5 min at 95°C.

3. Prepare the resolving gel monomer solution. Mix well and then add 25 μL of TEMED and 25 μL of a 10% w/v ammonium persulfate solution (see Note 8).
4. Pour the gel, leaving enough room for a stacking gel, and overlay with water by adding it slowly and evenly to avoid it mixing with the resolving gel solution. The gel should polymerize in about 20 min.
5. Prepare the stacking gel monomer solution. Mix well. Pour off the water that was overlaid on top of the resolving gel (see Note 9) and then add 25 μL of TEMED and 25 μL of a 10% w/v ammonium persulfate solution to the stacking gel solution, mix and pour the solution onto the top of the short plate. Insert the comb. The gel should polymerize in about 20 min.
6. Remove the comb and rinse the wells with milliQ-water or SDS-PAGE running buffer.
7. Assemble the gel electrophoresis unit, add SDS-PAGE running buffer to the upper and lower chambers of the gel unit, and load the samples (up to 20 μL for a 10 well comb) into the wells. Include pre-stained molecular mass markers in one lane.
8. Connect the gel unit to a power supply and run it at 100–200 V (constant voltage) until the dye front reaches the bottom of the gel.
9. Disassemble the gel unit and place the gel in fresh gel Coomassie stain solution for at least 30 min (see Note 10). Remove the staining solution and place into gel destain solution until the background of the gel is clear (see Note 11).

3.1.4 Forming Fibrils of Different Morphologies from Purified Crystallin Proteins

1. These methods are based on work described previously (5, 9, 10).

   Methods for forming fibrils from crystallins using trifluoroethanol are given in steps 2–7. Methods for forming fibrils from α-crystallin using guanidine hydrochlorid are given in steps 8–12.
2. Freeze-dried crystallin proteins (see Note 12) are reconstituted to 6 mg/mL in 10% v/v trifluoroethanol (TFE) in water, (pH 2.0).
3. Add EDTA to the solution to a final concentration of 1 mM by diluting from 100 μM EDTA stock solution.
4. Mix the samples by inverting them three times. Remove a 100 μL aliquot and immediately freeze at −20°C (see Note 13).
5. Incubate the samples at 60°C (see Note 14) for 18 h.
6. Remove a second 100 μL aliquot for fibril formation assessment, freeze to halt fibril formation, and store the sample at −20°C.

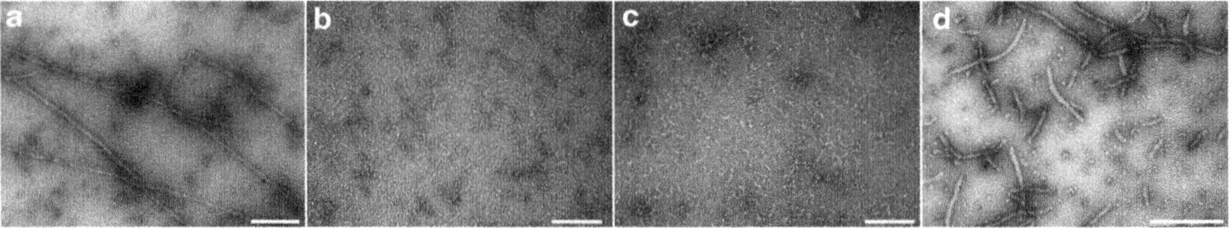

Fig. 3 Transmission electron micrographs of crystallin fibrils. Fibrils formed from (**a**) α-crystallin, (**b**) β-crystallin, and (**c**) γ-crystallin proteins using trifluoroethanol. (**d**) Fibrils formed from α-crystallin using guanidine hydrochloride. The scale bar in each panel represents 200 nm

7. Fibrillar samples may be used immediately or stored at −20°C (Fig. 3).
8. Reconstitute the freeze-dried α-crystallin to a concentration of 3 mg/mL in 0.1 M sodium phosphate buffer (pH 7.4) containing 1 M guanidine hydrochloride.
9. Remove a 100 μL aliquot and immediately freeze at −20°C (see Note 13).
10. Incubate samples at 60°C for 2 h.
11. Remove a second 100 μL aliquot for fibril formation assessment, freeze to halt fibril formation, and store at −20°C.
12. Fibrillar samples may be used immediately or stored at −20°C (Fig. 3).

3.2 Fibril Formation by the Casein Proteins

Recently, as part of a structure/function investigation of milk proteins, we identified that two of the four unrelated milk casein proteins, κ and α_{s2}, form amyloid fibrils spontaneously under physiological conditions (12–14). Importantly, their fibril formation is inhibited via a chaperone mechanism (12–15) by the action of the other casein proteins, β and α_{s1}, which contribute to the formation of casein micelles in milk. Under physiological conditions, κ-casein exists as an oligomeric (micelle-like) species in equilibrium with dissociated forms. The dissociated form of κ-casein, particularly the reduced, monomeric species, is highly amyloidogenic and aggregates through a unique fibril-forming mechanism (16, 17).

Fibrils formed from κ- and α_{s2}-casein are highly ordered and exhibit all the generic structural hallmarks of amyloid fibrils (13, 14). Our extensive studies with reduced and carboxymethylated (RCM) κ-casein have shown that it forms amyloid fibrils robustly and highly reproducibly at neutral pH (14, 18). Coupled to these characteristics, casein proteins are readily available, which makes them very attractive for use as a source of self-assembled bionanomaterials.

The following procedures are intended for the purification and preparation of RCM-κ-casein directly from milk for the production of amyloid fibrils. This involves the acid precipitation of caseins from raw (unpasteurized) milk, removal of the calcium-sensitive casein fraction (which comprises mostly α_{S1}-, α_{S2}- and β-casein) by a second precipitation step, reduction and carboxymethylation (RCM) of the subsequent calcium-soluble casein (or crude κ-casein), and, finally, the purification of RCM-κ-casein by ion-exchange chromatography.

3.2.1 Extracting κ-Casein from Milk

1. Prepare skim milk from fresh, raw milk by centrifuging the fresh milk at 2,000 × *g* for 10 min, collecting the supernatant.
2. Precipitate whole casein from the skim milk fraction (50 mL) at room temperature by adjusting the pH to 4.6 with 1 M HCl and stir the mixture for 30 min. Centrifuge at 2,000 × g for 5 min. Retain the precipitate.
3. Resuspend the protein precipitate up to a final volume of 25 mL by adding milliQ-water, then adding 1 M NaOH to bring the pH of the solution up to 7.5.
4. Acidify the mixture to pH 4.6 by again adding 1 M HCl and centrifuge for 5 min as before.
5. Resuspend the precipitate in 10 mL of milliQ-water, adjusting the pH to 6.5 with 1 M NaOH.
6. Precipitate the α_{S1}-, α_{S2}- and β-casein by adding approximately 1 mL of 8 M calcium chloride (to give a final concentration of 400 mM) and stirring the resulting mixture at 35°C for 1 h (see Note 15).
7. Centrifuge the mixture at 20,000 × *g* for 30 min. Discard the precipitate. The filtrate, which contains the κ-casein, is retained.

3.2.2 Reduction and Carboxymethylation of κ-Casein

1. To the filtrate, add 19.2 g of urea, 3.6 g of Tris–HCl, and 2.1 g of Tris-base, adjust the pH to 8.0, and bring the final volume of the solution to 40 mL (i.e., final concentrations of 8 M urea, 0.6 M Tris–HCl, and 0.4 M Tris-base).
2. Reduce the solution by adding 60 mg of DTT and stirring at room temperature for 1 h.
3. Alkylation is achieved by slowly adding 150 mg of IAA to the solution and stirring in the dark for 20 min.
4. Excess DTT, IAA and Tris buffer salts are removed by concentrating the sample to less than 1 mL using an ultrafiltration membrane or centrifugal device with a 30 kDa molecular mass cut-off. The sample is diluted to 50 mL with DEAE starting buffer and then again concentrated to 5 mL or less (see Note 16).
5. Transfer the sample directly to an anion-exchange column for further fractionation.

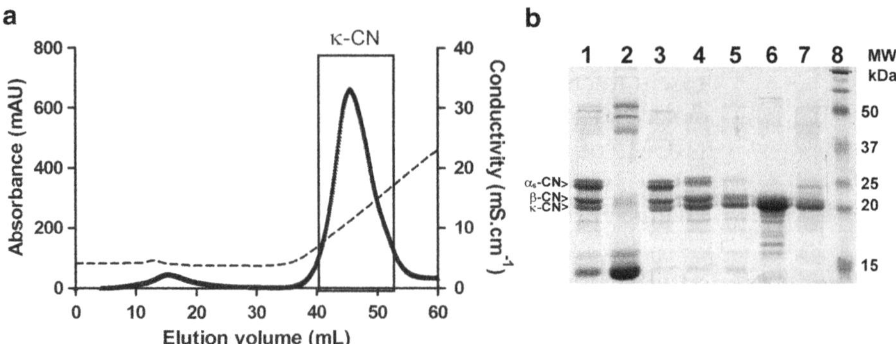

Fig. 4 Purification of RCM-κ-casein from whole milk. (**a**) RCM-κ-casein elutes as a broad peak centered at 45 mL following DEAE ion-exchange chromatography. The absorbance at 280 nm is shown on the *left-hand axis* (*solid line*) and conductivity (in mScm^{-1}) on the *right-hand axis* (*dotted line*). The fractions collected from the column are boxed and labelled (κ-CN). (**b**) SDS-PAGE of samples during the purification of RCM-κ-casein via the method described. *Lane 1* skim milk; *lane 2* filtrate at pH 4.6 (whey); *lane 3* pellet at pH 4.6 (casein); *lane 4* pellet in ~400 mM Ca^{2+} (all caseins); *lane 5* filtrate in ~400 mM Ca^{2+} (crude κ-casein); *lane 6* pooled fractions containing the peak from the DEAE column centered at 45 mL (*boxed*), i.e., purified RCM-κ-casein; *lane 7* commercially available κ-casein (Sigma); *lane 8* SDS protein markers

3.2.3 Purification of RCM-κ-Casein by Anion-Exchange Chromatography

1. As with Subheading 3.1.2, these directions assume the use of an AKTA FPLC system (GE Amersham, Sweden) equipped with a DEAE HiPrep 16/10 column. However, any fast-protein liquid chromatography system capable of gradient elution is suitable.
2. Prepare approximately 0.5 L of DEAE starting buffer and 0.25 L of DEAE eluting buffer. Filter and de-gas each buffer by passing them through a Millipore Steritop filter connected to vacuum and leave for 10–15 min.
3. Equilibrate the column with at least two column volumes (40 mL) of starting buffer at a flow rate of 2 mL/min.
4. Inject the sample (up to 5 mL) onto the column by loading it into a suitably sized sample loop. Perform the elution initially with starting buffer at 2 mL/min collecting 5 mL fractions. After one column volume, increase the concentration of eluting buffer at a linear gradient of approximately 0.4% per mL. The protein elutes as one broad peak centered at ~45 mL (Fig. 4).
5. Analyze each of the main fractions by SDS-PAGE using a 12% v/v gel. Collect and pool fractions containing the RCM-κ-casein (see Note 17).
6. The pooled samples are transferred to dialysis tubing with a molecular mass cut-off of 10 kDa and dialyzed extensively against milliQ-water (see Note 7).
7. The samples are frozen, then freeze-dried, and stored at −20°C.

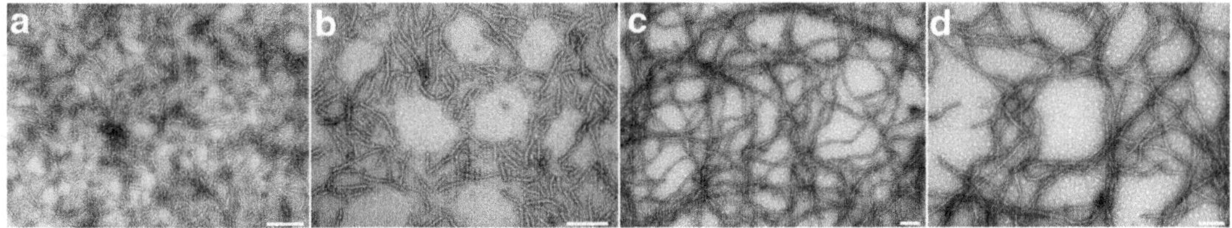

Fig. 5 Transmission electron micrographs of fibrils formed from RCM-κ-casein. Fibrils were formed at 37°C (**a**, **b**) and 50°C (**c**, **d**). The scale bar in each panel represents 200 nm

3.2.4 Forming Fibrils from RCM-κ-Casein

1. Freeze-dried RCM-κ-casein is reconstituted to 5 mg/mL in 50 mM phosphate buffer containing 0.04% w/v NaN_3, pH 7.0 (see Note 18).
2. Remove a 100 μL aliquot and immediately freeze at −20°C (see Note 19).
3. Incubate the sample at 37°C or 50°C for 24 h (see Note 20).
4. Remove a second 100 μL aliquot for fibril formation assessment and freeze to halt fibril formation. Fibrillar samples may be used immediately or stored at −20°C (Fig. 5).

3.3 Verifying the Formation of Fibrils by Crystallins and RCM-κ-Casein

All fibrils share a characteristic cross β-sheet array core structure, so called because individual fibrils are made up of sheets of β-strands which lie perpendicular to the core axis of the fibril (Fig. 1). These β-sheets stack together to form an individual protofilament. Mature fibrils are commonly composed of 2–6 protofilaments that plait together into rope-like fibers, 5–10 nm in diameter and up to a few micrometers in length. This underlying generic structure of fibrils has been resolved through techniques such as X-ray fiber diffraction, cryo-electron microscopy, and solid-state NMR spectroscopy; a cross formed by the meridional and equatorial reflections in X-ray fiber diffraction patterns being the definitive test for the presence of amyloid fibrils (Fig. 1b). However, these techniques are time-consuming and require specific expertise and facilities. Here, we outline two techniques that enable the presence of fibrils formed by crystallin and κ-casein proteins to be verified quickly and with equipment and facilities accessible to most researchers, i.e., via transmission electron microscopy to visualize the morphology of the fibrils and utilizing thioflavin T (ThT), a dye that fluoresces upon binding to the cross β-sheet array formed during fibril formation.

3.3.1 Examination of Fibrils Using Transmission Electron Microscopy

1. Remove a clean grid from the grid box using the reverse tweezers.
2. Dilute protein samples to ~0.5 mg/mL with buffer or milliQ-water.
3. Pipette 2 μL of the sample on to the carbon-coated side (see Note 21) of a single grid.

4. Rinse the grid by pipetting 10 μL of milliQ-water on to the grid. Remove the water by carefully touching the side of the grid with filter paper so that it "wicks" off.
5. Repeat step 3 twice more for a total of three rinses.
6. The sample is negatively stained by pipetting 10 μL of 2% w/v uranyl acetate solution on to the grid.
7. Remove the stain with filter paper as in step 3.
8. Allow the grid to air-dry for 5 min on the bench.
9. Store the grid in a grid box.
10. Microscopy training and guidelines on the use of the electron microscope should be observed.
11. Fibril samples can be analyzed under magnifications of 20,000–80,000 times (see Note 22) (see Figs. 3 and 5).
12. For all samples, multiple regions (at least three) on each grid should be examined to assess the structure of the deposited protein (see Note 23).

3.3.2 Monitoring Fibril Formation Using Thioflavin T

1. Thaw pre-incubation (control) and post-incubation (fibrillar) samples (see Note 24).
2. In triplicate, dilute 10 μL of samples into 1 mL of 50 mM Glycine-NaOH buffer (pH 9.0) containing 5 mM ThT.
3. Transfer the sample into a fluorescence cuvette and measure the ThT fluorescence of samples using a fluorimeter (see Note 25). Fibril formation is correlated with an increase in ThT fluorescence (due to the increase in β-sheet content of the sample).

4 Notes

1. Uranyl acetate is a water-soluble uranium compound that is often used as a negative stain in electron microscopy. Even with the relatively small amounts here, there are associated chemical and radiological hazards which require some basic safety precautions to be adopted when it is being used. Refer to the MSDS and place appropriate controls in place, with the emphasis being to avoid the possibility of inhalation or ingestion of the material.
2. Sheep or deer lenses may also be used.
3. An electric homogenizer, set on a slow speed, can be used for processing multiple lenses at once.
4. The amount of protein in the supernatant can be determined at this point using a suitable protein concentration assay such as the bicinchoninic acid (BCA) assay by diluting the sample 1:1,000 in the same buffer (i.e. 50 mM Tris–HCl, 100 mM

NaCl, 0.04% NaN_3, pH 7.4) or estimated by measuring the absorbance at 280 nm of a 1:1,000 dilution of the supernatant (one optical density unit is approximately 0.5 mg/mL of total lens soluble proteins) (11).

5. A small sample (20 μL) of crude homogenate should be retained for use on the SDS-PAGE gel if required.
6. Some preparations (particularly of older lenses) also contain a high molecular mass peak comprised of highly aggregated α-crystallin (>1 MDa in mass) which elutes in the void volume of the column, at approximately 90 mL.
7. Typically, samples are placed into 2–5 L of milliQ-water at 4°C with stirring for 2–3 days, with the milliQ-water being changed twice per day.
8. The ammonium persulfate solution should be prepared fresh daily.
9. The water can be removed from the gel by tipping the apparatus on its edge and using an adsorbent material (such as a filter paper) to "wick" it off.
10. The staining solution can be reused; however, the time taken for staining a gel will increase.
11. The destain solution can be filtered and reused.
12. This technique can be used for crystallin proteins derived from individual SEC peaks (as described in Subheading 3.1.2), combined SEC peaks or crude crystallin homogenate (5).
13. This sample is used as a control in further analyses (see Subheadings 3.3.1 and 3.3.2).
14. β-Crystallins and, to a lesser extent, γ-crystallins may form a white precipitate within the first 30 min of incubation at 60°C. This material consists of non-fibrillar aggregates (crystallin fibrils either remain within the solution or may, at very high concentrations, form a clear gel). If desired, the white precipitate can be removed by centrifugation at 13,000×g for 15 min. Then discard the pellet, return the supernatant to 60°C, and continue the incubation. Fibril formation conditions can be optimized by increasing or decreasing the protein concentration, EDTA concentration, incubation temperature, or incubation time.
15. Prior disulfide reduction of calcium-soluble casein aids fractionation of κ-casein by column chromatography.
16. Alternatively, the sample is immediately desalted against 50 mM ammonium bicarbonate using a gel filtration column and freeze-dried.
17. RCM-κ-casein migrates on SDS gels as a monomer with a molecular mass of ~19 kDa. Using this procedure, typically 80–100 mg can be obtained (from a theoretical maximum of 170 mg).

18. The absorbance at 280 nm of a 1.0 mg/mL solution of κ-casein is 0.95.
19. This sample is used as a control in further analyses.
20. Fibril formation conditions can be optimized by increasing the protein concentration, incubation temperature, or incubation time. Fibrils can be formed in buffers ranging from pH 6.0–8.0.
21. Check with the manufacturer, in most cases this will be the side which is more shiny.
22. Higher magnification may be possible; however, it becomes harder to produce quality images. We use a Philips Technai 100 transmission electron microscope, with an excitation voltage of 120 kV.
23. Where excessive tearing of the formvar grid coating occurs, dilute the samples further and prepare new grids. If no individual fibrils are observed, the fibril formation technique should be modified (such as by increasing or decreasing the protein concentration, EDTA concentration, incubation temperature, or incubation time).
24. Crystallin proteins and κ-casein bind ThT in their native state due to their high levels of β-sheet secondary structure. Therefore, control samples (non-incubated) are required to provide a baseline for the increase in ThT fluorescence that occurs due to fibril formation.
25. ThT fluorescence is measured by excitation of the sample at a wavelength of 440 nm and monitoring the emission at 490 nm.

Acknowledgments

This work was supported by grants from the Australian Research Council and Dairy Australia (DA). D.T. was supported by a postgraduate scholarship from DA and M.G. was supported by a postgraduate scholarship from Crop and Food Research New Zealand funded by the Foundation for Research Science and Technology.

References

1. Chiti F, Dobson CM (2006) Protein misfolding, functional amyloid, and human disease. Annu Rev Biochem 75:333–366
2. Ecroyd H, Carver JA (2008) Unraveling the mysteries of protein folding and misfolding. IUBMB Life 60:769–774
3. Jiménez JL, Nettleton EJ, Bouchard M, Robinson CV, Dobson CM, Saibil HR (2002) The protofilament structure of insulin amyloid fibrils. Proc Natl Acad Sci USA 99:9196–9201
4. Fowler DM, Koulov AV, Balch WE, Kelly JW (2007) Functional amyloid—from bacteria to humans. Trends Biochem Sci 32:217–224
5. Garvey M, Gras SL, Meehan S, Meade SJ, Carver JA, Gerrard JA (2009) Protein nanofibres of defined morphology prepared

from mixtures of crude crystallins. Int J Nanotech 6:258–273

6. Gras SL (2007) Amyloid fibrils: From disease to design. New biomaterial applications for self-assembling cross beta-fibrils. Aust J Chem 60:333–342
7. Cherny I, Gazit E (2008) Amyloids: not only pathological agents but also ordered nanomaterials. Angew Chem Int Ed Engl 47:4062–4069
8. Bloemendal H, de Jong W, Jaenicke R, Lubsen NH, Slingsby C, Tardieu A (2004) Ageing and vision: structure, stability and function of lens crystallins. Prog Biophys Mol Biol 86:407–485
9. Meehan S, Berry Y, Luisi B, Dobson CM, Carver JA, MacPhee CE (2004) Amyloid fibril formation by lens crystallin proteins and its implications for cataract formation. J Biol Chem 279:3413–3419
10. Meehan S, Knowles TP, Baldwin AJ, Smith JF, Squires AM, Clements P, Treweek TM, Ecroyd H, Tartaglia GG, Vendruscolo M, Macphee CE, Dobson CM, Carver JA (2007) Characterisation of amyloid fibril formation by small heat-shock chaperone proteins human alphaA-, alphaB- and R120G alphaB-crystallins. J Mol Biol 372:470–484
11. Horwitz J, Huang QL, Ding L, Bova MP (1998) Lens alpha-crystallin: chaperone-like properties. Methods Enzymol 290:365–383
12. Thorn DC, Ecroyd H, Carver JA (2009) The two-faced nature of milk casein proteins: amyloid fibril formation and chaperone-like activity. Aust J Dairy Technol 64:36–40
13. Thorn DC, Ecroyd H, Sunde M, Poon S, Carver JA (2008) Amyloid fibril formation by bovine milk alphaS2-casein occurs under physiological conditions yet is prevented by its natural counterpart, alphaS1-casein. Biochemistry 47:3926–3936
14. Thorn DC, Meehan S, Sunde M, Rekas A, Gras SL, MacPhee CE, Dobson CM, Wilson MR, Carver JA (2005) Amyloid fibril formation by bovine milk kappa-casein and its inhibition by the molecular chaperones alpha(S)- and beta-casein. Biochemistry 44:17027–17036
15. Morgan PE, Treweek TM, Lindner RA, Price WE, Carver JA (2005) Casein proteins as molecular chaperones. J Agric Food Chem 53:2670–2683
16. Ecroyd H, Thorn DC, Liu Y, Carver JA (2010) The dissociated form of kappa-casein is the precursor to its amyloid fibril formation. Biochem J 429:251–260
17. Ecroyd H, Koudelka T, Thorn DC, Williams DM, Devlin G, Hoffmann P, Carver JA (2008) Dissociation from the oligomeric state is the rate-limiting step in amyloid fibril formation by kappa-casein. J Biol Chem 283: 9012–9022
18. Carver JA, Duggan PJ, Ecroyd H, Liu Y, Meyer AG, Tranberg CE (2010) Carboxymethylated-kappa-casein: a convenient tool for the identification of polyphenolic inhibitors of amyloid fibril formation. Bioorg Med Chem 18:222–228

Chapter 7

Formation of Amphipathic Amyloid Monolayers from Fungal Hydrophobin Proteins

Vanessa K. Morris and Margaret Sunde

Abstract

The fungal hydrophobins are small proteins that are able to spontaneously self-assemble into amphipathic monolayers at hydrophobic:hydrophilic interfaces. These protein monolayers can reverse the wettability of a surface, making them suitable for increasing the biocompatibility of many hydrophobic nanomaterials. One subgroup of this family, the class I hydrophobins, forms monolayers that are composed of extremely robust amyloid-like fibrils, called rodlets. Here we describe protocols for the production and purification of recombinant hydrophobins and oxidative refolding to a biologically active, soluble, monomeric form. We describe methods to trigger self-assembly into the fibrillar rodlet state and techniques to characterize the physicochemical properties of the polymeric forms.

Key words Hydrophobins, Functional amyloid, Rodlet, Self-assembly, Amphipathic monolayer

1 Introduction

The known genomes of all filamentous fungi encode small proteins known as hydrophobins, which are characterized by the presence of high levels of hydrophobic amino acids and eight conserved cysteine residues (1, 2). These proteins are secreted by fungi and spontaneously self-assemble into polymeric structures at hydrophobic:hydrophilic interfaces, for example at the border between the air and aqueous growth medium (3). The assemblies are amphipathic and act to reduce the surface tension of the aqueous environment, which otherwise can be a barrier to the growth of hyphae up into the air and subsequent spore production. The hydrophobin monolayers also provide a water-resistant coating on fungal spores and on gas-exchange surfaces in lichens. The hydrophobic surface of the monolayers has a wettability that is even lower than Teflon®. Within the hydrophobin family two classes can be distinguished on the basis of

Juliet A. Gerrard (ed.), *Protein Nanotechnology: Protocols, Instrumentation, and Applications*, Methods in Molecular Biology, vol. 996, DOI 10.1007/978-1-62703-354-1_7, © Springer Science+Business Media New York 2013

the spacing of the cysteine residues and patterning of conserved residues and these underlie physical differences in the polymers formed by the two classes: class I hydrophobins form polymer films comprised of fibrillar structures, known as rodlets, which have dimensions of ~10 × 100–250 nm (4). The rodlets share many structural similarities with amyloid fibrils: they are insoluble polymers, bind the dyes Congo red and thioflavin T, exhibit a cross-β X-ray fiber diffraction pattern, and are long, straight, and unbranching. Class I hydrophobin rodlets can be considered to be functional amyloid, where the polymerized, fibrillar amyloid structure has biological activity and is beneficial to the fungus. Class II hydrophobin films are less robust and do not have a fibrillar morphology. However, both classes form amphipathic monolayers with similar surface activity (5).

Hydrophobins offer exciting possibilities in the area of nanobiotechnology because of their biocompatibility, molecular self-assembly properties, potential for chemical modification, and ability to assemble at either hydrophobic or hydrophilic surfaces and to reverse surface polarity. Many potential applications have been proposed, including increasing biocompatibility and wettability of materials and improving cell adhesion (1). The use of hydrophobins for biotechnology purposes has been constrained by the difficulty of producing and purifying these proteins in an active form and sufficient quantity and the lack of recombinant systems that would allow engineering of tailored functions.

The three-dimensional structures of two class II and one class I hydrophobin are now known. The structures of HFBI and HFBII from *Trichoderma reesei* have been determined by X-ray crystallography (6–8), and the solution structure of EAS from *Neurospora crassa* has been determined by heteronuclear nuclear magnetic resonance (NMR) spectroscopy (9). The hydrophobin fold consists of a small β-barrel, with a number of additional secondary structure elements accommodated on the periphery of the core (Fig. 1a). Both classes share the distinguishing feature that they have a relatively large exposed hydrophobic region on the surface (4). Recently there has been significant progress reported in the production and utilization of class II hydrophobins (10–12) and we have developed a system for the recombinant expression and in vitro oxidative refolding of class I hydrophobins (9, 13). Along with recombinant expression in bacteria, this has led to the possibility of engineering class I hydrophobins to manipulate polymerization and surface properties. We are now able to make milligram quantities of the class I hydrophobin EAS and variant proteins for biophysical studies and can produce ordered, oriented, amphipathic assemblies in vitro on hydrophobic and hydrophilic substrates. These monolayers are extremely robust and can only be depolymerized by treatment with strong acids. Self-assembly of the

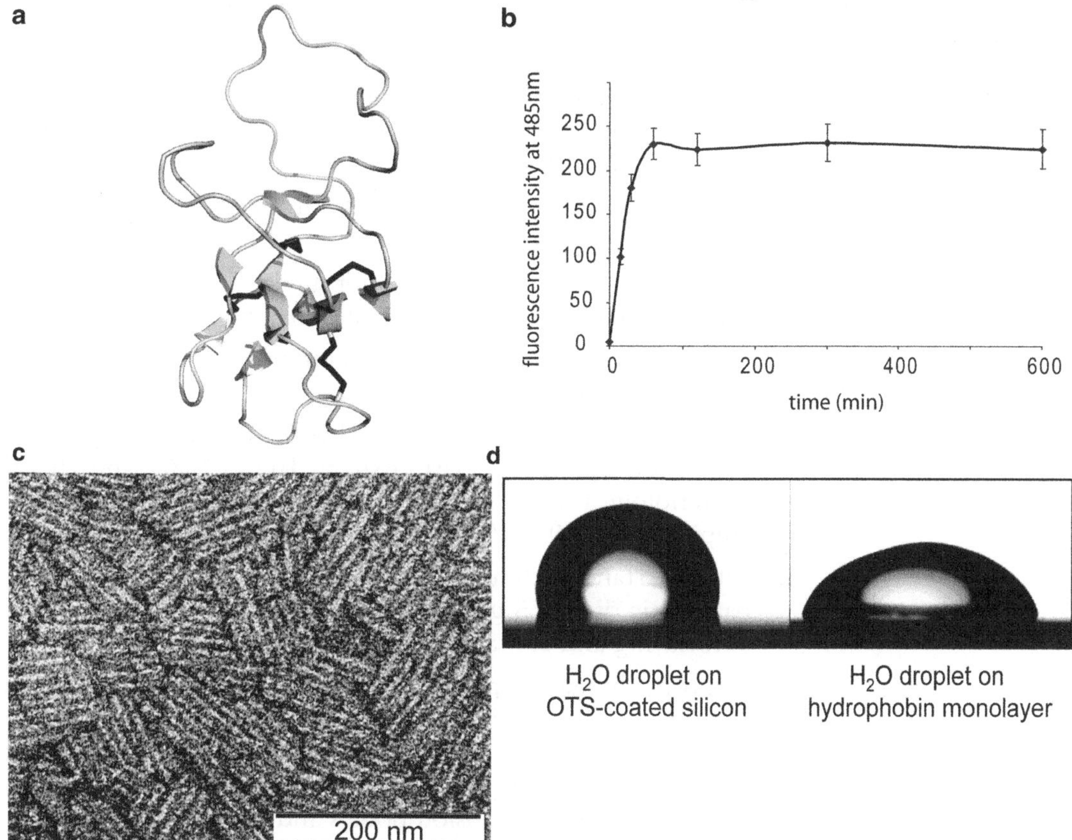

Fig. 1 Biophysical analysis of class I hydrophobins. (**a**) Cartoon representation of the structure of the soluble, monomeric form of the class I hydrophobin EAS from Neurospora crassa. This indicates that the protein is composed of a small β-barrel, constrained by four disulfides, and several long inter-cysteine loops that are accommodated on the periphery of the barrel. Figure produced from PDB 2FMC with the PyMOL molecular graphics system (15). (**b**) Time course of rodlet formation, induced by sample agitation, as measured by increase in thioflavin T fluorescence at 485 nm. This corresponds to fluorescence from ThT bound to the β-sheet core of the rodlets. (**c**) Negative stain transmission electron micrograph of EAS rodlets. (**d**) Profiles of water droplets on OTS-coated silicon and DewA hydrophobin-coated OTS-silicon. The spreading of the drop on the right indicates that the hydrophobin coating has increased the wettability of the hydrophobic OTS-coated silicon

class I hydrophobins into amyloid-like rodlets can be followed by the increase in rodlet-bound thioflavin T fluorescence and by negative stain transmission electron microscopy. The surface activity of the amphipathic hydrophobin monolayers can be characterized by contact angle measurements. These recombinant methods are also applicable to class II hydrophobin production and although the monolayers formed by class II hydrophobins are not amyloid-like, they are also amphipathic and can be characterized by contact angle measurements.

2 Materials

2.1 Protein Expression

1. pHUE expression plasmid engineered to express the desired hydrophobin as a fusion protein with an N-terminal His_6-ubiquitin tag (14).
2. Deubiquitylating enzyme produced from pUBP41 expression plasmid (14).
3. *Escherichia coli* BL21(DE3) cells or similar bacterial cells appropriate for high levels of expression of recombinant protein.
4. Ampicillin dissolved at 100 mg/ml in water and stored in single use aliquots at −20°C.
5. Luria-Bertani agar plates prepared with 10 g/l casein peptone pancreatic digest, 5 g/l yeast extract, 10 g/l NaCl, and 15 g/l bacteriological agar, sterilized by autoclaving and containing ampicillin at 100 μg/ml.
6. Luria-Bertani medium prepared with 10 g/l casein peptone pancreatic digest, 5 g/l yeast, and 10 g/l NaCl, sterilized by autoclaving and containing ampicillin at 100 μg/ml.
7. Isopropylthiogalactopyranoside (IPTG) dissolved at 1 M in water and stored in single use aliquots at −20°C.

2.2 Protein Purification and Refolding

The His_6-ubiquitin-hydrophobin fusion proteins are generally expressed in insoluble inclusion bodies and are purified by affinity chromatography using Ni-NTA agarose or equivalent matrix, under denaturing conditions. The following buffers are required for purification. In all cases, buffers can be prepared in advance and stored at room temperature but in the case of the lysis, guanidine denaturation, wash and elution buffers, β-mercaptoethanol should be added fresh immediately before use.

1. Ni-NTA agarose or equivalent affinity resin able to bind His-tagged proteins under denaturing conditions and in the presence of 10 mM β-mercaptoethanol.
2. Lysis buffer: 50 mM Tris–HCl, 150 mM NaCl, 4.8 mM β-mercaptoethanol, pH 8.0
3. Guanidine denaturation buffer: 10 mM Tris–HCl, 100 mM NaH_2PO_4, 6 M guanidine–HCl, 4.8 mM β-mercaptoethanol, pH 8.0.
4. Wash buffer: 10 mM Tris–HCl, 100 mM NaH_2PO_4, 8 M urea, 4.8 mM β-mercaptoethanol, pH 6.3.
5. Elution buffer: 10 mM Tris–HCl, 100 mM NaH_2PO_4, 8 M urea, 4.8 mM β-mercaptoethanol, pH 4.3.
6. Acetate refolding buffer: 50 mM sodium acetate, 10 mM GSH, 1 mM GSSG, pH 5.0.

7. Cleavage buffer: 50 mM Tris–HCl, 50 mM NaCl, 2 mM $CaCl_2$, pH 8.0.
8. Cleavage wash buffer: 50 mM Tris–HCl, 300 mM NaCl, 10 mM imidazole, 2 mM $CaCl_2$, pH 8.0.
9. Buffer A for rp-HPLC: 90% (v/v) water, 10% (v/v) methanol, 0.1% (v/v) trifluoroacetic acid.
10. Buffer B for rp-HPLC: 90% (v/v) acetonitrile, 10% (v/v) methanol, 0.1% (v/v) trifluoroacetic acid.

2.3 SDS-PAGE Analysis of Hydrophobins

1. 12% bis–tris polyacrylamide gel, pH 6.5–6.8 (prepared in-house), or 4–20% NuPAGE gradient gels (Novex, Invitrogen, VIC, Australia).
2. MES running buffer: 50 mM MES, 50 m Tris–HCl, 1 mM EDTA, 0.1% SDS, pH 7.5.
3. Soluble reducing agent (Invitrogen).
4. 4× LDS sample buffer (Invitrogen).

2.4 Analysis of Monomer Structure by NMR

1. NMR buffer: 20 mM sodium dihydrogen phosphate, 7% (v/v) D_2O, 40 μM DSS, pH 7.0 (not corrected for deuterium).

2.5 Rodlet Formation and Characterization by ThT Fluorescence

1. Thioflavin T (Sigma, NSW, Australia) at a concentration of 4 mM dissolved in water and stored at −20°C in single use aliquots, protected from light.
2. Vortex mixer or flat-bed shaker with attachments for small tubes or vials.
3. Wheaton glass vials, capacity 2 ml, with caps (Sigma).
4. Fluorimeter capable of excitation at 430 nm and measurement of emission at 485 nm.

2.6 Negative Stain TEM of Rodlets

1. Carbon-coated copper grids (200 mesh) with formvar or Pioloform™ support film.
2. 2% aqueous uranyl acetate solution, stored in dark bottle (BDH Chemicals, Poole, UK).
3. Filter paper.

2.7 Contact Angle Measurement

1. Silicon wafers (MMRC Pty Ltd. Mt Waverley, VIC, Australia).
2. 30% hydrogen peroxide solution (v/v).
3. 18.8 M sulfuric acid.
4. Octadecyltrichlorosilane (Aldrich, St. Louise, USA).
5. Drop shape analyzer with digital camera and associated software (Kruss DSA 10MK2, Hamburg, Germany).

3 Methods

3.1 Protein Expression

1. Transform *E. coli* BL21(DE3) cells with pHUE plasmid-encoding His_6-ubiquitin-hydrophobin fusion protein. Select transformed cells on LB_{amp} agar plates by growth at 37°C overnight.
2. Use single colony of transformed bacteria to inoculate a small volume of LB_{amp} medium and grow overnight to saturation.
3. Inoculate large volume LB_{amp} culture with overnight culture to give a starting OD_{600} of 0.05. Grow at 37°C with shaking until OD_{600} is ~0.7. Retain a sample of these preinduction cells for analysis by SDS-PAGE.
4. Add IPTG to a final concentration of 0.5 mM and grow for a further 2 h then harvest by centrifugation and store cell pellet at −20°C (see Note 1).

3.2 Protein Purification and Refolding

1. Resuspend bacterial cell pellets in lysis buffer (~10 ml/g of cells). Lyse by sonication on ice and separate soluble material from bacterial inclusion bodies and cell debris by centrifugation at 12,000 × *g* for 20 min (see Note 2).
2. Add β-mercaptoethanol to guanidine denaturation buffer immediately before use. Resuspend pellet in the same volume of guanidine denaturation buffer as used for lysis and solubilize proteins by stirring with a small magnetic stirrer for ~30 min, or until completely resuspended. A small amount of insoluble cell debris will remain after protein has been solubilized and this should be removed by a further centrifugation step at 12,000 × *g* for 20 min.
3. Transfer the supernatant, now containing the solubilized His_6-ubiquitin-hydrophobin fusion protein, into a clean tube.
4. Place clean, charged Ni-NTA beads in column (~2.5 ml per l of culture). Wash thoroughly with Milli-Q™ water (MQW) to remove 20% ethanol storage solution. Equilibrate with guanidine denaturation buffer by running 3 column volumes (CV) through the beads.
5. Incubate Ni-NTA agarose beads with the solubilized protein supernatant for 30–40 min with gentle agitation or allow the solution to flow slowly through the beads.
6. Wash the Ni-NTA agarose with 10 CVs of wash buffer to remove unbound proteins. Retain fractions for analysis by SDS-PAGE.
7. Elute fusion protein by addition of elution buffer to the matrix. Collect the eluted proteins in 1 CV fractions.
8. Analyze the fractions with SDS-PAGE to identify fractions containing significant concentrations of desired protein and

store at 4°C until required. Only urea-containing fractions can be analyzed by PAGE.

9. Prepare the acetate refolding buffer by weighing out the solid sodium acetate and adding it to 1 l of Milli-Q™ water. Weigh out reduced glutathione and oxidized glutathione and add these as solids to the liter of acetate while stirring to dissolve. Adjust pH to 5.0 and use buffer immediately.
10. Pool fractions for refolding and transfer to dialysis tubing. Dialyze against at least 25× volume of fresh dialysis buffer for 6–8 h at room temperature. Transfer tubing-containing protein into fresh refolding buffer and dialyze for a further 12–18 h at room temperature.
11. Transfer the dialysis tubing containing the protein into 1 l of cleavage buffer and dialyze for at least 4 h. Add deubiquitylating enzyme and incubate at 37°C for at least 2 h, or until cleavage is complete, as judged by SDS-PAGE (see Note 3).
12. Add NaCl and imidazole to the hydrophobin-containing solution to concentrations of 300 mM and 10 mM, respectively. Pass this solution back through Ni-NTA beads that have been pre-equilibrated with cleavage wash buffer, to remove the cleaved His_6-ubiquitin tag and collect the flow-through. Wash the resin with cleavage wash buffer and collect five 1 CV wash fractions. Analyze flow-through and wash fractions by SDS-PAGE.
13. Combine hydrophobin-containing samples and adjust the pH of this solution to pH 5 by addition of HCl.
14. Further purify the hydrophobin using a C_{18} reverse-phase column, running in 10% methanol, 0.1% trifluoroacetic acid (Buffer A for rp-HPLC), with a gradient of increasing acetonitrile (from Buffer B for rp-HPLC). Monitor absorbance at 215 and 280 nm and collect peak corresponding to hydrophobin (see Note 4). Freeze-dry and store lyophilized at −20°C.

3.3 SDS-PAGE Analysis of Hydrophobins

1. Prepare hydrophobin samples with 4× LDS sample buffer and addition of fresh reducing agent immediately before use and heat at 95°C for 5 min (see Note 5).
2. Analyze on 12% or 4–20% gradient polyacrylamide gels, running in MES buffer.
3. Stain and destain gel for visualization of protein bands as for conventional protein SDS-PAGE.

3.4 Analysis of Monomer Structure by NMR

1. Dissolve lyophilized hydrophobin sample from reverse-phase purification in NMR buffer at 50–100 μM and collect ^1H spectrum. Folded protein will display sharp, dispersed peaks in the amide region of the spectrum (10–6 ppm) and upfield-shifted methyl signals between 0 and 1 ppm (see Note 6).

3.5 Rodlet Formation by Class I Hydrophobins and Characterization by ThT Fluorescence

1. Prepare hydrophobin stock solution at 0.5–1 mg/ml in water in a glass vial.
2. Use this to prepare samples in glass vials with a final concentration of hydrophobin of 5–20 μM (see Note 7) in 40 μM ThT in 50 mM Tris–HCl, pH 8.0.
3. Set fluorimeter to excite at 435 nm and to collect an emission spectrum over 450–600 nm, or emission intensity at 485 nm, with slit widths of 5 nm.
4. Control samples will not be agitated.
5. Read fluorescence in all samples before agitation is started.
6. Induce rodlet formation by agitation of the sample, either on a vortex mixer or flat-bed shaker at room temperature. Optimum temperature for rodlet formation may vary for different hydrophobins.
7. Continue agitation until an increase in fluorescence is observed and reaches a maximum, as illustrated in Fig. 1b (see Note 8).

3.6 Negative Stain TEM of Class I Hydrophobin Rodlets

1. Prepare a sample of hydrophobin in filtered MQW at 10 μM.
2. Incubate a 20 μl drop of protein solution on Parafilm™ at room temperature for 10 min or under conditions where rodlets are known to form at the air:water interface (see Note 9).
3. Float grid on surface of drop for 1 min.
4. Wick off excess liquid by contact with filter paper.
5. Wash grid by incubation on successive drops of filtered MQW and removal of excess liquid at each stage by touching the edge of the grid to filter paper.
6. Float grid on the surface of a drop of 2 uranyl acetate stain for 10 min. Remove excess stain by touching edge of grid with filter paper.
7. Examine in a transmission electron microscope. Rodlets are usually associated laterally into bundles, with the length of the rodlets dependent on the protein and speed of the self-association (Fig. 1c).

3.7 Contact Angle Measurements with Class I and Class II Hydrophobins

1. Prepare silicon wafers for coating with octadecyltrichlorosilane (OTS) by first cleaning with piranha solution. In a fume hood and using appropriate protective clothing, produce piranha solution in a clean Pyrex™ dish by slowly adding 30% hydrogen peroxide solution to 18.8 M sulfuric acid (3:7 by volume). Heat this solution to 100–110°C and then immerse wafers for 10 min. Wash the wafers thoroughly with MQW water and then air-dry. Immerse cleaned wafers in a 5 mM solution of OTS in dichloromethane and incubate at room temperature for 30 min. Wash coated wafers with fresh dichloromethane, followed by ethanol and then MQW water.

Remove any excess OTS precipitate from the surface of the wafers by wiping with a tissue.

2. Incubate drop of hydrophobin solution (10 μM) on wafer for 30 min at room temperature (see Note 8).
3. Wash wafer thoroughly with MQW.
4. Place a 10 μl MQW drop on the hydrophobin-coated region of the wafer.
5. Collect digital image of drop profile and measure contact angle with drop shape analysis software, e.g., using Kruss DSA 10MK2 analyzer and associated software (Fig. 1d).

4 Notes

1. Exact conditions for maximum yield of fusion protein may vary between hydrophobin and may require optimization of expression time, expression temperature, and IPTG concentration.
2. Some His_6-ubiquitin-hydrophobin fusion proteins may be expressed in a soluble form in the cytoplasm. In particular, class II hydrophobins may be expressed in the soluble fraction. However, the hydrophobin component is not always correctly folded when it is in the soluble fraction. We generally find that the best approach is to denature and reduce the protein regardless of whether it is produced in the soluble fraction or in inclusion bodies and to carry out the purification under denaturing conditions, followed by oxidative refolding. If the protein appears to be in the soluble fraction, then whole cells are lysed by resuspension in guanidine denaturation buffer and stirring for 30 min at room temperature. Any remaining insoluble material and cell debris is removed by centrifugation, and the solubilized whole cell extract applied to Ni-NTA agarose.
3. The amount of deubiquitylating enzyme required must be determined for each fresh batch of enzyme.
4. The yield of correctly refolded hydrophobin is variable (~50–80%) and this is reflected in the HPLC elution trace. Usually there is more than one protein peak observed but only one peak corresponds to correctly folded protein that is able to self-assemble at hydrophobic–hydrophilic interfaces. Protein from all peaks should be collected and examined by ^{1}H NMR to determine whether it is folded and subjected to vortexing and incubation with ThT to determine whether it can form amyloid-like rodlets.
5. Hydrophobins do not stain very strongly with Coomassie Brilliant Blue but effective staining of hydrophobins after SDS-PAGE can be achieved if the disulfides are completely reduced immediately before electrophoresis. Fresh reducing agent must

be used and the liquid form of dithiothreitol from Invitrogen appears to be most effective.

6. The formation of four disulfides can also be used to monitor correct refolding of hydrophobins. This can be confirmed by high-resolution mass spectrometry since formation of the four disulfides results in loss of eight protons from the protein.
7. The optimal protein concentration for rodlet formation may differ for each hydrophobin and so a range of protein concentrations should be tested with each hydrophobin to determine the optimum.
8. The rate of rodlet formation is different for each hydrophobin and will need to be determined for each individual protein. We have observed a range of 5–240 min. The rate of rodlet formation may also be influenced by temperature and rate of agitation of the solution.
9. Hydrophobin rodlets formed by agitation of the solution may also be examined by negative stain transmission electron microscopy. Grids are floated on the surface of rodlet-containing samples for 10 min, and then excess solution is removed by wicking with filter paper before the grid is stained in the normal way. Rodlets may appear tangled or bundled when prepared in this way.

Acknowledgements

The authors would like to thank Dr. Ann Kwan for development of the recombinant expression and oxidative refolding protocol and for her advice and assistance. This work was supported by funding from the National Health and Medical Research Council of Australia (CDA402831) and the Australian Research Council (LP0776672 and DP0879121).

References

1. Linder MB, Szilvay GR, Nakari-Setala T, Penttila ME (2005) Hydrophobins: the protein-amphiphiles of filamentous fungi. FEMS Microbiol Rev 29:877–896
2. Wösten HAB (2001) Hydrophobins: multipurpose proteins. Ann Rev Microbiol 55: 625–646
3. Wosten HAB, de Vocht ML (2000) Hydrophobins, the fungal coat unravelled. Biochim Biophys Acta Rev Biomembr 1469:79–86
4. Sunde M, Kwan AH, Templeton MD, Beever RE, Mackay JP (2008) Structural analysis of hydrophobins. Micron 39:773–784
5. Wosten HA, de Vocht ML (2000) Hydrophobins, the fungal coat unravelled. Biochim Biophys Acta 1469:79–86
6. Hakanpaa J, Linder M, Popov A, Schmidt A, Rouvinen J, Linder MB, Szilvay GR, Nakari-Setala T, Penttila ME (2006) Hydrophobin HFBII in detail: ultrahigh-resolution structure at 0.75 A. Acta Crystallogr D Biol Crystallogr 62:356–367
7. Hakanpaa J, Paananen A, Askolin S, Nakari-Setala T, Parkkinen T, Penttila M, Linder MB, Rouvinen J (2004) Atomic resolution structure of the HFBII hydrophobin, a self-assembling amphiphile. J Biol Chem 279:534–539

8. Hakanpaa J, Szilvay GR, Kaljunen H, Maksimainen M, Linder M, Rouvinen J, Popov A, Schmidt A (2006) Two crystal structures of Trichoderma reesei hydrophobin HFBI–the structure of a protein amphiphile with and without detergent interaction. Protein Sci 15:2129–2140
9. Kwan AH, Winefield RD, Sunde M, Matthews JM, Haverkamp RG, Templeton MD, Mackay JP (2006) Structural basis for rodlet assembly in fungal hydrophobins. Proc Natl Acad Sci U S A 103:3621–3626
10. Wang X, Wang H, Huang Y, Zhao Z, Qin X, Wang Y, Miao Z, Chen Q, Qiao M (2010) Noncovalently functionalized multi-wall carbon nanotubes in aqueous solution using the hydrophobin HFBI and their electroanalytical application. Biosens Bioelectron 26:1104–1108
11. Zhao ZX, Qiao MQ, Yin F, Shao B, Wu BY, Wang YY, Wang XS, Qin X, Li S, Yu L, Chen Q (2007) Amperometric glucose biosensor based on self-assembly hydrophobin with high efficiency of enzyme utilization. Biosens Bioelectron 22:3021–3027
12. Zhao ZX, Wang HC, Qin X, Wang XS, Qiao MQ, Anzai JI, Chen Q (2009) Self-assembled film of hydrophobins on gold surfaces and its application to electrochemical biosensing. Colloids Surf B Biointerfaces 71(1):102–106
13. Kwan AH, Macindoe I, Vukasin PV, Morris VK, Kass I, Gupte R, Mark AE, Templeton MD, Mackay JP, Sunde M (2008) The Cys3-Cys4 loop of the hydrophobin EAS is not required for rodlet formation and surface activity. J Mol Biol 382(3):708–720
14. Catanzariti AM, Soboleva TA, Jans DA, Board PG, Baker RT (2004) An efficient system for high-level expression and easy purification of authentic recombinant proteins. Protein Sci 13:1331–1339
15. Schrodinger LLC (2010) The PyMOL molecular graphics system, Version 1.3r1.

Chapter 8

Proteins and Peptides as Biological Nanowires: Towards Biosensing Devices

Laura J. Domigan

Abstract

The current landscape of nanotechnology is such that attention is being given to those materials that self-assemble, as a mode of "bottom-up" fabrication of nanomaterials. The field of nanotubes and nanowires has long been dominated by carbon nanotubes and inorganic materials. However in more recent years, the search for materials with desirable properties, such as self-assembly, has unsurprisingly led to the biological world, where functional nanoscale biomolecular assemblies are in abundance.

Potential has been seen for a number of these assemblies to be translated into functional nanomaterials. The early days of bionanotechnology saw a lot of attention given to DNA molecules as nanowires, and proteins and peptides have now also been seen to have promise in this area. With most of the biological structures investigated having low conductivity in the native state, the use of biomolecules as templates for the formation of metallic and semiconductor nanowires has been the direction taken.

This chapter will discuss the use of various biomolecules and biomolecular assemblies as nanowires, with a particular emphasis on proteins, beginning with an introduction into the field of nanotubes and nanowires. Many applications are now recognized for nanowires, but for brevity, this chapter will focus solely on their use as biosensors, using glucose sensors as a case study.

Key words Nanotubes, Nanofibers, Nanowires, Nanosensors, Biosensors, Glucose sensing, Amyloid fibrils, Peptides, Actin, Microtubules, Collagen

1 Introduction

The term bionanotechnology refers to the interaction between the fields of biology and nanotechnology, with nanotechnology being defined as the creation, design, and manipulation of structures or particles with dimensions smaller than 100 nm (1). Once a field solely utilized by engineers and physical scientists, nanotechnology is now being used by the biological and medical research communities, and nanotechnology is also drawing inspiration, and in some cases raw materials, from the biological world.

In the search for new nanomaterials and components, a focus has been placed on those that self-assemble, with this "bottom-up" approach to manufacturing being recognized as the way of

Juliet A. Gerrard (ed.), *Protein Nanotechnology: Protocols, Instrumentation, and Applications*, Methods in Molecular Biology, vol. 996, DOI 10.1007/978-1-62703-354-1_8, © Springer Science+Business Media New York 2013

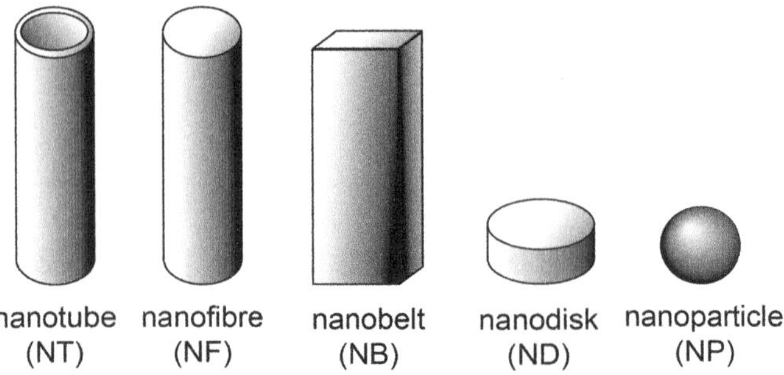

Fig. 1 A schematic representation of common nanostructures

manufacturing in the future (2). Traditional "top-down" fabrication of nanomaterials and components involves processes such as lithography to produce such structures by patterning; by contrast, "bottom-up" assembly refers to the creation of nanostructures via self-assembly from relatively simple building blocks, and biology can aid in the provision of these building blocks.

Nanostructures include nanotubes (NT), nanofibers (NF), nanobelts (NB), nanodisks (ND), and nanoparticles (NP), as shown in Fig. 1. The dimensions of these structures are usually less than 100 nm, for example, platinum nanoparticles are usually in the 6–10 nm diameter size range (3), and carbon nanotubes can vary from 0.4 to 100 nm diameter, with variable lengths (4, 5).

Many biological molecules and macromolecular assemblies exist on a very similar scale, for example, protein microtubules are 25 nm in diameter and several microns long (6), and adenovirus particles have diameters of approximately 100 nm (7), and so it is not surprising that biological molecules have attracted attention as new nanostructures.

The focus of this chapter is on the use of protein/peptide nanotubes (PNT) and nanofibers (PNF) as nanowires, or scaffolds for the formation of nanowires. Nanowires, be they biological, metallic, or semiconducting, have a range of different applications in the fields of electronics, optics, magnetic medium, thermoelectronics, and sensing. This range of applications is due to the unique properties that are displayed by nanowires. One of these applications is in biosensing. In this chapter, the role of nanowires in biosensing devices will be discussed, using glucose sensing as a case study.

2 Nanowires

Nanowires, along with nanotubes, nanobelts, and nanosprings, are a type of one-dimensional (1-D) nanostructure. 1-D nanostructures have been the subject of increasing research focus due to their unique properties, as when materials are reduced to the nanometer scale, new electrical, optical, and mechanical properties are often displayed. The high aspect ratio of 1-D nanostructures also provides a large surface-to-volume ratio, which can be beneficial to a number of applications, for example, the immobilization of enzymes for biosensing. There are also other properties that enhance these materials' desirability and applicability, such as control over dimensions and arrangement, and stability under a range of conditions, particularly those conditions that may be encountered in potential applications.

In general, 1-D nanostructures are formed by promoting the crystallization of solid-state structures along one direction, and this process can be carried out by a range of different mechanisms, with the main mechanisms identified by Wanekaya et al. (8) as:

1. The use of 1-D template structures to direct the formation of the nanostructure.
2. The use of the intrinsically anisotropic crystallographic structure of a solid to achieve 1-D growth.
3. The use of a liquid/solid interface to reduce the symmetry of a seed.
4. The use of appropriate capping agents to kinetically control the growth rates of various facets of a seed.

Nanowires can be formed from a number of different materials, with carbon nanotubes (CNT) and crystalline nanowires such as silicon nanowires (SiNW) gaining the most attention since their discovery in 1991 and 1998, respectively (9, 10). Carbon nanotubes can be either single-walled (SWNT) or multi-walled (MWNT). SWNTs are where there is a single graphite sheet which is wrapped into a cylindrical tube, and MWNTs are essentially made up of a number of SWNTs that are "concentrically nested like rings of a tree trunk" (11). Carbon nanotubes can be formed by a number of different methods such as chemical vapor deposition, carbon-arc discharge, and laser ablation. For a more comprehensive description of carbon nanotubes and their applications in nanotechnology, the reader is directed to a number of good reviews, as well as the previous edition of this book (11–14).

Silicon dominates the semiconductor industry and microelectronics, and since the creation of silicon "whiskers" with macroscopic dimensions in 1957 (15), the focus has been on getting this material smaller, with SiNW now existing with diameters of less than 10 nm (16). SiNW can be synthesized by both "top-down" and "bottom-up" approaches, such as e-beam lithography (EBL) and chemical vapor deposition, respectively. There are a number of good reviews available for more detail on SiNW and their use in nanotechnology (16–19).

Nanowires are also often formed from other metals, semiconductors, and inorganic compounds. Of interest to this text are nanowires formed from biological materials, specifically proteins and peptides.

3 Biological Nanowires

Not only are the dimensions of many biological macromolecules comparable to nanoscale building blocks, the presence of high-aspect-ratio elongated structures in biology means that a number of biological macromolecules and macromolecular assemblies have been investigated as nanowires (20). With most of the biological structures investigated having low conductivity in the native state, the use of biomolecules, such as nucleic acids, viruses, proteins, and peptides, as templates for the formation of metallic and semiconductor nanowires has been the direction taken. The different biomolecules that have been investigated for use as nanowires, along with their dimensions, sources, and with what they have been modified, are summarized in Table 1. Nucleic acids, viruses, and of course protein and peptide nanowires are discussed in further detail in the subsequent text.

3.1 Nucleic Acids

DNA molecules have been identified as interesting building blocks for nanotechnology due to their size, chemical robustness, and the ability to synthesize large amounts for relatively low costs. In terms of size, the diameters of ssDNA and dsDNA are approximately 1 nm and 2 nm, respectively, and the length of DNA can be easily tuned, from nanometers to microns, using a range of techniques such as PCR, DNA ligation, and enzymatic digestion, with the length per nucleoside subunit 0.34 nm (60).

In regard to using the DNA molecule itself as a nanowire, the inherent conductivity of DNA is in debate, with current values in the nanoAmp range recorded under applied potentials up to 6 V for DNA networks of 100 nm wires (61), and studies also reporting that DNA is insulating at lengths greater than 40 nm, where currents remained below the noise level of approximately 1 pA under applied potentials up to 10 V for bundles of approximately ten DNA molecules (62).

What has been of more interest, due to these poor electrical characteristics, is the use of DNA molecules as templates for self-assembly of nanowires. This work began in 1998, when Braun created the first conductive nanowire using DNA as a template for the formation of conductive silver nanowires (21). Since this pioneering study, DNA nanowires have been formed via modification with silver (21), gold (22, 63), platinum (24, 25), palladium (26, 27), and copper (28).

3.2 Viruses

Whole biological organisms have also attracted attention as nanowires, such as the tobacco mosaic virus (TMV). TMV is a rod-shaped RNA virus, which is formed from the self-assembly of 2,130 coat protein subunits which are arranged helically with a coil of RNA that spans the length of the virus. The TMV is uniform in

Table 1
Summary of the variety of different biomolecules and biomolecular assemblies that have been investigated as nanowires, or as templates for nanowire formation

Biomolecule/biomolecular assembly	Dimensions (diameter/length)	Source	Modified nanowires (biomolecule as a template)	References
(A) Nucleic acids				
DNA	1–2 nm/tunable in micron range	PCR	Au, Ag, Pt, Pd, Cu	(21–28)
(B) Viruses				
Tobacco mosaic virus (TMV)	18 nm, 4 nm (central channel)/300 nm		CdS, PbS, Pd, FeO, Si, Ni, Co, Ag, Au, Pt, CoPt, $FePt_3$	(29–33)
(C) Proteins				
Collagen	1.5 nm/300 nm	Self-assembly from proteins produced by recombinant expression	Au, ZnO	(34–36)
Actin	8 nm/up to 10 μm		Au	(37)
Microtubules	25 nm/several microns	Isolated from porcine brain tissue	Ag, Au, Co, FeO, Ni, Pd,	(38–40)
Amyloid fibrils	10–20 nm/up to several microns	Self-assembly from proteins produced by recombinant expression, and from crude waste materials	Conducting polymers (PEDOT-S, PTAA, PPF), H_2PtCl_6, Ag, Fe^{3+}	(41–50)
(D) Peptides				
Peptide nanotubes	1–100 nm/up to several microns in length	Self-assembly from engineered peptide solutions	Au, Ag, Cu, Pt, Zns	(51–56)
Amyloid-like peptide nanotubes, e.g., FF peptide	1–100 nm/up to several microns in length		Au, Ag, Co_3O_4	(57–59)

diameter and length, existing with dimensions of 18 nm and 300 nm, respectively. The TMV also has a 4 nm hollow core, and has chemically distinct internal and external surfaces, meaning that it may be modified on both its internal and external surfaces.

Nanowires have been formed using TMV as a template with surface modification of the TMV by cadmium sulfide, lead sulfide, iron oxide, silica (29), platinum, palladium, and gold (30, 31). The central channel of TMV has also been used as a template, with 3 nm nickel and cobalt nanowires being produced this way (32), as well as a variety of other nanowires (31, 33).

Other viruses have also been used for this purpose, such as the M13 bacteriophage, which was used as a scaffold for the synthesis of ZnS, CdS, CoPt, and FePt nanowires (64).

4 Protein Nanowires

There are a number of naturally occurring fibrous protein structures, and globular proteins can also be induced to form fibrous structures, as in the case of amyloid fibrils. The morphology and dimensions of these protein structures make them excellent candidates for the formation of self-assembling 1-D nanostructures.

Protein structures have the advantage of chemical and biological diversity, as well as providing building blocks with heterogeneity. The amino acid building blocks also give the opportunity to exploit the chemistry of the various side chains for derivatization and functionalization, be it for metallic modification or the immobilization of enzymes. The option for genetic control of the primary sequence further enhances this property, and via recombinant expression it is possible to easily generate large amounts of a protein of interest, making them a readily available source. Additional to these properties, fibrous proteins have a higher relative stability as compared to globular proteins, and unique mechanical properties (65).

The use of proteins as biological nanowires has been explored in some detail, with a variety of different protein structures used. Discussed in this section will be a range of protein structures that have been used for the creation of protein nanowires, namely, collagen, actin, microtubules, and amyloid fibrils. Peptide nanowires will be covered in Subheading 5. These protein/peptide nanotubes (PNTs), and nanofibers (PNFs), have been used as template molecules in a number of different ways. Figure 2 shows a schematic representation of the three most common ways: by modification with metallic nanoparticles (NPs), conducting polymers (CPs), and metal ions. Of course, there are a number of different methods via which these modifications can take place, examples of which are given in the text.

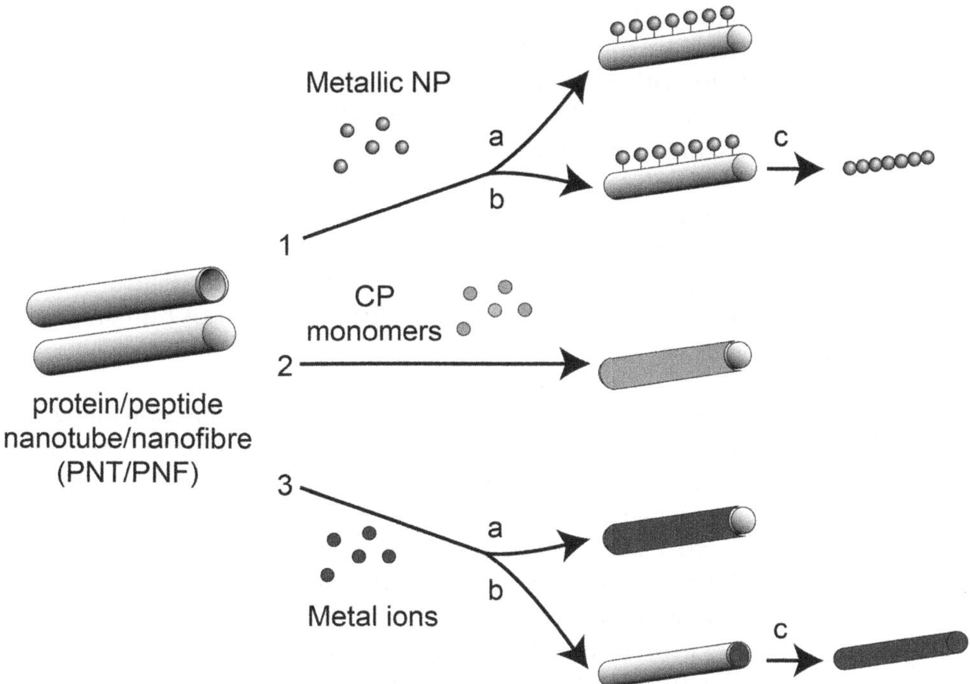

Fig. 2 Schematic showing the most common ways in which protein/peptide nanotubes/nanofibers (PNTs/PNFs) are used for the templating of conductive nanowires. (*1*) The PNT/PNF is modified with metallic nanoparticles (NPs). This may occur such that the PNT/PNF itself is decorated with metallic NPs (**a**), or with the PNT/PNF being used as a "sacrificial" template for the formation of a chain of metallic NPs (**b**). (*2*) The PNT/PNF is modified with conducting polymer (CP) monomers, with the polymerization reaction being carried out in the presence of the PNT/PNF, forming a coating of conducting polymer on the surface of the PNT/PNF. (*3*) The PNT/PNF is modified with metal ions. (**a**) A coating of metal may be formed on the exterior of the PNT/PNF, or (**b**) on the interior of a PNT. In some cases, (**c**) the PNT can then be removed, leaving a thin metallic nanowire

4.1 Collagen

As the major component of the extracellular matrix of bone, cartilage, skin, blood vessels, and corneas, collagen is one of the most abundant proteins in higher organisms. The collagen triple helix is approximately 300 nm long and 1.5 nm in diameter, and collagens consist of one or more of these triple helices (66). It is this hierarchical design, along with the relative dimensions of the building molecules, that gives collagen its superior mechanical properties, creating a tough and robust material (67).

Collagen has been modified to form both gold and zinc oxide nanowires (34–36). Bai et al. used a collagen-like triple helix, which is formed from a polypeptide that was genetically engineered and contains a fragment from the natural collagen sequence (35). Recombinant technology was used to create a template for nanowire formation that was monodisperse and easily mineralized with metal ions (Fig. 2, 3a). The triple helix was incubated with trimethylphosphine gold chloride ([$AuPMe_3Cl$]) for 4 days and then reduced by hydrazine hydrate for 1 day at 4 °C, which resulted in the formation of Au crystals on the helix. Precoating of the triple helix with a Au-mineralizing peptide, which has a high affinity for organic Au salts, was also carried out, via which a more uniform coating was obtained.

With collagen being using largely in tissue engineering applications, the application of metallically modified collagen has also been investigated here. Orza et al. constructed a matrix of gold-coated collagen nanofibers and demonstrated how these could be used for the growth and differentiation of human adult stem cells (36).

4.2 Actin

Filamentous actin (F-actin), formed by the polymerization of monomers of globular actin (G-actin), has a diameter of 8 nm and exists in lengths up to 10 μm (68). In vivo, the polymerization and depolymerization of actin is a dynamic equilibrium; however, in vitro, actin can be stabilized against depolymerization (69), and this is essential for the use of actin as a filamentous building block, or as a nanowire.

Protein-metal nanowires have been formed based on the use of G-actin as a molecular building block, with the filaments being coated with gold nanoparticles (37), the process shown in Fig. 2, 1a. The method for this modification involved firstly the preparation of an F-actin filament by the polymerization of G-actin monomers in the presence of ATP, NaCl, and $MgCl_2$. The F-actin filament was then reacted with 1.4 nm Au-NPs that were modified with a single *N*-hydroxysuccinimide active ester, which react with surface amide groups on the F-actin filament, covalently attaching the Au-NPs. The F-actin filament was then dissociated back to the monomeric units by removal of ATP, Mg^{2+}, and K^{+}, by dialysis, and an average loading of one NP per G-actin monomer was seen. The Au-NP modified G-actin was then repolymerized and the resulting filaments were subjected to catalytic enlargement of the Au-NP to produce a continuous gold wire.

4.3 Microtubules

Microtubules are proteinaceous cylindrical structures which are assembled from tubulin protofilaments, resulting in a structure with an external diameter of 25 nm and several microns in length (6). In vivo, they are part of the cytoskeleton, and are involved in processes such as motility and intracellular transport. Tubulin, from which microtubules are formed, can also form other structures, such as sheets, spirals, and rings. Nanowires of nickel, cobalt, silver, gold, and iron oxide have all been formed around the microtubule template (38–40).

4.4 Amyloid Fibrils

Amyloid fibril is the name given to the insoluble fibrous quaternary structure formed by the assembly of normally soluble protein or peptide monomers into intermolecularly hydrogen bonded β-sheets

(70). The name "amyloid" was given based on the first identification of these insoluble deposits (71), where they were named so due to them being able to be stained with iodine, making them amylose-like. In later years, protein components were identified (72), but the name amyloid has remained.

Amyloid fibrils are best known for their role in a number of diseases, such as Alzheimer's disease, type II diabetes, Parkinson's disease, and Creutzfeldt-Jakob disease (70). In these diseases, commonly occurring proteins misfold forming insoluble fibrillar aggregates. It is now known that amyloid fibrils can form from a variety of different protein and peptide solutions, and this, as well as a number of other desirable properties (2), has shifted research into amyloid fibrils from being solely into their involvement in disease, to include their use as novel bionanomaterials. It has also recently been shown that large quantities of amyloid fibrils can be produced from waste materials, namely, eye lens proteins (73), an important development, as if amyloid fibrils are to be used successfully in bionanotechnology, there is a need for production of amyloid fibrils from a readily available and inexpensively sourced protein (74).

The basic unit of amyloid fibrils is known as a protofilament, and amyloid fibrils consist of a number of these, typically 2–6, each of which is about 2–5 nm in diameter (75). This protofilament structure of amyloid fibrils is seen to be conserved across a range of proteins (75), although the way the protofilaments associate with each other can differ. In some cases, the protofilaments twist together to form rope-like fibrils 7–13 nm wide (75, 76), while in other cases, the protofilaments associate laterally to form long ribbons that are 2–5 nm thick and up to 30 nm wide (77).

Over the past decade, there has been an increasing interest in the use of amyloid fibrils in bionanotechnology for a range of different applications due to them possessing a number of desirable properties that recommends their use in this field (74, 78–82). Some of these properties are those previously described generic properties of fibrous proteins whereas others are specific to the amyloid fibril structure. The topography of amyloid fibrils also recommends their use as nanomaterials. They naturally exist in the nanometer scale, with widths of generally 7–10 nm and lengths of up to a few microns (2). These lengths are usually highly heterogeneous, although methods to control the length distribution of amyloid fibrils are emerging (83).

When using biological molecules in the assembly of nanostructures it is important to have a high level of longevity in the structure formed (2). Because of their highly ordered β-sheet structure, amyloid fibrils are stable at pH and temperature extremes (41), as well as at high pressure (84). This observed stability is particularly relevant, due to the similarities in these conditions to those likely to be encountered in an industrial setting, where these nanostructures may be manufactured or used (85).

The mechanical properties of amyloid fibrils are also of interest in relation to their use as a nanomaterial. Amyloid fibrils have been shown to have a strength comparable to steel and are flexible, with a stiffness similar to that of silk (86). These mechanical properties are overall similar to those seen in spider silk, which could be expected as both are β-sheet-rich protein fibrils (86). The major difference is that while spider silk requires complex biological machinery to reach its final form, amyloid fibrils self-assemble in solution without the need for other factors.

Research has looked at amyloid fibrils as both unmodified and modified nanowires, beginning with the work of Schiebel et al. in 2003, where amyloid-like fibers formed from the N-terminal and middle region (NM region) of Sup25p, a prion determinant from *Saccharomyces cerevisiae*, were shown to act as insulators with high resistance and, after modification via gold toning, to conduct with low resistance of $R = 86\ \Omega$ (Fig. 3) (85). Another study on unmodified fibrils formed from an elastin-related polypeptide showed these amyloid fibrils to be able to sustain electrical conduction, with current values in the range of several nanoAmps at 0–2 V (87); these however, are still low current values in comparison to other nanowires such as carbon nanotubes.

Amyloid fibrils have also been used as nanowire templates, formed via the fibrils undergoing modification by conducting polymers as well as other metallic modifications. Insulin amyloid fibrils have been used as scaffolds for the formation of conducting polymer nanowires, with insulin amyloid fibrils being modified with the conducting polymers poly(thiophene acetic acid) (PTAA), alkoxysulfonate PEDOT (PEDOT-S), and poly(propylene fumarate) (42, 44, 45). Figure 2, 2, represents this type of modification.

Insulin amyloid fibrils have also been modified with the platinum complex, hexachloroplatinic acid (H_2PtCl_6) (46), and silver nanoparticles (47), as well as being used as a sacrificial template for the formation of gold nanowires (48, 88). This method is represented schematically in Fig. 2, 1b. In more detail, insulin fibrils were deposited onto soda lime glass substrates, and after 30 s rinsed with water and dried with nitrogen gas. 12 nm Au-NPs were then deposited onto the insulin fibril network, and after a 10-min incubation, rinsed with water and dried again, to remove any unbound Au-NPs. Insulin fibrils were then removed by exposing samples to low-pressure air plasma, which left chains of Au-NPs on the substrate. Amyloid fibrils formed from α-synuclein and lysozyme have also been used as templates, with fibrils modified with iron (49) and silver, respectively (50).

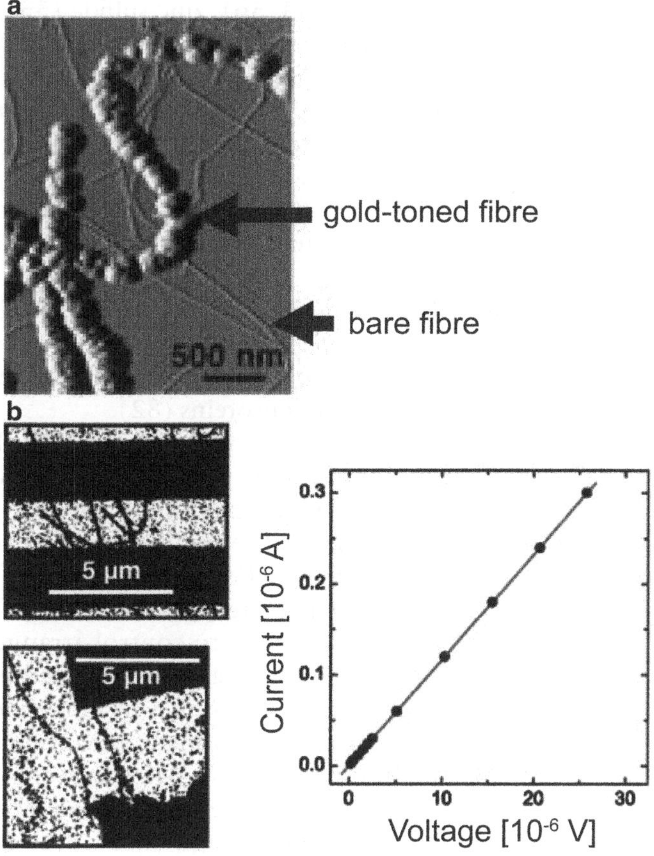

Fig. 3 The pioneering work of Scheibel et al., where amyloid-like fibers formed from the NM region of Sup25p were shown to act as insulators with high resistance and, after modification via gold toning, to conduct with low resistance. (**a**) Non-modified and gold-toned fibers as imaged by AFM. (**b**) Electrical conductivity of gold-toned fibers was investigated by deposition on patterned electrodes (*left*). The fibers exhibited linear I–V curves, demonstrating ohmic conductivity with low resistance of $R = 86\ \Omega$

5 Peptide Nanowires

Peptides have been identified as excellent building blocks for nanotechnology, due to chemical variety, biocompatibility, and their ability to spontaneously associate into a range of structures such as nanofibrils, nanotapes, nanoribbons, nanospheres, and nanotubes (89). The elongated structures achievable have variable dimensions, some with diameters as small as 1 nm, and others up to hundreds of nanometers (89).

Self-assembling peptides have also been investigated as nanowires, mainly as template molecules. Like proteins, peptides offer surfaces which can be functionalized, which has seen the creation of gold (51–53), silver (53, 57), copper (54), cobalt oxide (59), platinum (53, 56), zinc sulfide (55), and even trilayered metal-insulator-metal coaxial cables (58). The method used by Reches and Gazit was particularly elegant, with ionic silver added to PNTs in solution, upon which silver NPs formed within the tubes which developed to a more uniform wire within the PNT after reduction with citric acid (57). The PNT was then removed by proteolytic lysis with proteinase K, resulting in individual silver nanowires, approximately 20 nm in diameter (Fig. 2, 3b).

5.1 Amyloid-Inspired Peptides

Peptide design is based on the predicted interactions between the side chains, which has resulted in inspiration from amyloid fibrils, with a number of self-assembling peptides being fragments of amyloid-forming proteins (82).

The peptide that has attracted the most attention in this field over the last few years is the dipeptide diphenylalanine (FF). This short aromatic peptide is the key recognition sequence of the Alzheimer's disease β-amyloid peptide and forms discrete and hollow nanotubes in solution (57). Aside from being modified with metals, as mentioned above, Gazit and coworkers have also shown the ability to control arrangement of these peptide nanotubes, with vertical arrays created by vapor deposition (90), and alignment of nanotubes in a magnetic field (91). Some investigation has also been made into their conductive properties in the unmodified state. By immobilizing FF peptide nanotubes between electrodes using dielectrophoresis, Castillo et al. created current–voltage (I–V) curves for small bundles of nanotubes (92). The current transmitted through the immobilized nanotubes after an applied potential of 0–3 V was in the picoAmp range, confirming that the peptide nanotubes had high resistance.

6 Nanosensors

Biosensing is the detection and quantification of biological and chemical species, and it is critical to many areas of health care and life science. It involves the transduction of a signal that is associated with the selective recognition of a species of interest (93). As efficient biosensing requires direct interaction with biomolecules of dimensions in the nanoscale, it is predictable that this field requires techniques and probing tools that exist on a similar scale (12), and so, nanostructures, such as nanowires, offer the opportunity to develop novel sensors.

Types of biosensing devices based on nanowires include single nanowire field-effect transistors (FETs), porous nanowire films, and nanoelectrode arrays. For this chapter, glucose sensing has been chosen as a case study, and examples of nanowire-based glucose sensors designed in each of these styles will be given.

7 Case Study: Glucose Sensing

Glucose sensing has a very important role in both the diagnosis and management of diabetes, as well as in the food industry. In diabetes, careful monitoring of blood and urine glucose levels is necessary to correctly diagnose and manage the disease. Glucose monitoring is used in the food industry during fermentation processes, as the amount of glucose greatly influences the quality of food products.

Since the creation of the first glucose monitor in 1971, this has been a field that has grown rapidly due to the high demand, with glucose sensors dominating the biosensor market (94). There are a number of different types of glucose sensors, with detection methods based on fluorescence, electrochemistry, and spectroscopy, among many others. For the purpose of this chapter, the focus is on those sensors which use enzyme electrodes, a mode of detection which was first created in 1962 (95).

Enzyme electrodes function by using enzymes that catalyze oxidation-reduction (redox) reactions, and so, donate and accept electrons. This electron movement from the enzyme may then be used to produce a concentration-dependent current or voltage, which may be measured using electrodes. In glucose sensing, the enzyme glucose oxidase (GOx) is widely used. GOx catalyzes the conversion of glucose to gluconolactone. GOx-modified electrodes can be used in glucose sensing either by measuring oxidation and reduction of a mediator molecule, e.g., ferrocene, or by measuring peroxide amperometrically using a platinum electrode. The schematic below shows how these two scenarios work (Fig. 4).

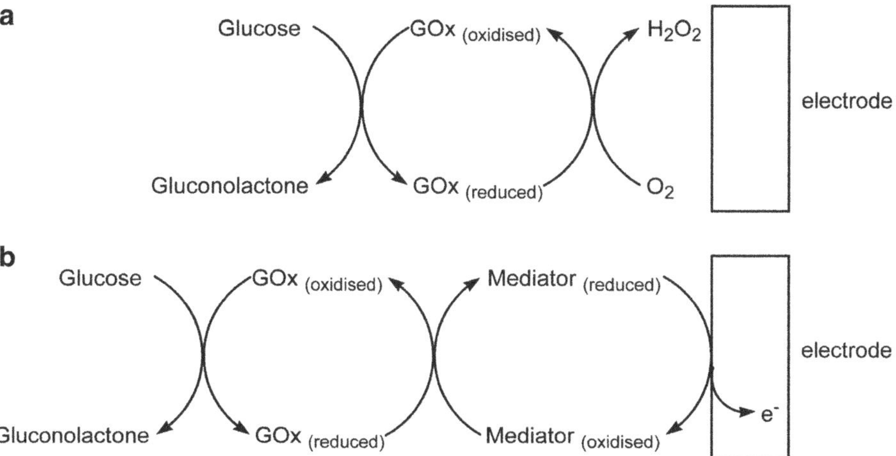

Fig. 4 Schematic showing the functioning of glucose oxidase (GOx) enzyme electrodes as glucose sensors. These can work in one of two ways; (**a**) through the amperometric detection of hydrogen peroxide using a platinum electrode, and (**b**) by measuring the oxidation and reduction of a mediator molecule

Single nanowire FETs function by exhibiting a conductivity change along the nanowire in response to variations in the electric field or potential at the surface (93). The standard setup involves a semiconductor nanowire connected to metal source and drain electrodes though which current flows. This current flow between the source and drain electrodes, via the nanowire, is switched on and off by a third electrode, the gate. The surface of the nanowire is often modified with receptor groups that recognize specific molecules, i.e., the analyte of interest. The binding of the analyte of interest results in an increase/decrease in the conductance of the nanowire and this is the signal that is associated with the presence of this analyte.

Besteman et al. created the first biosensor based on an individual SWNT (Fig. 5) (96). Semiconducting SWNT were grown with CVD on silicon oxide wafers, and GOx immobilized on the SWNT using a linking molecule, pyrenebutanoic acid succinimidyl ester, that binds the SWNT through van der Waals coupling with a pyrene group, and the enzyme via an amide bond. By AFM, it was ascertained that the SWNT had a density of approximately one GOx molecule per 12 nm. Gold electrodes were deposited on top of SWNTs with e-beam lithography, to ensure good contact between the electrode and the SWNT. The GOx-coated semiconducting SWNT was found to act as reversible pH sensors, and also glucose sensors, showing an increase in conductivity upon addition of glucose.

Porous nanowire films can exist as thin films which are deposited on top of more conventional electrodes, or as a 3-D gel. This type of setup favors an enzyme-based biosensing system, as the nanowire film can also function as an immobilization matrix for the enzyme of interest, in this case GOx. This type of biosensor was created by Huang et al. in their study involving Mn_2O_3-Ag nanofibers (97). The nanofibers were fabricated by electrospinning followed by calcinations, after which the nanofibers were dispersed in Nafion copolymer solution. GOx was then added to this solution, and the Mn_2O_3-Ag-Nafion-GOx mixture deposited onto a glassy carbon electrode (GCE). The modified electrode was then exposed to glutaraldehyde vapor for GOx cross-linking. By this method, an amperometric glucose sensor with a fast response time, good sensitivity, and limit of detection was created. The use of the nanofibers in this case added a high specific surface area, good biocompatibility, and numerous efficient electron transfer pathways.

A number of glucose sensors have been created based on nanoelectrode arrays, such as arrays of carbon nanotubes. For example, Liu et al. immobilized GOx onto aligned CNTs and achieved direct electron transfer from the redox center of GOx to the CNT (98). This was done by covalently attaching flavin adenine dinucleotide (FAD), the redox active center of GOx, to the ends of tubes and

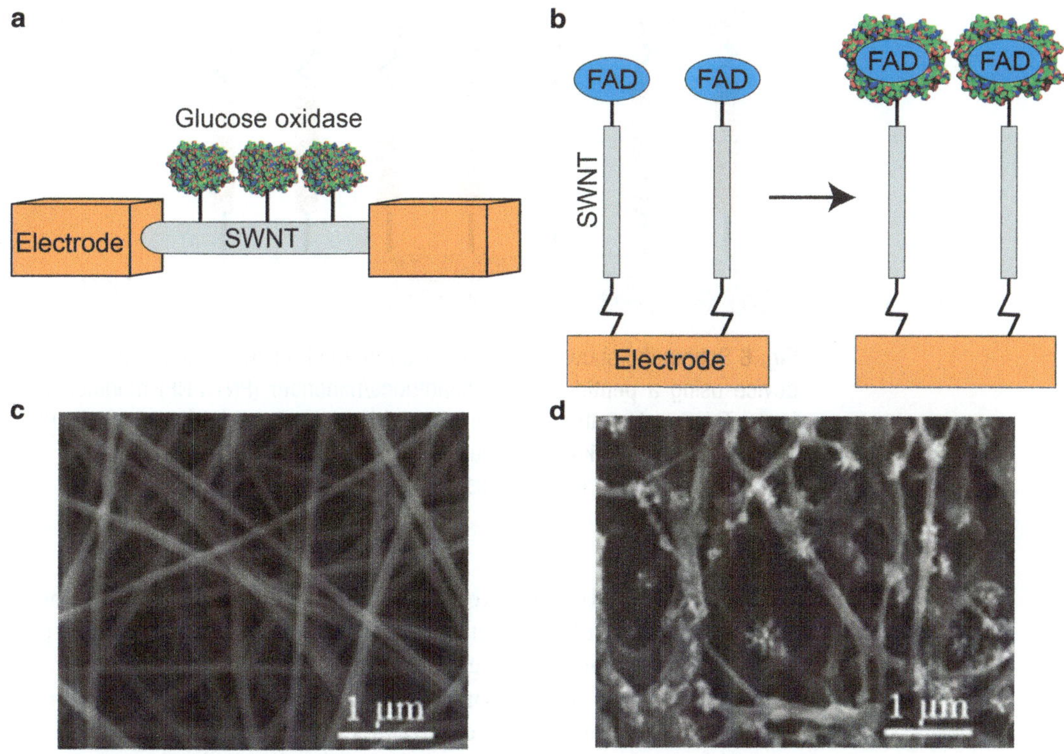

Fig. 5 Examples of different types of nanowire-based glucose sensors. (**a**) Schematic representation of the work of Besteman et al., where a glucose oxidase-modified single-walled carbon nanotube (SWNT) acted as a single-molecule biosensor. Schematic adapted from (96). (**b**) Liu et al. achieved direct electron transport via covalent attachment of FAD, the redox center of glucose oxidase, to vertically aligned SWNTS, creating a nanoelectrode array-based sensor. Schematic adapted from (98). (**c**, **d**) A porous nanowire film glucose sensor was formed using Mn_2O_3-Ag nanofibers. Images show the nanofibers before (**c**) and after (**d**) calcinations. This type of sensor creates a large surface area for enzyme immobilization, in this case, immobilization of glucose oxidase. Images reproduced with permission from (97) (Copyright © 2011 WILEY-VCH Verlag GmbH & Co. KGaA, Weinheim)

refolding the enzyme around the FAD cofactor. A similar study was also carried out by Willner and coworkers (99), although without the comparison with wild-type immobilized GOx. It was seen that the rate constant for electron transfer improved upon direct attachment of the nanotubes to FAD (98).

8 Protein/Peptide "Nanowires" in Biosensing

Protein and peptide nanofibers and nanotubes have been shown to have low conductivity in the unmodified state, making their use as nanowires restricted to that as template molecules. However, their high aspect ratios and the availability of a variety of functional groups do make them attractive enzyme immobilization scaffolds,

Fig. 6 Schematic showing a generalized method for the creation of biosensing device using a peptide/protein nanotubes/nanofiber (PNT/PNF)-modified electrode. This involves the immobilization of the enzyme of interest to PNTs/PNFs, often via a cross-linking molecule, which are attached to an appropriate electrode. The order that this process is carried out in varies

which can then be used to modify electrodes, creating a type of porous "nanowire" film-sensing device, with the advantage of proteins/peptides showing high biocompatibility.

Yemini et al. first reported this type of novel electrochemical biosensing platform, with electrodes modified by diphenylalanine peptide nanotubes showing enhanced reversibility and anodic/cathodic peak currents for the redox couple $Fe(CN)_6^{4-}/Fe(CN)_6^{3-}$ (100). They then went on to show that the PNT-modified gold electrodes could also be used as a biosensing device for glucose sensing, by covalently attaching glucose oxidase to the PNT (101). A similar device was created by Chen and coworkers, with PNTs formed from the ionic-complementary peptide, EFK16-11 (102, 103). A sensing device for hydrogen peroxide detection was also created in this way, with horseradish peroxidase (HRP) covalently immobilized to a collagen-like peptide-modified electrode (104).

Although there were some differences in the methods used in each of these examples, a general method for the creation of this type of biosensing device is shown in the schematic in Fig. 6. The material chosen for the electrode surface will likely be influenced by the chemistry of the nanostructure used, as for longevity and stability in this type of device it is necessary to have strong interactions between the PNT and the electrode surface. Hydrophobic and hydrophilic interactions may be enough, as in the case of the amphiphilic peptides EFK16-11 and EAK16-11, which adsorb well on mica (105), and highly ordered pyrolytic graphene (HOPG), respectively (Fig. 7) (103). In other cases, it may be necessary to engineer an attachment site, for example, thiol groups were incorporated into FF peptides using Traut's reagent (101).

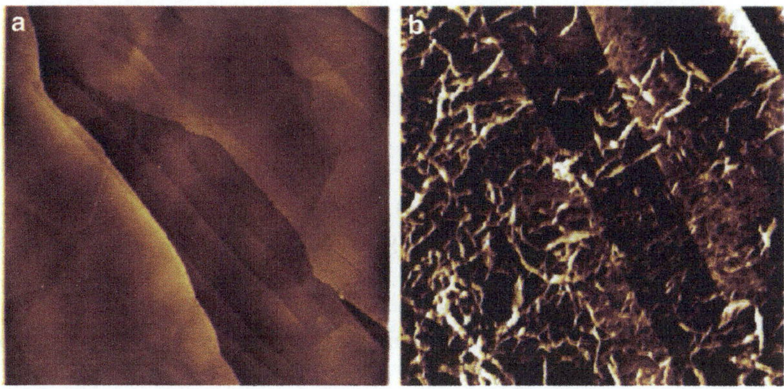

Fig. 7 An example of the use of PNTs as an enzyme immobilization scaffold for biosensing. Tapping mode AFM images of (**a**) a bare HOPG surface, and (**b**) a PNT (EFK16-11)-modified HOPG surface. Scan size is 6 μm^2 (103) (Reprinted with permission from Qian, Z., Khan, M. A., Mikkelsen, S., and Chen, P. (2009)) Improved enzyme immobilization on an ionic-complementary peptide-modified electrode for biomolecular sensing, Langmuir 26, 2176–2180 (Copyright 2011 American Chemical Society)

The enzyme of interest to the biosensing application must be immobilized on the PNT in some way. Conventional methods for enzyme immobilization on electrode surfaces include adsorption, cross-linking, electrochemical copolymerization, entrapment in polymeric gels or carbon paste, and covalent attachment (106). Some limitations of enzyme immobilization on electrode surfaces have been identified by Wang et al. as low quantities of attached enzymes, partial loss of enzymatic activity, nonbiocompatible immobilization compounds, low stability, and a time-consuming immobilization process. It is thought that the use of biomaterials as immobilization matrixes can help to overcome some of these problems.

In the examples provided, the enzymes of interest were immobilized onto the PNT by cross-linkers, namely, glutaraldehyde (GA) and carbodiimide (EDC), which are both commonly used cross-linking molecules. As peptides and proteins provide the versatility of the different amino acid side chains, there are a number of different options when it comes to the selection of a cross-linking molecule (107). Cross-linking sites may be engineered so as to provide a scenario closer to that shown in the schematic in Fig. 1; however, in many cases the end product is a mat of nanotubes with multiple enzyme immobilization sites as shown in Fig. 7. There are also methods available to limit the degree of nanotube-nanotube cross-linking, therefore increasing the surface area available for enzyme loading (103).

9 Conclusions

Potential has been seen in a number of biological molecules for their use as template molecules for the formation of conductive nanowires, with proteins and peptides attracting large amounts of attention. With the focus on nanoscale biosensing technologies increasing, it is only a matter of time before biological nanowires are used in this way, although further research, particularly into methods to control their growth and arrangement, is needed before this, as well as their use in other applications, can become a reality.

References

1. Chan WCW (2006) Bionanotechnology progress and advances. Biol Blood Marrow Transplant 12:87–91
2. Gras SL (2007) Amyloid fibrils: from disease to design. New biomaterial applications for self-assembling cross-beta fibrils. Aust J Chem 60:333–342
3. El-Sayed MA (2001) Some interesting properties of metals confined in time and nanometer space of different shapes. Acc Chem Res 34:257–264
4. Tang ZK, Zhang L, Wang N, Zhang XX, Wen GH, Li GD, Wang JN, Chan CT, Sheng P (2001) Superconductivity in 4 angstrom single-walled carbon nanotubes. Science 292:2462–2465
5. Ding R, Lu G, Yan Z, Wilson M (2001) Recent advances in the preparation and utilization of carbon nanotubes for hydrogen storage. J Nanosci Nanotechnol 1:7–29
6. Nogales E (2001) Structural insights into microtubule function. Annu Rev Bioph Biom 30:397–420
7. Perez JM, Simeone FJ, Saeki Y, Josephson L, Weissleder R (2003) Viral-induced self-assembly of magnetic nanoparticles allows the detection of viral particles in biological media. J Am Chem Soc 125:10192–10193
8. Wanekaya AK, Chen W, Myung NV, Mulchandani A (2006) Nanowire-based electrochemical biosensors. Electroanalysis 18:533–550
9. Iijima S (1991) Helical microtubules of graphitic carbon. Nature 354:56–58
10. Yu DP, Bai ZG, Ding Y, Hang QL, Zhang HZ, Wang JJ, Zou YH, Qian W, Xiong GC, Zhou HT et al (1998) Nanoscale silicon wires synthesized using simple physical evaporation. Appl Phys Lett 72:3458
11. Baughman RH, Zakhidov AA, De Heer WA (2002) Carbon nanotubes–the route toward applications. Science 297:787
12. Li J, Ng HT, Chen H (2005) Carbon nanotubes and nanowires for biological sensing. Methods Mol Biol 300:191–224
13. Tasis D, Tagmatarchis N, Bianco A, Prato M (2006) Chemistry of carbon nanotubes. Chem Rev 106:1105–1136
14. Schnorr JM, Swager TM (2011) Emerging applications of carbon nanotubes. Chem Mater 23(3):646–657
15. Treuting R, Arnold S (1957) Orientation habits of metal whiskers. Acta Metall 5:598
16. Schmidt V, Wittemann JV, Senz S, Gösele U (2009) Silicon nanowires: a review on aspects of their growth and their electrical properties. Adv Mater 21:2681–2702
17. Law M, Goldberger J, Yang P (2004) Semiconductor nanowires and nanotubes. Annu Rev Mater Res 34:83–122
18. Wan Y, Sha J, Chen B, Fang Y, Wang Z, Wang Y (2009) Nanodevices based on silicon nanowires. Recent Pat Nanotechnol 3:1–9
19. Bandaru P, Pichanusakorn P (2010) An outline of the synthesis and properties of silicon nanowires. Semicond Sci Technol 25:024003
20. Gazit E (2007) Use of biomolecular templates for the fabrication of metal nanowires. FEBS J 274:317–322
21. Braun E, Eichen Y, Sivan U, Ben-Yoseph G (1998) DNA-templated assembly and electrode attachment of a conducting silver wire. Nature 391:775–778
22. Patolsky F, Weizmann Y, Lioubashevski O, Willner I (2002) Au-nanoparticle nanowires based on DNA and polylysine templates. Angew Chem Int Ed Engl 41:2323–2327

23. Kim H, Roh Y, Hong B (2010) Selective alignment of gold nanowires synthesized with DNA as template by surface-patterning technique. IEEE T Nanotechnol 9:254–257
24. Ford WE, Harnack O, Yasuda A, Wessels JM (2001) Platinated DNA as precursors to templated chains of metal nanoparticles. Adv Mater 13:1793–1797
25. Seidel R, Ciacchi LC, Weigel M, Pompe W, Mertig M (2004) Synthesis of platinum cluster chains on DNA templates: conditions for a template-controlled cluster growth. J Phys Chem B 108:10801–10811
26. Richter J, Seidel R, Kirsch R, Mertig M, Pompe W, Plaschke J, Schackert HK (2000) Nanoscale palladium metallization of DNA. Adv Mater 12:507–510
27. Richter J, Mertig M, Pompe W, Mönch I, Schackert HK (2001) Construction of highly conductive nanowires on a DNA template. Appl Phys Lett 78:536
28. Monson CF, Woolley AT (2003) DNA-templated construction of copper nanowires. Nano Lett 3:359–363
29. Shenton W, Douglas T, Young M, Stubbs G, Mann S (1999) Inorganic–organic nanotube composites from template mineralization of tobacco mosaic virus. Adv Mater 11:253–256
30. Lim JS, Kim SM, Lee SY, Stach EA, Culver JN, Harris MT (2010) Quantitative study of Au (III) and Pd (II) ion biosorption on genetically engineered tobacco mosaic virus. J Colloid Interface Sci 342:455–461
31. Dujardin E, Peet C, Stubbs G, Culver JN, Mann S (2003) Organization of metallic nanoparticles using tobacco mosaic virus templates. Nano Lett 3:413–417
32. Knez M, Bittner AM, Boes F, Wege C, Jeske H, Maiβ E, Kern K (2003) Biotemplate synthesis of 3-nm nickel and cobalt nanowires. Nano Lett 3:1079–1082
33. Tsukamoto R, Muraoka M, Seki M, Tabata H, Yamashita I (2007) Synthesis of CoPt and FePt3 nanowires using the central channel of tobacco mosaic virus as a biotemplate. Chem Mater 19:2389–2391
34. Bai H, Xu F, Anjia L, Matsui H (2009) Low temperature synthesis of ZnO nanowires by using a genetically-modified collagen-like triple helix as a catalytic template. Soft Matter 5:966–969
35. Bai H, Xu K, Xu Y, Matsui H (2007) Fabrication of Au nanowires of uniform length and diameter using a monodisperse and rigid biomolecular template: collagen-like triple helix. Angew Chem Int Ed Engl 46:3319–3322
36. Orza A, Soritau O, Olenic L, Diudea M, Florea A, Ciuca D, Mihu C, Casciano D, Biris A (2011) Electrically conductive gold-coated collagen nanofibers for placental-derived mesenchymal stem cells enhanced differentiation and proliferation. ACS Nano 5:4490–4503
37. Patolsky F, Weizmann Y, Willner I (2004) Actin-based metallic nanowires as bio-nanotransporters. Nat Mater 3:692–695
38. Behrens S, Wu J, Habicht W, Unger E (2004) Silver nanoparticle and nanowire formation by microtubule templates. Chem Mater 16:3085–3090
39. Zhou JC, Gao Y, Martinez-Molares AA, Jing X, Yan D, Lau J, Hamasaki T, Ozkan CS, Ozkan M, Hu E et al (2008) Microtubule-based gold nanowires and nanowire arrays. Small 4:1507–1515
40. Zhou JC, Wang X, Xue M, Xu Z, Hamasaki T, Yang Y, Wang K, Dunn B (2010) Characterization of gold nanoparticle binding to microtubule filaments. Mater Sci Eng C 30:20–26
41. Scheibel T (2005) Protein fibers as performance proteins: new technologies and applications. Curr Opin Biotech 16:427–433
42. Herland A, Thomsson D, Mirzov O, Scheblykin IG, Inganäsa O (2008) Decoration of amyloid fibrils with luminescent conjugated polymers. J Mater Chem 18:126–132
43. Herland A, Björk P, Hania PR, Scheblykin IG, Inganäs O (2007) Alignment of a conjugated polymer onto amyloid-like protein fibrils. Small 3:318–325
44. Hamedi M, Herland A, Karlsson R, Inganas O (2008) Electrochemical devices made from conducting nanowire networks self-assembled from amyloid fibrils and alkoxysulfonate PEDOT. Nano Lett 8:1736–1740
45. Tanaka H, Herland A, Lindgren LJ, Tsutsui T, Andersson MR (2008) Enhanced current efficiency from bio-organic light-emitting diodes using decorated amyloid fibrils with conjugated polymer. Nano Lett 8:2858–2861
46. Tang Q, Solin N, Lu J, Inganäs O (2010) Hybrid bioinorganic insulin amyloid fibrils. Chem Commun 46:4157–4159
47. Leroux F, Gysemans M, Bals S, Batenburg KJ, Snauwaert J, Verbiest T, Van Haesendonck C, Van Tendeloo G (2010) Three-dimensional characterization of helical silver nanochains mediated by protein assemblies. Adv Mater 22:2193–2197
48. Hsieh S, Hsieh C (2010) Alignment of gold nanoparticles using insulin fibrils as a sacrificial biotemplate. Chem Commun 46:7355–7357

49. Choi YS, Kim J, Bhak G, Lee D, Paik SR (2011) Photoelectric protein nanofibrils of alpha-synuclein with embedded iron and phthalocyanine tetrasulfonate. Angew Chem 123:6194–6198
50. Malisauskas M, Meskys R, Morozova-Roche LA (2008) Ultrathin silver nanowires produced by amyloid biotemplating. Biotech Progr 24:1166–1170
51. Djalali R, Chen Y, Matsui H (2003) Au nanocrystal growth on nanotubes controlled by conformations and charges of sequenced peptide templates. J Am Chem Soc 125:5873–5879
52. Djalali R, Chen Y, Matsui H (2002) Au nanowire fabrication from sequenced histidine-rich peptide. J Am Chem Soc 124:13660–13661
53. Kasotakis E, Mossou E, Adler-Abramovich L, Mitchell E, Forsyth V, Gazit E, Mitraki A (2009) Design of metal-binding sites onto self-assembled peptide fibrils. Biopolymers 92:164–172
54. Banerjee IA, Yu L, Matsui H (2003) Cu nanocrystal growth on peptide nanotubes by biomineralization: size control of Cu nanocrystals by tuning peptide conformation. Proc Nat Acad Sci 100:14678
55. Banerjee IA, Yu L, Matsui H (2005) Room-temperature wurtzite ZnS nanocrystal growth on Zn finger-like peptide nanotubes by controlling their unfolding peptide structures. J Am Chem Soc 127:16002–16003
56. Song Y, Challa SR, Medforth CJ, Qiu Y, Watt RK, Peña D, Miller JE, van Swol F, Shelnutt JA (2004) Synthesis of peptide-nanotube platinum-nanoparticle composites, Chem Commun 0:1044–1045
57. Reches M, Gazit E (2003) Casting metal nanowires within discrete self-assembled peptide nanotubes. Science 300:625–627
58. Carny O, Shalev DE, Gazit E (2006) Fabrication of coaxial metal nanocables using a self-assembled peptide nanotube scaffold. Nano Lett 6:1594–1597
59. Ryu J, Kim SW, Kang K, Park CB (2009) Synthesis of diphenylalanine/cobalt oxide hybrid nanowires and their application to energy storage. ACS Nano 4:159–164
60. Gu Q, Cheng C, Gonela R, Suryanarayanan S, Anabathula S, Dai K, Haynie DT (2006) DNA nanowire fabrication. Nanotechnology 17:R14
61. Cai L, Tabata H, Kawai T (2000) Self-assembled DNA networks and their electrical conductivity. Appl Phys Lett 77:3105–3106
62. Storm A, van Noort J, de Vries S, Dekker C (2001) Insulating behavior for DNA molecules between nanoelectrodes at the 100 nm length scale. Appli Phys Lett 79:3881–3883
63. Kim HJ, Roh Y, Hong B (2006) Controlled gold nanoparticle assembly on DNA molecule as template for nanowire formation. J Vac Sci Technol A 24:1327
64. Mao C, Solis DJ, Reiss BD, Kottmann ST, Sweeney RY, Hayhurst A, Georgiou G, Iverson B, Belcher AM (2004) Virus-based toolkit for the directed synthesis of magnetic and semiconducting nanowires. Science 303:213
65. Wong Po Foo C, Kaplan DL (2002) Genetic engineering of fibrous proteins: spider dragline silk and collagen. Adv Drug Del Rev 54:1131–1143
66. Kadler KE, Holmes DF, Trotter JA, Chapman JA (1996) Collagen fibril formation. Biochem J 316:1
67. Buehler MJ (2006) Nature designs tough collagen: Explaining the nanostructure of collagen fibrils. Proc Nat Acad Sci 103:12285–12290
68. Kabsch W, Vandekerckhove J (1992) Structure and function of actin. Annu Rev Biophys Biomol Struct 21:49–76
69. Cooper JA (1987) Effects of cytochalasin and phalloidin on actin. J Cell Biol 105:1473–1478
70. Chiti F, Dobson CM (2006) Protein misfolding, functional amyloid, and human disease. Annu Rev Biochem 75:333–366
71. Virchow R (1854) Weitere mittheilungen über das vorkommen der pflanzlichen cellulose beim menschen. Virchows Arch 6:268–271
72. Friedreich N, Kekulé A (1859) Zur amyloidfrage. Virchows Arch 16:50–65
73. Garvey M, Gras S, Meehan S, Meade S, Carver J, Gerrard J (2009) Protein nanofibres of defined morphology prepared from mixtures of crude crystallins. Int J Nanotechnol 6:258–273
74. Waterhouse SH, Gerrard JA (2004) Amyloid fibrils in bionanotechnology. Aust J Chem 57:519–523
75. Serpell LC, Sunde M, Benson MD, Tennent GA, Pepys MB, Fraser PE (2000) The protofilament substructure of amyloid fibrils. J Mol Biol 300:1033–1039
76. Goeden-Wood NL, Keasling JD, Muller SJ (2003) Self-assembly of a designed protein polymer into beta-sheet fibrils and responsive gels. Macromolecules 36:2932–2938
77. Saiki M, Honda S, Kawasaki K, Zhou D, Kaito A, Konakahara T, Morii H (2005) Higher-order molecular packing in amyloid-like fibrils constructed with linear arrangements of hydrophobic and hydrogen-bonding side-chains. J Mol Biol 348:983–998
78. Mankar S, Anoop A, Sen S, Maji SK (2011) Nanomaterials: amyloids reflect their brighter side, Nano Rev 2

79. Gras S (2009) Surface- and solution- based assembly of amyloid fibrils for biomedical and nanotechnology applications. Adv Chem Eng 35:161–209
80. Gras SL (2007) Protein misfolding: a route to new nanomaterials. Adv Powder Technol 18:699–705
81. Hamada D, Yanagihara I, Tsumoto K (2004) Engineering amyloidogenicity towards the development of nanofibrillar materials. Trends Biotechnol 22:93–97
82. Cherny I, Gazit E (2008) Amyloids: Not only pathological agents but also ordered nanomaterials. Angew Chem Int Ed Engl 47:4062–4069
83. Domigan LJ, Healy JP, Meade S, Blaikie RJ, Gerrard JA (2011) Controlling the dimensions of amyloid fibrils: Towards homogenous components for bionanotechnology. Biopolymers 97:123–133
84. Dirix C, Meersman F, MacPhee CE, Dobson CM, Heremans K (2005) High hydrostatic pressure dissociates early aggregates of TTR 105-115, but not the mature amyloid fibrils. J Mol Biol 347:903–909
85. Scheibel T, Parthasarathy R, Sawicki G, Lin X-M, Jaeger H, Lindquist SL (2003) Conducting nanowires built by controlled self-assembly of amyloid fibers and selective metal deposition. Proc Nat Acad Sci 100:4527–4532
86. Smith JF, Knowles TPJ, Dobson CM, MacPhee CE, Welland ME (2006) Characterization of the nanoscale properties of individual amyloid fibrils. Proc Nat Acad Sci 103:15806–15811
87. del Mercato LL, Pompa PP, Maruccio G, Torre AD, Sabella S, Tamburro AM, Cingolani R, Rinaldi R (2007) Charge transport and intrinsic fluorescence in amyloid-like fibrils. Proc Nat Acad Sci 104:18019–18024
88. Hsieh C, Hsieh S (2011) Nanoparticle chain formation on functional surfaces using insulin fibrils as a structure directing agent. J Mater Chem 21:16900–16904
89. Valéry C, Artzner F, Paternostre M (2011) Peptide nanotubes: molecular organisations, self-assembly mechanisms and applications. Soft Matter 7:9583–9594
90. Adler-Abramovich L, Aronov D, Beker P, Yevnin M, Stempler S, Buzhansky L, Rosenman G, Gazit E (2009) Self-assembled arrays of peptide nanotubes by vapour deposition. Nature Nanotechnol 4:849–854
91. Hill RJA, Sedman VL, Allen S, Williams P, Paoli M, Adler-Abramovich L, Gazit E, Eaves L, Tendler SJB (2007) Alignment of aromatic peptide tubes in strong magnetic fields. Adv Mater 19:4474–4479
92. Castillo J, Tanzi S, Dimaki M, Svendsen W (2008) Manipulation of self-assembly amyloid peptide nanotubes by dielectrophoresis. Electrophoresis 29:5026–5032
93. Patolsky F, Zheng G, Lieber CM (2006) Nanowire-based biosensors. Anal Chem 78:4260–4269
94. Oliver N, Toumazou C, Cass A, Johnston D (2009) Glucose sensors: a review of current and emerging technology. Diabetic med 26:197–210
95. Clark LC Jr, Lyons C (1962) Electrode systems for continuous monitoring in cardiovascular surgery. Anna NY Acad Sci 102:29–45
96. Besteman K, Lee J-O, Wiertz FGM, Heering HA, Dekker C (2003) Enzyme-coated carbon nanotubes as single-molecule biosensors. Nano Lett 3:727–730
97. Huang S, Ding Y, Liu Y, Su L, Filosa R Jr, Lei Y (2011) Glucose biosensor using glucose oxidase and electrospun Mn2O3-Ag nanofibers. Electroanalysis 23:1912–1920
98. Liu J, Chou A, Rahmat W, Paddon-Row MN, Gooding JJ (2005) Achieving direct electrical connection to glucose oxidase using aligned single walled carbon nanotube arrays. Electroanalysis 17:38–46
99. Patolsky F, Weizmann Y, Willner I (2004) Long-range electrical contacting of redox enzymes by SWCNT connectors. Angew Chem Int Ed Engl 43:2113–2117
100. Yemini M, Reches M, Gazit E, Rishpon J (2005) Peptide nanotube-modified electrodes for enzyme-biosensor applications. Anal Chem 77:5155–5159
101. Yemini M, Reches M, Rishpon J, Gazit E (2005) Novel electrochemical biosensing platform using self-assembled peptide nanotubes. Nano Lett 5:183–186
102. Yang H, Fung SY, Sun W, Mikkelsen S, Pritzker M, Chen P (2008) Ionic-complementary peptide-modified highly ordered pyrolytic graphite electrode for biosensor application. Biotechnol Progr 24:964–971
103. Qian Z, Khan MA, Mikkelsen S, Chen P (2009) Improved enzyme immobilization on an ionic-complementary peptide-modified electrode for biomolecular sensing. Langmuir 26:2176–2180
104. Yemini M, Xu P, Kaplan DL, Rishpon J (2006) Collagen-like peptide as a matrix for enzyme immobilization in electrochemical biosensors. Electroanalysis 18:2049–2054

105. Yang H, Fung S-Y, Pritzker M, Chen P (2007) Surface-assisted assembly of an ionic-complementary peptide: Controllable growth of nanofibers. J Am Chem Soc 129:12200–12210
106. Spahn C, Minteer SD (2008) Enzyme immobilization in biotechnology. Rec Patents Eng 2:195–200
107. Guisan JM (2006) Immobilization of enzymes and cells. Humana, Totowa, NJ

Chapter 9

Nanotechnology with S-Layer Proteins

Bernhard Schuster and Uwe B. Sleytr

Abstract

Nanosciences are distinguished by the cross-fertilization of biology, chemistry, material sciences, and solid-state physics and hence open up a great variety of new opportunities for innovation. The technological utilization of self-assembly systems, wherein molecules spontaneously associate under equilibrium conditions into reproducible supramolecular aggregates, is one key challenge in nanosciences for life and nonlife science applications. The attractiveness of such processes is due to their ability to build uniform, ultrasmall functional units and the possibility to exploit such structures at meso- and macroscopic scale very frequently by newly developed techniques and methods. By the utilization of crystalline bacterial cell-surface proteins (S-layer proteins) innovative approaches for the assembly of supramolecular structures and devices with dimensions of a few to tens of nanometers have been developed. S-layers have proven to be particularly suited as building blocks in a molecular construction kit involving all major classes of biological molecules. The controlled immobilization of biomolecules in an ordered fashion on solid substrates and their controlled confinement in definite areas of nanometer dimensions are key requirements for many applications including the development of bioanalytical sensors, biochips, molecular electronics, biocompatible surfaces, and signal processing between functional membranes, cells, and integrated circuits.

Key words Surface layers, S-layers, Two-dimensional protein crystals, Biomimetics, Self-assembly, Nanotechnology, Nanobiotechnology, Nanoparticle, Construction kit, Supported lipid membranes

1 Introduction

Many prokaryotic organisms have regular arrays of (glyco)proteins on their outermost surface (1–5). These monomolecular crystalline surface layers, termed S-layers (6, 7), are found in members of nearly every taxonomical group of walled bacteria and cyanobacteria and represent an almost universal feature of archaeal cell envelopes (Fig. 1). Since the biomass of prokaryotic organisms surpass the eukaryotic biomass and S-layer proteins make up approximately 10% of the cell proteins, they represent one of the most abundant biopolymers on earth. S-layers are generally composed of a single protein or glycoprotein species with a molecular mass of 40,000–230,000 and exhibit either oblique (p1, p2), square (p4), or hexagonal (p3, p6) lattice symmetry with unit cell dimensions in the

Juliet A. Gerrard (ed.), *Protein Nanotechnology: Protocols, Instrumentation, and Applications*, Methods in Molecular Biology, vol. 996, DOI 10.1007/978-1-62703-354-1_9,

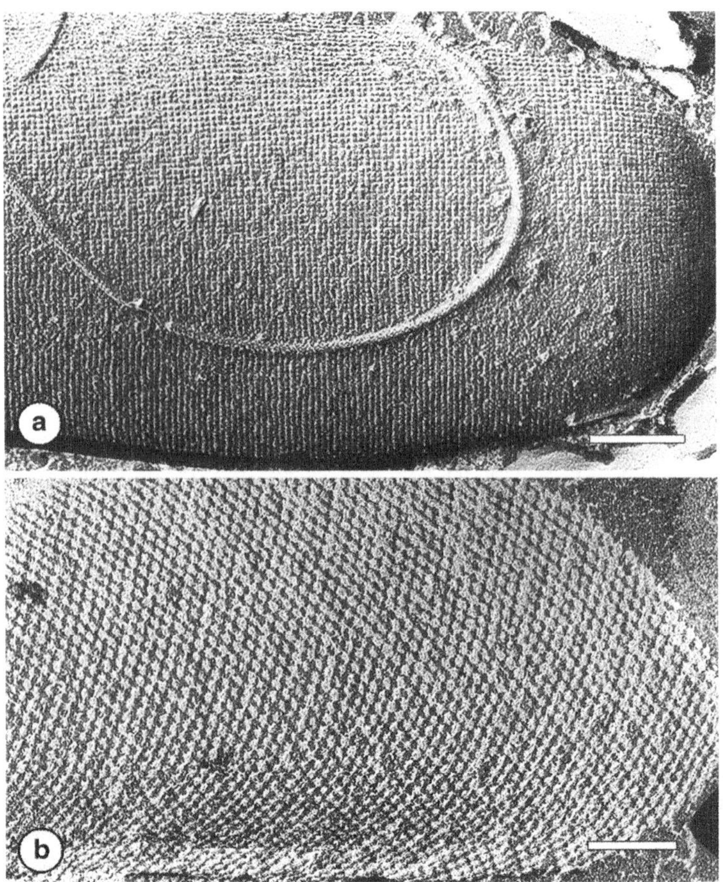

Fig. 1 Electron micrographs of freeze-etching preparations of whole cells of (**a**) *Lysinibacillus sphaericus* CCM 2177, showing a square S-layer lattice and (**b**) *Thermoplasma thermohydrosulfuricus* revealing a hexagonally ordered array. Bar in (**a**) 200 nm and in (**b**) 100 nm (Reprinted from ref. 58 with permission from the publisher. © 2001, Elsevier Science)

range of 3–30 nm (see Fig. 2). One morphological unit consists of one, two, three, four, or six identical subunits, respectively. The monomolecular arrays are generally 5–10 nm thick and show pores of identical size (diameter 1.5–8 nm) and morphology. In most S-layers the outer face is less corrugated than the inner face. Moreover S-layers are highly anisotropic structures regarding the net charge and hydrophobicity of the inner and outer surface (8–10). Due to the crystalline character of S-layers functional groups (e.g., carboxyl-, amino-, hydroxyl groups) are repeated with the periodicity of the protein lattice (11, 12).

Since S-layers possess a high degree of structural regularity, these crystalline arrays are excellent models for studying the dynamic aspects of assembly of a supramolecular structure in vivo and in vitro (13–19). Moreover the use of S-layers has provided

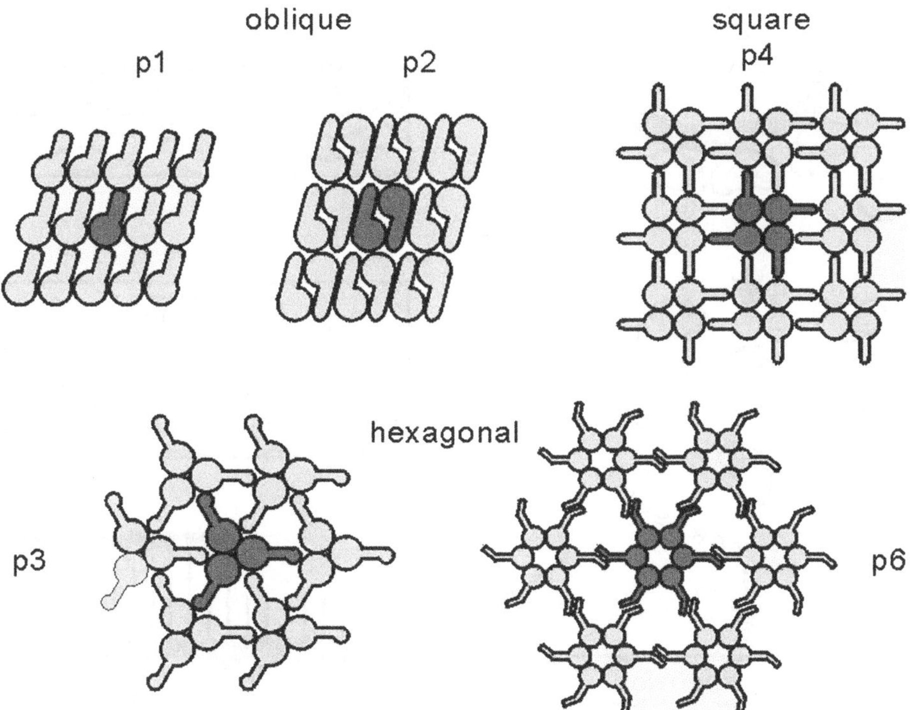

Fig. 2 Schematic representation of S-layer lattice types grouped according to the possible two-dimensional space group symmetries. Morphological units were chosen arbitrarily and are shown in dark gray (Reprinted from ref. 27 with permission from the publisher. © 2003, Wiley-VCH)

innovative approaches for the assembly of supramolecular structures and devices (20). S-layers have proven to be particularly suited as building blocks and patterning elements in a biomolecular construction kit involving all major classes of biological and chemically synthesized molecules or nanoparticles and quantum dots (21, 22). In this context one of the most important properties of isolated S-layer (glyco)protein subunits is their capability to reassemble into monomolecular arrays in suspension, at the air interface, on a solid surface, on floating lipid monolayers (see Fig. 3), and on liposomes or particles (2, 3, 23–29).

An important line of development in S-layer-based technologies is presently directed towards the genetic manipulation of S-layer proteins (22, 23). These strategies open new possibilities for the specific tuning of their structure and function. S-layer proteins incorporating specific functional domains of other proteins while maintaining the self-assembly capability will lead to new ultrafiltration membranes, affinity structures, enzyme membranes, metal precipitating matrices, microcarriers, biosensors, diagnostics, biocompatible surfaces, and vaccines (21–23, 29).

Although up to now most S-layer technologies developed concerned life sciences, an important emerging field of future

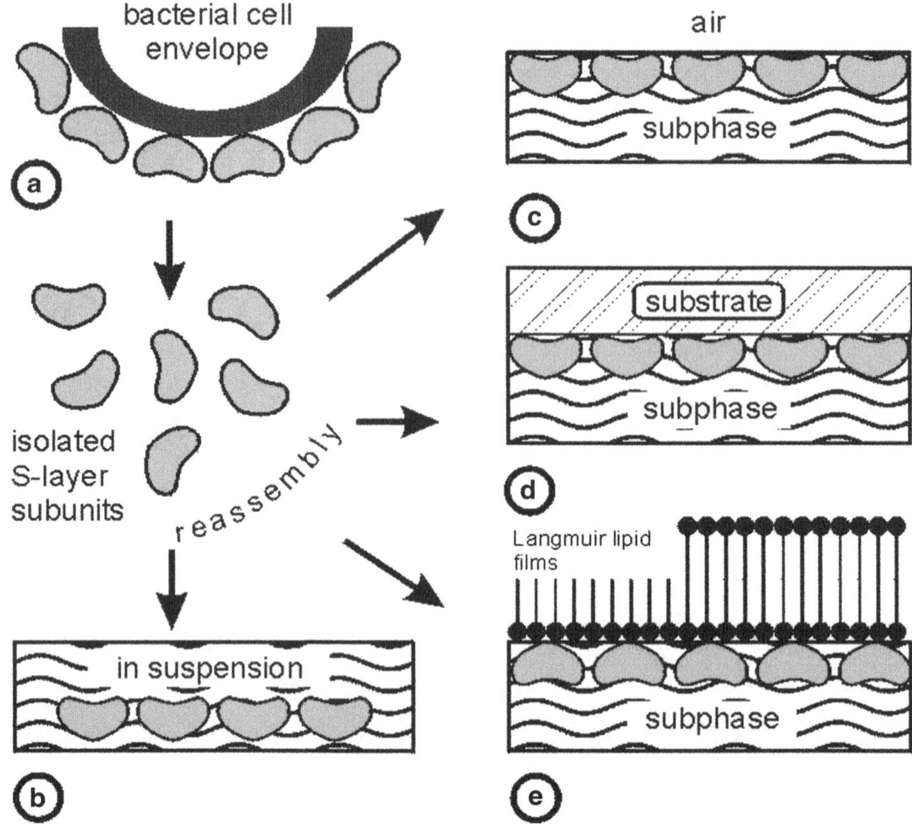

Fig. 3 Schematic illustration of the recrystallization of isolated S-layer subunits into crystalline arrays (**a**). The self-assembly process can occur in suspension (**b**), at the air–liquid interface (**c**), on solid supports (**d**), and on Langmuir lipid films (**e**) (Reprinted from ref. 24 with permission from the publisher. © 1999, Wiley-VCH)

applications relates to nonlife sciences. Native or genetically modified S-layers recrystallized on solid supports can be used as patterning elements for accurate spatial positions of nanometer scale metal particles or as a matrix for chemical deposition of metals as required for molecular electronics and nonlinear optics (20).

2 Materials

2.1 Bacterial Strain, Growth in Continuous Culture, and Isolation

2.1.1 Bacterial Strains

1. *Geobacillus stearothermophilus* PV72, kindly provided by F. Hollaus (Österreichisches Zuckerforschungsinstitut, Tulln, Austria).
2. *Lysinibacillus sphaericus* CCM 2177 (formerly *Bacillus sphaericus* CCM 2177) obtained from the Czech Collection of Microorganisms (CCM), Brno, Czech Republic.

2.1.2 Growth in Continuous Culture

1. SVIII medium contained per liter: 10 g peptone, 5 g yeast extract, 5 g lab lemco, 1.2 g $K_2HPO_4.3H_2O$, 0.1 g $MgSO_4{\cdot}7H_2O$, 0.6 g sucrose.
2. Bioreactor type Biostat E (Braun, Melsungen, Germany), or other bioreactor.
3. Mass flow controller (Brooks, Veenendaal, The Netherlands), or other mass flow controller.
4. 1 M NaOH.
5. 2 M H_2SO_4.
6. pH and redox probes of gel paste-type and an amperometric probe (Ingold, Urdorf, Switzerland), or other equivalent probes.
7. Spectrophotometer (model 25, Beckmann, Fullerton, CA), or other spectrophotometer.
8. Sodium dodecyl sulfate (SDS)-polyacrylamide gel electrophoresis (PAGE) apparatus (Bio-Rad, Hercules, CA), or other SDS-PAGE apparatus.
9. Densitometer (Elscript 400AT/SM; Hirschmann, Germany), or other densitometer.
10. Centrifuge (Sepatech 17RS, Heraeus, Hanau, Germany), or other centrifuge.
11. Buffer A: 50 mM Tris–HCl, pH 7.2.

2.1.3 Preparation of Cell Wall Fragments

1. Ultrasonic treatment (Ultrasonics Sonicator W-385, Farmingdale, NY) or other tabletop.
2. Centrifuge (JA-HS, Beckmann), or other centrifuge.
3. 0.75% Triton X-100 (Serva, Heidelberg, Germany), dissolved in buffer A.

2.1.4 Isolation of S-Layer Proteins

1. 5 M guanidine hydrochloride (GHCl; Fluka) in buffer A.
2. Ultracentrifuge (Beckmann L5-65), or other ultracentrifuge.
3. 10 mM $CaCl_2$ in distilled water (Fluka).
4. Dialysis tube, cutoff: 12–16 kDa, pore size 25 Å (Biomol, Hamburg, Germany).
5. Spectrophotometer (Hitachi U 2000, Tokyo, Japan), or other spectrophotometer.

2.2 S-Layer Proteins on Solid Supports

2.2.1 Solid Supports

1. Silicon nitride and silicon wafers (100 orientation, p-type, boron-doped, resistivity 25–45 Ω cm, native oxide layer; MEMC, Italy or Wacker Chemitronic, Burghausen, Germany).
2. Metallic wafers: gold-coated supports (Pharmacia, Peapack, New Jersey) and evaporated titanium, aluminum, palladium.

3. Polymers: polyester, polypropylene, poly(ethylene terephthalate), poly(methacrylic acid methyl ester), polycarbonate (Wettlinger Kunststoffe, Wien, Austria).
4. Glass slides (Assistant Micro Slides Elka, No 2400, Sondheim, Germany), mica (Dumico, Rotterdam, The Netherlands), highly oriented pyrolytic graphite (HOPG; SPI Supplies, West Chester, PA, USA).

2.2.2 Cleaning and Modification of Solid Supports

1. Solvents (acetone, propan-2-ol, ethanol, ammonia (29%), hydrogen peroxide (30%), hydrogen chloride (37%), and dried toluene), Milli-Q water, and N_2-gas.
2. Silanes: octadecyltrichlorosilane, (3-methacryloyloxypropyl)-trimethoxysilane, trimethoxysilane, decyldimethylsilane, hexamethyldisilane, 2-aminopropyltrimethoxysilane, 3-mercaptopropyltrimethoxysilane (ABCR, Karlsruhe, Germany).
3. Plasma cleaner (Gala Instruments, Bad Schwalbach, Germany) and O_2 gas. Contact angle measurements (Kruess contact angle measurement system Easy Drop DSA 15, Kruess, Hamburg, Germany).

2.2.3 Crystallization of S-Layer Proteins at Solid Supports

1. Buffer B: 1 mM citrate buffer, adjusted to pH 4.0 with NaOH or HCl.
2. Buffer C: 10 mM $CaCl_2$ in 0.5 mM Tris–Tris–HCl, pH 9 with NaOH or HCl.
3. pH meter.
4. 0.1 mg/mL *G. stearothermophilus* PV72 (Subheading 2.1.1) in buffer B.
5. 0.1 mg/mL *L. sphaericus* CCM 2177 (Subheading 2.1.1) in buffer C.
6. Rotator (Reax2, Heidolph, Schwabach, Germany).

2.2.4 Atomic Force Microscopy

1. Digital Instruments Nanoscope IIIa (Santa Barbara, CA) or other was used with an E-scanner (nominal scan size, 12 μm) or a J-scanner (nominal scan size, 130 μm).
2. Standard 200 μm long oxide-sharpened silicon nitride cantilevers (NanoProbes, Digital Instruments) with a nominal spring constant of 0.06 nm^{-1}.

2.3 Patterning of Crystalline S-Layer Proteins by an Excimer Laser and Soft Lithography

1. S-layer protein SbpA of *L. sphaericus* CCM 2177 (Subheading 2.1.1).
2. Buffer C (Subheading 2.2.3).
3. Cleaning and characterization of solid supports (Subheading 2.2.2).

4. A 100 nm thick chromium coating on quartz glass consisting of lines and squares (feature sizes ranging from 200 nm to 1,000 nm) with different line-and-space ratios.
5. ArF excimer laser (model EMG 102E, Lambda Physik, Göttingen, Germany).
6. Silicon mold master: 4-in. silicon wafers, photoresist (Clariant AZ 9260; Microchemicals, Ulm, Germany), photolithography.
7. Poly(dimethylsiloxane) (PDMS; Sylgard 184, DOW Corning, Midland, MI).
8. Oven, exsiccator.
9. Atomic force microscopy (Subheading 2.2.4).
10. Epifluorescence microscopy.
    (a) Buffer D: 0.1 M $NaHCO_3$–Na_2CO_3 buffer, adjusted to pH 9.2 with NaOH or HCl.
    (b) Fluorescence marker (fluorescein isothiocyanate, FITC), and DMSO (both Sigma-Aldrich, Wien, Austria).
    (c) Fluorescence microscope (ECLIPSE TE 2000-S, Nikon, Tokyo, Japan).

2.4 Formation of Nanoparticle Arrays

2.4.1 Preparation of Supports

1. Standard formvar and carbon-coated electron microscope grids (Groepl, Tulln, Austria).
2. Coating of grids with SiO_2 by evaporation (EPA 100, Leybold-Heraeus, Köln, Germany).
3. SiO_2-coated grids were O_2-plasma treated in a plasma cleaner (Subheading 2.2.2).

2.4.2 Electrostatic Binding of Nanoparticles to S-Layers

1. 0.1 mg/mL S-layer protein SbpA of *L. sphaericus* CCM 2177 in buffer C (Subheading 2.2.3) and O_2-treated SiO_2-coated grids.
2. Nanoparticles: citrate-stabilized gold nanoparticles (mean diameter of 5 nm; Sigma-Aldrich, Wien, Austria) and amino-modified cadmium selenide (CdSe) nanoparticles (mean diameter of 4 nm; University of Hamburg, Germany).

2.4.3 Transmission Electron Microscopy

1. Negative staining: uranyl acetate (2.5% in Milli-Q water, Merck).
2. Transmission electron microscope (TEM; FEI Tecnai G^2 20, FEI, Eindhoven, The Netherlands).

2.5 S-Layer-Supported Lipid Membranes

2.5.1 Painted and Folded Membranes

1. 1,2-Diphytanoyl-*sn*-Glycero-3-Phosphocholine (DPhyPC; Avanti, Alabaster, AL).
2. Hexadecane, *n*-decane, *n*-hexane, and pentane.
3. Chloroform and ethanol.

4. Electrolyte: 0.01 M to 1 M KCl or NaCl in Milli-Q-water and if desired 10 mM $CaCl_2$ (all chemicals obtained from Merck).
5. Painted membranes: homemade Teflon chamber with a drilled orifice, 0.9 mm in diameter, which divided the two compartments with a volume of about 12 mL, each (for more details see refs. 30–33).
6. Copper wire (∅ ~1 mm) covered by a Teflon (polytetrafluoroethylene) tube and bent in "L"-shape to form some kind of brush.
7. Folded membranes: homemade Teflon chamber with a Teflon film (25 μm thick; Goodfellow, Cambridge, England) which divided the two compartments with a volume of about 3.5 mL, each. Into the Teflon film a hole, ~140 μm in diameter, was punched by a perforating tool (syringe needle which has been sharpened inside and outside; for further details see refs. 30, 31, 34).
8. Two 1 mL single-use syringes (B. Braun, Melsungen, Germany), or others. One syringe is connected by plastic tubes (∅ ~ 1 mm) to the *cis*-compartment and the other one to the trans-compartment, respectively.

2.5.2 Technical Equipment

1. Patch clamp amplifier (EPC 10, HEKA, Lamprecht, Germany), or others with corresponding software (Patchmaster, HEKA).
2. Two silver/silver chloride (Ag/AgCl) electrodes (see Note 1).
3. Vibration isolation (LW3030, Newport, Darmstadt, Germany) with a Faraday cage on the top.

3 Methods

3.1 Bacterial Strain, Growth in Continuous Culture, and Isolation

3.1.1 Growth in Continuous Culture

1. *Geobacillus stearothermophilus* PV72 (35, 36) was grown on 50 mL of SVIII medium (37) in a 300 mL shaking flasks at 57°C to mid-logarithmic growth.
2. 200 mL of this suspension was used as the inoculum for 5 L of SVIII medium sterilized in a bioreactor. Before inoculation, 20 mL of a sterile glucose solution (6 g of glucose in total) was added.
3. Cultivation was performed at 57°C and at a stirring speed of 300 rpm. In continuous culture, the dilution rate was kept at 0.1 h^{-1}. The rate of aeration was 0.5 L of air per min. The pH value of the culture was kept at 7.2 ± 0.2 by addition of either 1 M NaOH or 2 M H_2SO_4. Aeration rate was controlled by a mass flow controller. Redox potential was measured by a platinum contact redox probe. The partial oxygen pressure was

monitored with an amperometric probe. The cell density was measured at 600 nm in a spectrophotometer. In principle, *Lysinibacillus sphaericus* CCM 2177 (38) was cultivated under the same conditions, but as this organism is a mesophilic one, the temperature was lowered to 32°C (see Note 2).

4. For controlling the homogeneity of the culture, 10 mL samples were taken from the bioreactor at different times. Aliquots were plated on SVIII agar, and the grown biomass (at 57°C for 18 h) was used for SDS-PAGE (35). The gel system consisted of a 4% stacking gel and a 10% separation gel.
5. Single-cell colonies grown on SVIII agar plates were subjected to SDS-PAGE for final identification. The relative amounts from both types of S-layer proteins were estimated from SDS-gels by densitometric evaluation.
6. For biomass harvesting, the culture suspension from the overflow of continuous culture was collected in heat-sterilized bottles at 2–4°C. Cells were separated from spent medium by continuous centrifugation at 16,000 × *g* at 4°C, washed with buffer A, and stored at −20°C.

3.1.2 Preparation of Cell Wall Fragments

1. The frozen biomass (100 g) was suspended in 350 mL buffer A. The suspension was separated into three parts and the cells were broken by ultrasonic treatment for 2 min at maximal output. To avoid autocatalytic processes all preparation steps have to be done on ice at 4°C.
2. The intact and broken cells were separated by centrifugation at 48,000 × *g* for 10 min. The upper, lighter pellet was detached and collected. The lower, darker pellet was again suspended in buffer A, treated with ultrasonic, and sedimented. This procedure was repeated four times.
3. In order to remove contaminating plasma membrane fragments the crude cell wall preparations (collected pellets) were extracted with 250 mL 0.75% Triton X-100 (dissolved buffer A) and stirred 10 min at room temperature (RT; 22 ± 2°C).
4. The cell wall fragments were sedimented at 48,000 × *g* for 10 min. The extraction step was repeated three times.
5. The pellet was frozen in aliquots at −20°C.

3.1.3 Isolation of S-Layer Proteins

1. Cell wall fragments (2 mg) were suspended in 30 mL GHCl-solution in buffer A and stirred at RT for 30 min.
2. The suspension was sedimented at 90,000×*g* and 4°C for 45 min in an ultracentrifuge.
3. The supernatant was dialyzed either against a $CaCl_2$-solution (*G. stearothermophilus* PV72) or against distilled water (*L. sphaericus* CCM 2177) for three times at least 2 h each, at 4°C (see Note 3).

4. The S-layer self-assembly products (see Note 4) were sedimented for 15 min at 40,000×*g* at 4°C. The supernatant containing single subunits and oligomeric precursors was stored at 4°C and used within 5 days.
5. For determination of the protein concentration the measured adsorption at 280 nm was multiplied by 1.75 and 1.64 for the S-layer protein SbsB of *G. stearothermophilus* PV72 and SbpA of *L. sphaericus* CCM 2177, respectively.
6. The protein solutions were adjusted to a concentration of 1 mg protein per mL and used for all recrystallization experiments described further in Fig. 3.

3.2 S-Layer Proteins on Solid Supports

3.2.1 Preparation of Solid Supports

1. Silicon wafers were immersed in hot acetone followed by rinsing in propan-2-ol and finally washed with ethanol and Milli-Q water. The advancing contact angle of water on the clean silicon surface was 65°.
2. In order to increase the hydrophilicity of the substrates, the silicon wafers were treated in an oxygen (O_2) plasma (leaning time, 20 s; plasma pressure, 0.01 bar; power density, 70%; high purity grade O_2).
3. The plasma-treated silicon substrates with an advancing contact angle of water of 5° were used immediately for the recrystallization studies.
4. Other solid supports (e.g., metals, polymers, glass) were only rinsed with ethanol and Milli-Q water before use.
5. Silanization procedures (solution or vapor phase) using different silanes were applied to obtain silicon or glass substrates with more hydrophobic surfaces (39, 40). The substrates were cleaned in a solution containing 1:1:5 parts of ammonia (29%), hydrogen peroxide (30%), and Milli-Q water at 80°C for 10 min.
6. The silicon or glass substrates were treated with 1:1:6 parts of concentrated hydrogen chloride (37%), hydrogen peroxide (30%), and deionized water at 80°C for 15 min. Finally the substrates were rinsed thoroughly with Milli-Q water and dried in a stream of nitrogen gas.
7. For silanization out of a solution the substrates were further rinsed with acetone and dried toluene.
8. The supports were put into anhydrous toluene containing 1% silane.
9. Silanization, e.g., with decyldimethylsilane (DMS) was carried out for 30 min to 2 h with mild shaking at RT.
10. The silanized supports were rinsed with toluene, methanol, and Milli-Q water.

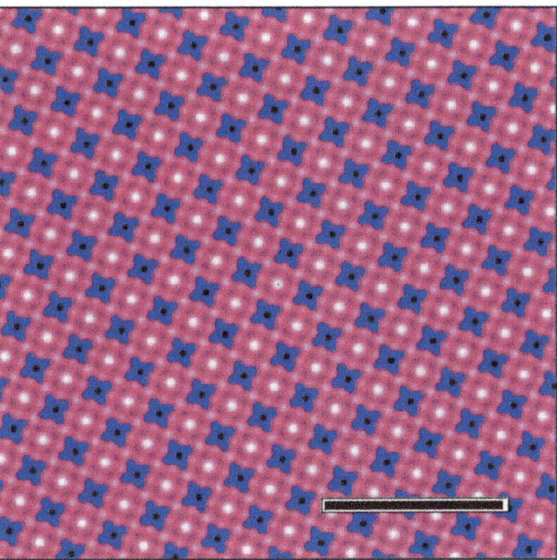

Fig. 4 Scanning force microscopy image of the S-layer protein SbpA from *L. sphaericus* CCM 2177 recrystallized on a silicon wafer. Image was recorded in contact mode in a liquid cell (bar, 50 nm) (Reprinted from ref. 27 with permission from the publisher. © 2003, Wiley-VCH)

11. Silanization from vapor-phase was performed with silanes of shorter chain lengths (e.g., hexamethyldisilane; HMDS).
12. Supports were baked with some drops of silane in an airtight glass vessel at 60°C for 2 h and finally rinsed with methanol.

3.2.2 S-Layer Protein Recrystallization at Solid Supports

1. For recrystallization of the S-layer protein SbsB of *G. stearothermophilus* PV72 and SbpA of *L. sphaericus* CCM 2177, buffer B and buffer C were used, respectively. The protein concentration in all experiments was 0.1 mg/mL.
2. Recrystallization on solid supports was carried out either in rotating Eppendorf tubes which had been previously filled with the protein solution or in glass wells. In the latter case, the substrates were placed onto the air/liquid interface.
3. After a recrystallization time of 4 h at RT the supports were removed by tweezers, washed, and stored in Milli-Q water (4°C).

3.2.3 Atomic Force Microscopy

1. Scanning was carried out in contact mode in a liquid cell filled with a 100 mM NaCl solution (Fig. 4). The applied force was kept to a minimum during scanning to prevent modification of the sample surface by the tip. Scan speed was approx. 6 Hz. Images were flattened line by line during recording using the software of the microscope.
2. Atomic force microscopy studies showed crystalline domains with average diameters of 10–20 μm for SbsB and of

Table 1
Substrate-induced recrystallization properties of the S-layer proteins SbsB (*Geobacillus stearothermophilus* PV72) and SbpA (*Lysinibacillus sphaericus* CCM 2177)[a]

Supports	Surface and the modifications	SbsB	SbpA
Metallic supports	Gold	+	+
	Titanium	+	+
	Aluminum	n	+
	Palladium	+	+
Silicon wafers (100 orientation, p-type)	Si (native oxide layer)	+	+[b]
	Si (native oxide layer) cleaned[c]	–	–
	Si (native oxide layer) O_2-plasma treated	–	+
	Si_3N_4	+	+
Silanized silicon wafers	Octadecyltrichlorosilane	+	+
	(3-methacryloyloxypropyl)-trimethoxysilane	+	+
	Trimethoxysilane	+	+
	Decyldimethylsilane	+	+
	Hexamethyldisilane	+	+
	2-aminopropyltrimethoxysilane	+	+
	3-mercaptopropyltrimethoxysilane	+	+
Polymers	Polyester	+	n
	Polypropylene	+	+
	Poly(ethylene terephthalate)	n	+
	Poly(methacrylic acid methyl ester)	n	+
	Polycarbonate	+	n
Others	Glass	+	+
	Mica	–	+
	Highly oriented pyrolytic graphite (HOPG)	–	+

[a]+, crystallization; –, no crystallization; n, not tested
[b]Very large crystalline domains
[c]Cleaning: acetone, propan-2-ol, ethanol, and Milli-Q water as described in (Subheading 3.2)

0.1–10 μm for SbpA, when crystallized on a variety of solid supports (see Table 1). In particular, SbsB generated crystalline monolayers only on hydrophobic solid supports, whereas SbpA formed extended crystalline domains at hydrophilic surfaces, but only small patches on hydrophobic ones.

3.3 Patterning of Crystalline S-Layer Proteins

3.3.1 Excimer Laser Patterning

1. Silicon wafers were cleaned with several solvents and O_2-plasma treated (Subheading 3.2).
2. Recrystallization of isolated S-layer protein on the silicon wafer was carried out as previously described (Subheading 3.2).
3. Prior to irradiation the recrystallized S-layer was carefully dried in a stream of high-purity nitrogen gas in order to remove

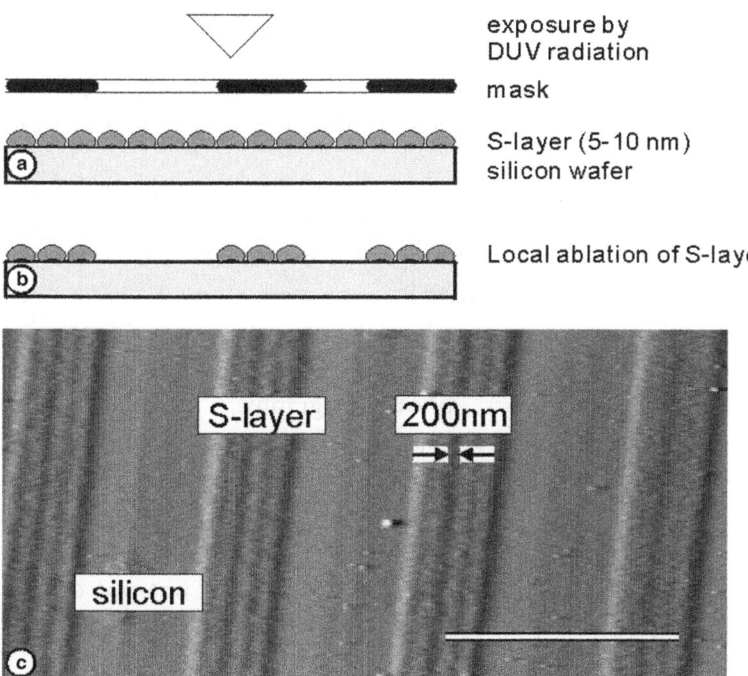

Fig. 5 Schematic drawing of the patterning of S-layers by exposure to deep ultraviolet radiation. (**a**) A pattern is transferred onto the S-layer by exposure to ArF excimer laser radiation through a microlithographic mask. (**b**) The S-layer is specifically removed from the silicon surface in the exposed areas. (**c**) Scanning force microscopical image of a patterned S-layer on a silicon wafer. Bar corresponds to 3 μm (Modified after ref. 24 with permission from the publisher. © 1999, Wiley-VCH)

excess water not required for maintaining the structural integrity of the protein lattice (see Note 5).

4. The lithographic mask was brought into direct contact with the S-layer-coated silicon wafer (Fig. 5).
5. The whole assembly was irradiated by the ArF excimer laser (see Note 6) in a series of one to five pulses with an intensity of about 100 mJ/cm^2 per pulse (pulse duration 8 ns, 1 pulse per second).
6. The mask was removed and the S-layer-coated silicon wafer immediately immersed in buffer.

S-layers which had been patterned by ArF excimer laser radiation may also be used as high-resolution etching masks in nano/microlithography (39, 40). This application requires enhancement of the patterned protein layer by electro-less metallization prior to subsequent reactive ion etching. Since S-layers are only 5–10 nm thick and thus much smaller than conventional resists (500–1,000 nm mean thickness) proximity effects are strongly reduced yielding a considerable improvement in edge resolution. Further on, for the development of bioanalytical sensors, patterned S-layers may also be used as electrode structures for binding biologically active molecules at specified target areas.

3.3.2 Soft Lithography Patterning

A well-known soft lithography technique, micromolding in capillaries (MIMIC) (41, 42), can be used for patterning and self-assembly of two-dimensional S-layer protein arrays on silicon supports.

1. For mold formation, 6 μm high mesa-structure mold masters were fabricated in photoresist on 4-in. silicon wafers using photolithography.
2. Poly(dimethylsiloxane), PDMS, was used to generate the molds from the masters (see Note 7) (43, 44). Ten parts of the silicone elastomer and one part of the curing agent (w:w) were mixed and degassed in an exsiccator.
3. The PDMS solution was put on the master laying in a Petri dish, and again, the solution was degassed until no bubbles were observed.
4. The PDMS mold was backed at 50°C for at least 4 h and subsequently removed from the master and cut to a proper size. Microchannels were formed when the recessed grooves in the PDMS mold were brought into conformal contact with the planar support, typically a native oxide-terminated silicon wafer.
5. The microchannels were filled from one end with protein solution (0.1 mg/mL SbpA in buffer C) by capillary action. The silicon supports (solvent cleaned) were O_2-plasma treated before the application of the mold in order to increase the wettability of the surface and to improve channel filling (see Note 8).
6. After self-assembly and crystallization of the S-layer protein (30 min to 24 h), the PDMS mold was removed under Milli-Q water, leaving the patterned S-layer arrays on the support.
7. The patterning was detected either by atomic force microscope (Subheading 3.2) or by epifluorescence microscopy.
8. For fluorescence microscopical detection the protein structures were labeled with fluoresceinisothiocyanate (FITC; see Note 9). The solid-supported S-layer patterns were incubated with the FITC suspension (1 mg FITC in 100 μl DMSO, diluted with 2 mL buffer D) for 1 h at RT in the dark.
9. After labeling the samples were washed with buffer D and finally, the patterning was investigated by epifluorescence microscopy (Fig. 6).

The MIMIC technique can be utilized for lateral patterning of simple and moderately complex crystalline S-layer arrays ranging in

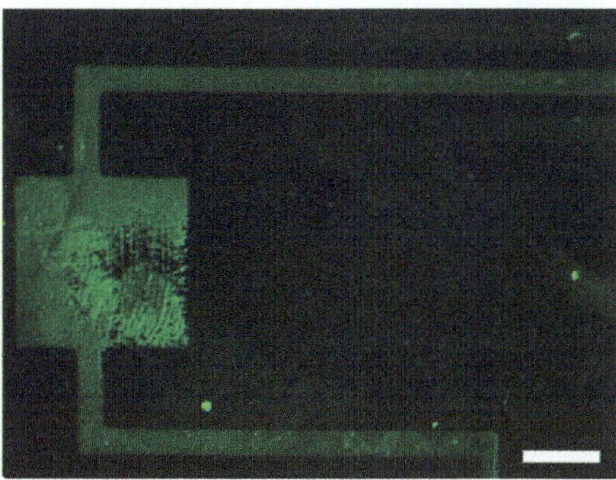

Fig. 6 Fluorescence image of the FITC-labeled S-layer protein SbpA patterned at a plasma-treated native silicon oxide support using a poly(dimethylsiloxane) (PDMS) mold. The bar represents 50 μm (Reprinted from ref. 55 with permission from the publisher. © 2005, Springer, GmbH)

critical dimension from submicron to hundreds of microns. Furthermore, the native chemical functionality of the S-layer protein is completely retained as demonstrated by attachment of human IgG antibody and subsequent binding of anti-human IgG antigen on the patterned S-layer substrates (45). This versatile MIMIC patterning technique can also be combined with immobilization techniques (Subheading 3.4), e.g., for controlled binding of nanoparticles with well-defined locations and orientations.

3.4 Formation of Nanoparticle Arrays

3.4.1 Preparation of Supports

1. To obtain comparable surface properties to silicon wafers, standard Formvar and carbon-coated electron microscope grids were coated with a 1–10 nm thick layer of SiO_2 by evaporation.
2. O_2-plasma treatment was carried out as described before (Subheading 3.2).

3.4.2 S-Layer Recrystallization

1. A solution of SbpA of *L. sphaericus* CCM 2177 (0.1 mg SbpA per mL buffer C) was filled in glass wells.
2. The SiO_2-coated grids were placed horizontally at the liquid/air interface and removed after 4 h. In most cases, there is not only a crystalline SbpA layer on the grid but also adsorbed self-assembly products.
3. The S-layer protein-coated grids were washed and stored in Milli-Q water at 4°C.

3.4.3 Nanoparticles

1. Citrate-stabilized gold nanoparticles with a mean diameter of 5 nm were negatively charged. The amino-modified, positively charged CdSe nanoparticles were prepared according to the literature (46–50) (see Note 10).

2. For non-covalent, electrostatic binding of nanoparticles to S-layer lattices, SbpA-coated grids (with or without attached S-layer self-assembly products, see Note 4) were incubated in the nanoparticle solution for 1 h at RT and washed with Milli-Q water.

3.4.4 Transmission Electron Microscopy

Transmission electron microscopy analysis was performed either on negatively stained but most frequently on untreated preparations. The structural (lattice constants, symmetries) and chemical diversity (surface-active functional groups) of S-layer proteins allow the formation of nanocrystal superlattices with a spatially controlled packing. Due to electrostatic interactions, anionic citrate-stabilized gold nanoparticles (5 nm in diameter) formed a superlattice at those sites where the inner face of the S-layer lattice was exposed. On the contrary, cationic semiconductor nanoparticles (such as amino-functionalized CdSe particles) formed arrays on the outer face of the solid-supported S-layer lattices (51).

3.5 S-Layer-Supported Lipid Membranes

3.5.1 Formation of Painted Lipid Membranes

1. Lipid membranes (Fig. 7a) were made from a 1% (wt/wt) solution of DPhyPC in *n*-decane (52, 53). The stock solution was stored at −20°C.
2. The orifice was pre-painted with DPhyPC, dissolved in chloroform (10 mg/mL), and dried with nitrogen for at least 20 min.
3. The compartments were filled with the electrolyte (12 mL each).
4. The *cis*-cell was grounded; the *trans*-cell was connected by another Ag/AgCl-electrode to the patch clamp amplifier.
5. A drop of lipid mixture was put on the Teflon brush and was stroked up the orifice. Membrane formation should be seen immediately (see Note 11).
6. Thinning of the membranes was followed by measuring the capacitance of the lipid membrane.
7. After a constant capacitance was reached (takes ~20–40 min) experiments to study the intrinsic parameters of the lipid membrane have been performed.

3.5.2 Formation of Folded Lipid Membranes

1. DPhyPC was dissolved in *n*-hexane/ethanol (9:1). The stock solution was stored at −20°C at a concentration of 5 mg lipid per mL.
2. At least 30 min before the formation of the membrane, the aperture was preconditioned with a small drop of hexadecane/pentane (1:10) (Fig. 7b). Both compartments were filled to just below the aperture with electrolyte (54, 55).
3. A volume of 2 μl of the lipid stock solution was spread on the aqueous surface of each compartment, and the solvent was

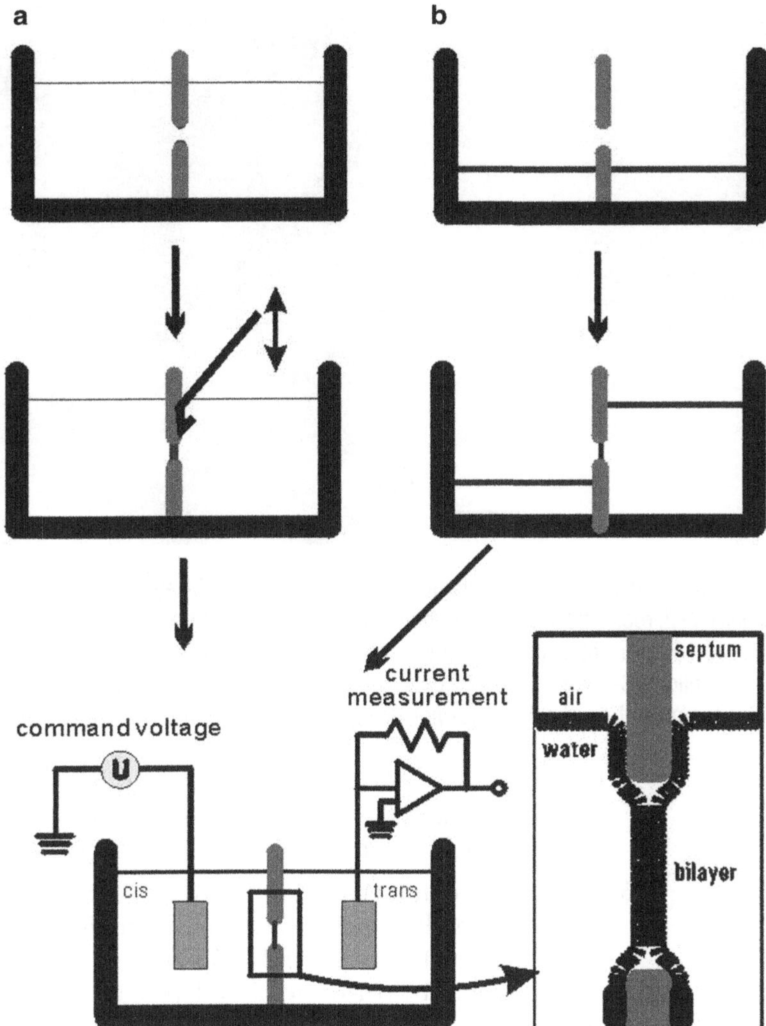

Fig. 7 Schematic illustration of the formation of (**a**) a painted and (**b**) a folded lipid membrane. On the lower, *left* part a schematic illustration of the set-up is given (not drawn to scale); the insert shows a schematic drawing of the bilayer lipid membrane (Reprinted from ref. 55 with permission from the publisher. © 2005, Springer, GmbH)

allowed to evaporate for at least 20 min. Raising the level of the electrolyte within the compartments to above the aperture by means of the syringes led to formation of a lipid membrane which was checked by measuring its conductance and capacitance (see Note 12).

4. The current response from a given voltage function was measured to provide the capacitance and conductance of the lipid membranes (56, 57). A triangular voltage function (+40 mV to −40 mV, 20 ms) may be used to determine the capacitance of the lipid membrane. The specific capacitance is about 0.4 to

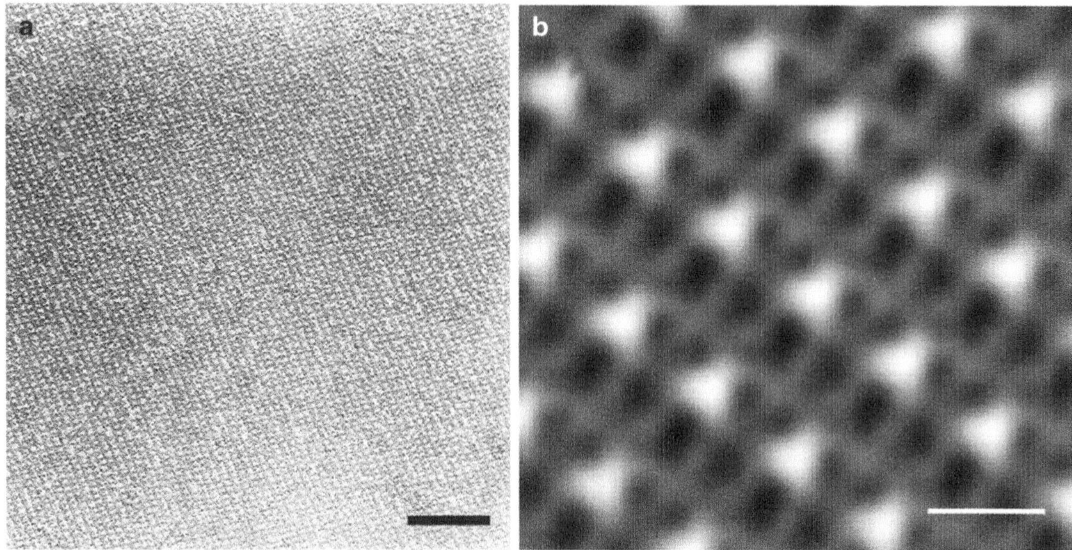

Fig. 8 (**a**) Electron micrograph of negatively stained preparation of the S-layer protein SbsB isolated from *G. stearothermophilus* PV72, recrystallized on a monolayer made of DPhPC/hexadecylamine (molar ratio 10:4). The bar corresponds to 100 nm. (**b**) Computer image reconstitution of the transmission electron microscopy image showing of the oblique S-layer lattice of SbsB. The bar corresponds to 10 nm (Reprinted from ref. 64 with permission from the publisher. © 2002, Elsevier Science)

0.5 μF/cm² and 0.6 to 0.8 μF/cm² for the painted and folded membranes, respectively (30, 32, 58–62) (see Note 13).

5. Membrane conductance is usually < 10^{-8} S/cm². The settings of the two built-in Bessel filters of the amplifier for the current-monitor signal were 10 and 1.5 kHz, respectively. All experiments should be performed at RT (see Note 14).
6. After each experiment, the Teflon aperture was cleaned extensively with chloroform, methanol, and ethanol and finally rinsed with Milli-Q-water.

3.5.3 Recrystallization of the S-Layer Proteins SbpA and SbsB

1. After forming the painted or folded lipid membrane, the S-layer solution was carefully injected into the *trans*-compartment to a final protein concentration of 0.1 mg/mL.
2. The same volume of buffer was added to the *cis*-compartment.
3. According to our experience the recrystallization process of S-layer subunits on lipid membranes was generally completed within 3 h (63).
4. If the lipid membrane should be supported by the S-layer protein SbpA, 10 mM $CaCl_2$ has to be added to the electrolyte to make recrystallization possible. On the other hand, no $CaCl_2$ is needed for the recrystallization of the S-layer SbsB protein of *G. stearothermophilus* PV72 (Fig. 8).

5. The closely attached S-layer lattice did not affect the specific capacitance, whereas the resistance of the membranes increased slightly (60, 64–67).
6. Recrystallization of the S-layer protein can be investigated by transition electron microscopy (Subheading 2.4.3) on deposited negatively stained preparations or by atomic force microscopical investigations (Subheading 2.2.4) of the lipid-coated polymer septum.

The advantage of S-layer-supported lipid membranes is the enhanced long-term stability (66, 67), the stability against voltage ramps even up to 500 mV and more (68), the increased bending stiffness (69), and thus, the higher robustness against hydrostatic pressure gradients (63, 64). Thus, it might be possible to distinguish at mechano-sensitive ion channels (70–72), reconstituted in S-layer-supported lipid membranes, between the curvature-induced mechanical activation and the flow-induced activation (64). In addition, the tightly attached S-layer lattice allows complete reconstitution of membrane-active peptides (66, 73) but also of complex membrane proteins like α-hemolysin (65, 67, 68).

4 Notes

1. For chlorination immerse silver wire as anode in a 0.1 M HCl solution and pass a current of 10 mA for 5 min through the wire.
2. *L. sphaericus* CCM 2177 tends to make the medium alkaline, and thus, one has to be prepared to add H_2SO_4 early enough to maintain a pH value of 7.2.
3. The S-layer-containing solutions should be dialyzed against large volumes, usually 3 L of distilled water, with or without $CaCl_2$ is taken at each dialysis step. Attention should be applied to cool down the distilled water to 4°C before the dialysis is performed.
4. Single isolated S-layer subunits from many prokaryotic organisms have shown the ability to assemble into regular lattices identical to those observed on intact cells upon removal of the disrupting agents used for their isolation, e.g., upon dialysis. The S-layer self-assembly processes lead to the formation of flat sheets, open-ended cylinders, or spheres. Ionic strength, temperature, protein concentration, and polymer associated with S-layers can determine both the rate and extent of assembly (for review see refs. 25–28).
5. Excess water has to be carefully removed before patterning in order to prevent interference fringes caused by the water film.

6. S-layer protein is completely removed by ArF ($\lambda = 193$ nm) irradiation at a dosage of 100–200 mJ/cm^2.
7. Poly(dimethylsiloxane) (PDMS), a hydrophobic elastomer, is well known for absorbing organic solvents (74). However, although it is not investigated in detail, organic components might also dissolve away from the PDMS hence leading to no reproducible measurements, in particular when working with biomolecules like phospholipids (Erik Reimhult, personal communication).
8. Rapid filling of the micron-scale channels may be followed with an optical microscope and capillaries may be filled even when the solution entered from both ends of the mold.
9. FITC binds to the free amino groups of the S-layer protein.
10. The amino-modified CdSe nanoparticles were prepared by an organometallic synthesis using a mixture of highly boiling primary aminoalkanes and trioctylphosphine (TOP) as the coordinating solvent. The CdSe nanocrystals from about 1 ml freshly prepared sol were precipitated by adding a small amount of methanol. After removal of the supernatant, the particles were transferred in 5 ml aqueous solution of 20 mM *N,N*-dimethyl-mercaptoethylammonium chloride and 1 mM 2-(butylamino)-ethanethiol in the case of an additional functionalization with a secondary amine. Five minutes of ultrasonic treatment led to an optically clear solution.
11. You have to push the Teflon brush very tightly against the septum when the lipid is stroked up the orifice. If the membrane ruptures, try it again with the Teflon brush without dropping new lipid on it.
12. If no membrane formation can be achieved, remove the lipid of the air/water interface with a suction pump and try it again with a smaller amount of lipid. In addition, be very careful that all solutions and the electrolytes are free of any dust or other contaminants.
13. The dielectric constant for lipid membranes is taken as $\varepsilon = 2.1$, corresponding to the average dielectric constant of a long-chain hydrocarbon (32).
14. If particularly the humidity is too high or the weather is sultry, membrane formation is very rare and the membranes are usually not very stable.

Acknowledgement

The research was funded by the Austrian Science Fund (FWF): P20256-B11.

References

1. Sleytr UB, Messner P, Pum D et al. (1996) Crystalline bacterial cell surface proteins. Academic Press, R.G. Landes Company, Austin, USA
2. Sleytr UB, Huber C, Pum D et al (2007) Nanobiotechnology with S-layer proteins. FEMS Microbiol Lett 267:131–144
3. Sleytr UB, Egelseer EM, Ilk N et al (2007) S-layers as basic building block for a molecular construction kit. FEBS J 274:323–334
4. Messner P, Schäffer C, Egelseer EM et al (2010) Occurrence, structure, chemistry, genetics, morphogenesis, and functions of S-layers. In: König H, Claus H, Varma A (eds) Prokaryotic cell wall compounds—structure and biochemistry. Springer, Berlin, Germany
5. Sleytr UB, Messner P (2009) Crystalline bacterial cell surface layers (S-layers). In: Schaechter M (ed) Encyclopedia of microbiology, 3rd edn. Elsevier, Oxford
6. Sleytr UB (1978) Regular arrays of macromolecules on bacterial cell walls: structure, chemistry, assembly and function. Int Rev Cytol 53:1–64
7. Sleytr UB, Beveridge TJ (1999) Bacterial S-layers. Trends Microbiol 7:253–260
8. Pum D, Sára M, Sleytr UB (1989) Structure, surface charge, and self-assembly of the S-layer lattice from *Bacillus coagulans* E38-66. J Bacteriol 171:5296–5303
9. Sára M, Pum D, Sleytr UB (1992) Permeability and charge-dependent adsorption of the S-layer lattice from *Bacillus coagulans* E38-66. J Bacteriol 174:3487–3493
10. Moreno-Flores S, Kasry A, Butt HJ et al (2008) From native to non-native two-dimensional protein lattices through underlying hydrophilic/hydrophobic nanoprotrusions. Angew Chem Int Ed 47:4707–4710
11. Messner P, Pum D, Sleytr UB (1986) Characterization of the ultrastructure and the self assembly of the surface layer (S-layer) of *Bacillus stearothermophilus* strain NRS 2004/3a. J Ultrastruct Mol Struct Res 97: 73–88
12. Sára M, Sleytr UB (1987) Charge distribution on the S-layer of *Bacillus stearothermophilus* NRS 1536/3c and the importance of charged groups for morphogenesis and function. J Bacteriol 169:2804–2809
13. Györvary ES, Stein O, Pum D et al (2003) Self-assembly and recrystallization of bacterial S-layer proteins at silicon supports imaged in real time by atomic force microscopy. J Microsc 212:300–306
14. Diederich A, Sponer C, Pum D et al (1996) Reciprocal influence between the protein and lipid components of a lipid-protein membrane model. Coll Surf B: Biointerfaces 6:335–346
15. Sleytr UB, Györvary ES, Pum D (2003) Crystallization of S-layer protein lattices on surfaces and interfaces. Prog Organ Coat 47:279–287
16. Pum D, Sàra M, Schuster B et al (2006) Bacterial surface layer proteins: a simple but versatile biological self-assembly system in nature. In: Chen J, Jonoska N, Rozenberg G (eds) Nanotechnology: science and computation. Springer, Berlin, Heidelberg, Germany
17. Lopez AE, Moreno-Flores S, Pum D et al (2010) Surface dependence of protein nanocrystal formation. Small 6:396–403
18. Pum D, Tang J, Hinterdorfer P et al (2010) S-layer protein lattices studied by scanning force microscopy. In: Kumar CSSR (ed) Nanomaterials for the life sciences, vol. 7. Biomimetic and bioinspired nanomaterials. Wiley-VCH, Weinheim, Germany
19. Sleytr UB (1975) Heterologous reattachement of regular arrays of glycoproteins on bacterial surfaces. Nature 257:400–402
20. Pum D, Sleytr UB (2009) Protein-based nanobioelectronics. In: Offenhäusser A, Rinaldi R (eds) Nanobioelectronics for electronics, biology, and medicine. Springer, Berlin, Germany
21. Sleytr UB, Egelseer EM, Ilk N et al (2010) Nanobiotechnological applications of S-layers. In: König H, Claus H, Varma A (eds) Prokaryotic cell wall components—structure and biochemistry. Springer, Heidelberg, Germany
22. Egelseer EM, Ilk N, Pum D et al (2010) Nanobiotechnological applications of S-layers. In: Flickinger MC (ed) Encyclopedia of industrial biotechnology: bioprocess, bioseparation, and cell technology. Wiley, Weinheim, Germany
23. Ilk N, Egelseer EM, Ferner-Ortner J et al (2008) Surfaces functionalized with self-assembling S-layer fusion proteins for nanobiotechnological applications. Colloids Surf A: Physicochem Eng Aspects 321:163–167
24. Sleytr UB, Messner P, Pum D et al (1999) Crystalline bacterial cell surface layers (S layers): from supramolecular cell structure to biomimetics and nanotechnology. Angew Chemie Int Ed 38:1034–1054
25. Sleytr UB, Sára M, Pum D (2000) Crystalline bacterial cell surface layers (S-layers): a versatile self-assembly system. In: Ciferri A (ed) Supramolecular polymerization Marcel Dekker. Basel, New York
26. Sleytr UB, Sára M, Pum D et al (2001) Molecular nanotechnology and nanobiotechnology with two-dimensional protein crystals (S-layers). In: Rosoff M (ed) Nano-surface chemistry. Marcel Dekker, New York, Basel
27. Sleytr UB, Sára M, Pum D et al (2003) Self assembly protein systems: microbial S-layers.

In: Steinbüchel A, Fahnestock S (eds) Biopolymers, vol 7. Wiley-VCH, Weinheim, Germany
28. Sleytr UB, Sára M, Pum D et al (2005) Crystalline bacterial cell surface layers (S-layers): a versatile self-assembly system. In: Ciferri A (ed) Supramolecular polymers, 2nd edn. CRC Press, Taylor & Francis Group, Boca Raton, FL
29. Sára M, Egelseer EM, Huber C et al (2006) S-layer proteins: potential applications in nano(bio)technology. In: Rehm B (ed) Microbial bionanotechnology: biological self-assembly systems and biopolymer-based nanostructures. Horizon Scientific Press, Hethersett, Norwich, UK
30. Hanke W, Schlue WR (1993) Planar lipid bilayers: methods and applications. In: Sattelle DB (ed) Biological techniques series. Academic Press, London, UK
31. Alvarez O (1986) How to set up a bilayer system. In: Miller C (ed) Ion channel reconstitution. Plenum Press, New York
32. Benz R, Fröhlich O, Läuger P et al (1975) Electrical capacity of black lipid films and lipid bilayers made from monolayers. Biochim Biophys Acta 394:323–334
33. Winterhalter M (2000) Black lipid membranes. Curr Opin Coll Interface Sci 5:250–255
34. Montal M (1974) Formation of bimolecular membranes from lipid monolayers. Methods Enzymol B 32:545–554
35. Messner P, Hollaus F, Sleytr UB (1984) Paracrystalline cell wall surface layers of different Bacillus stearothermophilus strains. Int J Syst Bacteriol 34:202–210
36. Sleytr UB, Sára M, Küpcü Z et al (1986) Structural and chemical characterization of S-layers of selected strains of *Bacillus stearothermophilus* and *Desulfotomaculum nigrificans.* Arch Microbiol 146:19–24
37. Bartelmus W, Perschak F (1957) Schnellmethode zur Keimzahlbestimmung in der Zuckerindustrie. Z Zuckerind 7:276–281
38. Pum D, Sleytr UB (1995) Anisotropic crystal growth of the S-layer of *Bacillus sphaericus* CCM 2177 at the air/water interface. Colloids Surf A 102:99–104
39. Pum D, Stangl G, Sponer C et al (1997) Deep ultraviolet patterning of monolayers of crystalline S-layer protein on silicon surfaces. Colloids Surf B 8:157–162
40. Pum D, Stangl G, Sponer C et al (1997) Patterning of monolayers of crystalline S-layer proteins on a silicon surface by deep ultraviolet radiation. Microelectron Eng 35:297–300
41. Xia Y, Whitesides GM (1998) Soft lithography. Angew Chem Int Ed 37:550–575
42. Michel B, Bernard A, Bietsch A et al (2001) Printing meets lithography: soft approaches to high resolution patterning. IBM J Res Dev 45:697–719
43. Kumar A, Biebuyck HA, Whitesides GM (1994) Patterning self-assembled monolayers: applications in materials science. Langmuir 10:1498–1511
44. Kim E, Xia Y, Whitesides GM (1995) Making polymeric microstructures: capillary micromolding. Nature 376:581–584
45. Györvary ES, O'Riordan A, Quinn AJ et al (2003) Biomimetic nanostructure fabrication: non-lithographic lateral patterning and self-assembly of functional bacterial S-layers at silicon supports. Nano Lett 3:315–319
46. Talapin DV, Rogach AL, Kornowski A et al (2001) Highly luminescent monodisperse CdSe and CdSe/ZnS nanocrystals synthesized in a hexadecylamine—trioctylphosphine oxide—trioctylphospine mixture. Nano Lett 1:207–211
47. Talapin DV, Rogach AL, Mekis I et al (2002) Synthesis and surface modification of amino-stabilized CdSe, CdTe and InP nanocrystals. Colloids Surf A 202:145–154
48. Rosenthal SJ, McBride J, Pennycook SJ et al (2007) Synthesis, surface properties, composition and structural characterization of CdSe, core/shell and biologically active nanocrystals. Surf Sci Rep 62:111–157
49. Asokan S, Krueger KM, Colvin VL et al (2007) Shape-controlled synthesis of CdSe tetrapods using cationic surfactant ligands. Small 3: 164–1169
50. Sperling RA, Parak WJ (2010) Surface modification, functionalization and bioconjugation of colloidal inorganic nanoparticles. Philos T Roy Soc A 368:333–1383
51. Györvary ES, Schroedter A, Talapin D et al (2003) Formation of nanoparticle arrays on S-layer protein lattices. J Nanosci Nanotechnol 4:15–120
52. Mueller P, Rudin DO, Tien HT et al (1962) Reconstitution of cell membrane structure in vitro and its transformation into excitable systems. Nature 194:979–981
53. Fettiplace R, Gordon LGM, Hladky SB et al (1975) Techniques in formation and examination of black lipid bilayer membranes. In: Korn ED (ed) Methods of membrane biology, vol 4. Plenum Press, New York
54. Montal M, Mueller P (1972) Formation of bimolecular membranes from lipid monolayers and a study of their electrical properties. Proc Natl Acad Sci U S A 69:3561–3566
55. Schuster B, Györvary ES, Pum D et al (2005) Nanotechnology with S-layer proteins. In: Vo-Dinh T (ed) Protein nanotechnology: protocols, instrumentation and application, book series: methods molecular biology, vol 300. Humana Press, Totowa, NJ

56. Darszon A (1983) Strategies in the reassembly of membrane proteins into lipid bilayer systems and their functional assay. J Bioenerg Biomembr 15:321–334
57. Schindler H (1989) Planar lipid-protein membranes: strategies of formation and of detection dependencies of ion transport functions on membrane conditions. Methods Enzymol 171:225–253
58. Sleytr UB, Sára M, Pum D et al (2001) Characterization and use of crystalline bacterial cell surface layers. Progr Surf Sci 68:231–278
59. Tien HT, Ottova AL (2001) The lipid bilayer concept and its experimental realization: from soap bubbles, kitchen sink, to bilayer lipid membranes. J Membr Sci 189:83–117
60. Schuster B, Sleytr UB (2009) Composite S-layer lipid structures. J Struct Biol 168:207–216
61. Schuster B, Sleytr UB (2005) 2D-protein crystals (S-layers) as support for lipid membranes. In: Tien TH, Ottova A (eds) Advances in planar lipid bilayers and liposomes, vol 1. Elsevier Science, Amsterdam, The Netherlands
62. Schuster B (2005) Biomimetic design of nanopatterned membranes. NanoBiotechnology 1:153–164
63. Schuster B, Sleytr UB, Diederich A et al (1999) Probing the stability of S-layer-supported planar lipid membranes. Eur Biophys J 28:583–590
64. Schuster B, Sleytr UB (2002) The effect of hydrostatic pressure on S-layer supported lipid membranes. Biochim Biophys Acta 1563:29–34
65. Schuster B, Sleytr UB (2002) Single channel recordings of α-hemolysin reconstituted in S-layer-supported lipid bilayers. Bioelectrochemistry 55:5–7
66. Schuster B, Pum D, Sleytr UB (1998) Voltage clamp studies on S-layer-supported tetraether lipid membranes. Biochim Biophys Acta 1369:51–60
67. Schuster B, Pum D, Braha O et al (1998) Self-assembled α-hemolysin pores in an S-layer-supported lipid bilayer. Biochim Biophy Acta 1370:280–288
68. Schuster B, Pum D, Sára M et al (2001) S-layer ultrafiltration membranes: a new support for stabilizing functionalized lipid membranes. Langmuir 17:500–503
69. Hirn R, Schuster B, Sleytr UB et al (1999) The effect of S-layer protein adsorption and crystallization on the collective motion of a lipid bilayer studied by dynamic light scattering. Biophys J 77:2066–2074
70. Chang G, Spencer RH, Lee AT et al (1998) Structure of the MscL homolog from mycobacterium tuberculosis: a gated mechanosensitive ion channel. Science 282:2220–2226
71. Jones SE, Naik RR, Stone MO (2000) Use of small fluorescent molecules to monitor channel activity. Biochem Biophys Res Co 279:208–212
72. Booth IR, Louis P (1999) Managing hypoosmotic stress: aquaporins and mechanosensitive channels in Escherichia coli. Curr Opin Microbiol 2:166–169
73. Schuster B, Weigert S, Pum D et al (2003) New method for generating tetraether lipid membranes on porous supports. Langmuir 19:2392–2397
74. Malmstadt N, Nash MA, Purnell RF et al (2006) Automated formation of lipid-bilayer membranes in a microfluidic device. Nano Lett 6:1961–1965

Part II

New Proteins

Chapter 10

Stimuli-Responsive Peptide Nanostructures at the Fluid–Fluid Interface

Chun-Xia Zhao and Anton P.J. Middelberg

Abstract

The self-organization of peptide-based nanostructures at a confined fluid–fluid interface, for example, the air–water or oil–water interface, is important in the context of stabilizing macroscopic soft-matter foams and emulsions. The unique ability to design interfacial nanostructures by controlling the subtle cooperativity that drives peptide self-assembly, and the ability to switch molecular cooperativity by facile triggers such as pH, opens new vistas for controlling macroscopic soft matter in industries as diverse as healthcare and industrial processing. Here we describe research aimed at developing new understanding into soft-matter formation and control, through variation of peptide sequence and bulk conditions. Macroscopic foaming and microfluidic emulsification studies prove particularly useful in visualizing and hence understanding the synergistic link between molecular design, mesoscopic interfacial properties, and bulk soft-matter stability.

Key words Peptide, Emulsion, Foam, Biosurfactant, Surfactant, Switch, Responsive, Nanomaterial, Protein, Interfacial rheology

1 Introduction

Peptide self-assembly is a burgeoning research field finding increasing scientific and application focus (1, 2). Much activity to date has centered on the design of three-dimensional structures, such as gels and scaffolds for cell organization, and the design of nanostructures in bulk, for example, fibrils and tapes, or modification of the two-dimensional solid–liquid interface (for example, in biosensor applications). In contrast, the area of nanostructure formation confined at a fluid–fluid interface, for example air–water or oil–water, has been a relatively neglected area of research.

Several fundamental investigations of peptide organization have been undertaken at the air–water and oil–water interface. Recent studies have revealed the importance of molecular cooperativity and in particular electrostatics (3), molecular orientation (4), peptide sequence (5), and secondary structure (6) on the resulting nanostructures. Insight at the air–water interface has been aided by

Juliet A. Gerrard (ed.), *Protein Nanotechnology: Protocols, Instrumentation, and Applications*, Methods in Molecular Biology, vol. 996, DOI 10.1007/978-1-62703-354-1_10,

the relative ease of, for example, neutron reflectivity (refer to a recent review by Zhao et al. (7)), while the oil–water interface is being explored with molecular dynamics simulation (8). Considering that scientific understanding of peptide self-assembly complexity at the fluid–fluid interface is in its early stages, it is unsurprising that there are few examples of translation of knowledge to facilitate the practical control of macroscopic foams and emulsions. Foams and emulsions comprise a high content of air–water and oil–water interface, respectively, and have global economic importance across diverse industrial sectors ranging from medicine and personal care to industrial processing.

One class of peptide materials has recently been demonstrated to offer a high degree of foam and emulsion macroscopic control. Interfacial studies (9) of a helix-forming sequence, Lac21 (10), showed it possesses high interfacial affinity and good solubility, giving the practical advantage that peptide self-assembly occurs from the bulk aqueous phase. Designed peptides using the same basic helix motif have subsequently been shown to provide unique control over foam and emulsion formation and dissipation, in response to facile triggers such as pH (11, 12). Macroscopic control occurs through changes in interfacial properties including through the formation and dissipation of a cohesive network having mechanical properties similar to collagen (13). A peptide network having high mechanical strength is able to slow the kinetics of foam (12) and emulsion (14) coalescence, and under extreme mechanical cohesiveness, emulsion droplets exhibit capsule-like behavior.

2 Materials

2.1 Acid Clean

1. 5% detergent solution (Decon 90, Decon Laboratories Ltd, Hove, East Sussex, UK).
2. 30% hydrogen peroxide.
3. 98% sulfuric acid.
4. 1,000 ml beakers.
5. Two spill trays.
6. Magnetic stirrer.
7. Protection: gloves, lab coat, closed-in shoes and safety glasses, protective apron, PVC sleeve protectors, and full face shield.

2.2 Peptides and Metal Ions

1. AM1: This 21-residue peptide AM1 (Ac-MKQLADS LHQLARQ VSRLEHA-$CONH_2$) forms cohesive, mechanically strong interfacial films via cross-linking of interfacially adsorbed peptide molecules with metal ions (11).
2. AFD4: This peptide surfactant AFD4 (Ac-MKQLADS LHQLAHK VSHLEHA-$CONH_2$) permits intra- as well as

intermolecular bridging by metal ions. AFD4 includes additional histidine residues having appropriate spacing to permit metal ion chelation by successive turns of a peptide α-helix (15).

3. AM1 (MW 2473) and AFD4 (MW 2435) are custom synthesized by chemical synthesis with purity >95% by RP-HPLC. Peptide concentration is determined by quantitative amino acid analysis (e.g., by the Australian Proteome Analysis Facility, Sydney, NSW, Australia).
4. 100 mM zinc sulfate ($ZnSO_4$).
5. 100 mM zinc chloride ($ZnCl_2$).
6. 100 mM cobalt chloride ($CoCl_2$).
7. 100 mM nickel nitrate ($Ni(NO_3)_2$).

2.3 Foam Formation and Switching

1. Buffer solution: 25 mM sodium 4-(2-hydroxyethyl)-1-piperazine ethanesulfonate (HEPES), pH 7.4.
2. Peptide solutions: 1 mL solutions of 0.15 mg/mL (AM1, AFD4) prepared in 25 mM HEPES buffer (pH 7.4) (see Note 1).
3. pH shifting solutions: 1.6 M H_2SO_4 and 1 M NaOH.
4. Chelating agent: 100 mM sodium ethylenediaminetetraacetate (EDTA) at pH 8.
5. Metal ions: 100 mM $ZnSO_4$ solution.
6. Syringe: 60 mL plastic syringe.
7. Syringe pump: a syringe pump.
8. Foam preparation apparatus: A custom built glass foaming apparatus (shown in Fig. 1) consists of a glass tube (10 cm × 1 cm diameter), open at the top and fitted with a porous glass frit at the bottom. Below the glass frit are an air inlet connection and a valve for draining liquid from the tube. The air inlet is connected by plastic tubing to an air filled 60 mL syringe, mounted on the syringe pump (12).

2.4 Emulsion Formation and Switching

1. Buffer solution: 25 mM sodium 4-(2-hydroxyethyl)-1-piperazine ethanesulfonate (HEPES), pH 7.4.
2. Metal ions: 100 mM $ZnSO_4$ solution.
3. Peptide solutions: 2.8 mL solutions of 0.15 mg/mL (AM1, AFD4) prepared in 25 mM HEPES buffer (pH 7.4), 250 μM $ZnSO_4$, pH 7.4 (see Note 1).
4. Oil phase: toluene.
5. Visualization dye for emulsion switching: Sudan III (1-(4-(phenylazo)phenylazo)-2-naphthol) (50 μM) is dissolved in the oil phase; methylene blue (3,7-bis(dimethylamino)phenazathionium chloride) (10 μM) is included in the aqueous phase.
6. pH shifting solutions: 1.9 M H_2SO_4.

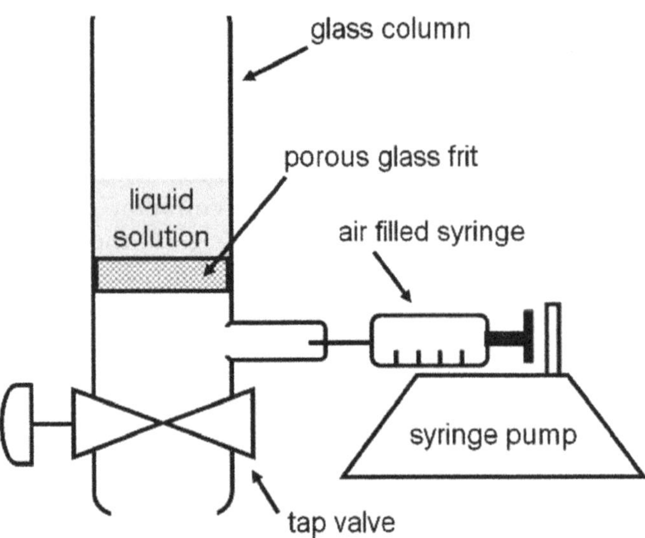

Fig. 1 Schematic of apparatus used for foam formation (12) (Reproduced by permission of The Royal Society of Chemistry (RSC))

7. Chelating agent: 100 mM sodium ethylenediaminetetraacetate (EDTA) at pH 7.6.
8. Emulsification apparatus: a rotor-stator homogenizer (Ystral X10).

2.5 Emulsion Formation in Microfluidic Devices, with Switching

1. Continuous phases: 100 μM AM1 and 100 μM $ZnSO_4$ in 25 mM HEPES buffer pH 7.0.
2. Dispersed phase: dodecane.
3. Switching solution: 100 mM sodium ethylenediaminetetraacetate EDTA at pH 7.4.
4. Microfluidic device (14) (shown in Fig. 2): A T-junction microfluidic device made of PMMA (polymethylmethacrylate) was purchased from Epigem Ltd (UK). The channel surface is hydrophilic by coating an aromatic epoxy. The microchannel is 100 μm deep. The parallel channel is 100 μm wide, and the perpendicular channel is 50 μm wide. The expansion channel is 500 μm wide.
5. Syringe pumps: three motor-driven syringe pumps (PHD 2000 Harvard, Instech).
6. Syringes: three SGE glass syringes, 1 mL, 500 μL, and 100 μL.
7. Microscope: an optical microscope (Eclipse 50i, Nikon).
8. Picture capture: Droplet pictures are recorded by high-speed video camera (Powershot A640, Canon).

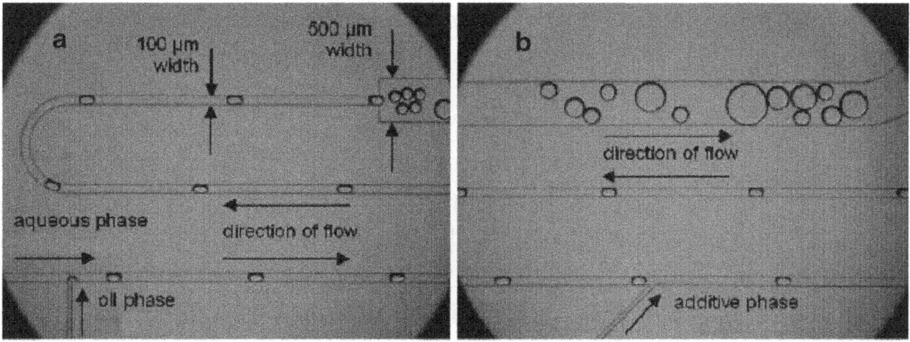

Fig. 2 Microfluidic chips used for emulsion droplet formation and switching. Photomicrographs of the microfluidic chip show (**a**) dodecane droplet generation at the T-junction, droplet collisions in the 500 µm expansion channel leading to extensive coalescence, and (**b**) entry of the additive phase at the downstream Y-junction (14) (Copyright Wiley-VCH Verlag GmbH & Co. KGaA. Reproduced with permission)

2.6 Double Emulsion Formation in Microfluidic Devices

1. Continuous phases: 100 µM AFD4 peptide solution in Milli-Q water in the presence of 200 µM $ZnSO_4$, pH 7.0; 1.0 mM sodium dodecyl sulfate (SDS) solution.
2. Dispersed phase: ternary solvent system Miglyol 812–ethanol–water at the volume ratio of (1:1:0.04) in or without the presence of Span 80; the heavy phase of the ternary system sunflower oil–ethanol–water at the volume ratio of (1:2:0.04) with 20 mM Span 80.
3. Microfluidic device: a T-junction microfluidic device as described in Section 2.5.
4. Syringe pumps: two motor-driven syringe pumps (PHD 2000 Harvard, Instech).
5. Syringes: two SGE glass syringes, 1 mL and 500 µL.
6. Microscope: an optical microscope (Eclipse 50i, Nikon).
7. Picture capture: droplet pictures recorded by video camera (Powershot A640, Canon).

3 Methods

3.1 Glassware Acid Clean (see Note 2*)*

1. Soak glassware in about 5% (v/v) Decon-90 in water for a minimum of 3 h, then rinse six times with Milli-Q water to prevent carry-over of Decon-90 residues or other organic residues that could react with peroxide-sulfuric acid (controls: gloves, lab coat, closed-in shoes, and safety glasses).
2. Collect hydrogen peroxide solution 30% (v/v) stored in a plastic bag in the refrigerator and 98% sulfuric acid solution stored in an acid cabinet.

3. Put on all the personal protective equipment (PPE) including protective apron, PVC sleeve protectors, and full face shield (see Note 3).
4. Put glassware to be cleaned in a spill tray containing 1–2 cm water in the bottom of a tray in the fume hood (see Note 4).
5. Prepare the solution for acid cleaning of glassware, comprising 1 part 30% (v/v) hydrogen peroxide, 1 part 98% (w/v) sulfuric acid, freshly mixed ("piranha solution").
6. Piranha solution is prepared by slowly adding sulfuric acid to a beaker containing hydrogen peroxide with continuous magnetic stirring slowly and continuously (see Note 5).
7. Piranha solution becomes quite hot. Pour Piranha solution into glassware to be cleaned, completely filling the glassware, and allow to stand in contact with glass for up to 15 min (see Note 6).
8. After soaking for 15 min, drain the cleaned glassware completely and fill with water as soon as possible to remove the risk of contact with undiluted piranha solution.
9. After soaking with piranha solution, glassware is rinsed with Milli-Q water at least 10 times.
10. Dispose of solutions in accordance with local waste chemical management procedures.

3.2 Surface Tension or Interfacial Tension Measurement by Drop Shape Analysis

The surface tension and interfacial tension are measured with a Krüss Drop Shape Analysis System DSA10 (Krüss GmbH, Hamburg, Germany).

3.2.1 Surface Tension Measurement

1. Set up the DSA10; input the known parameters including temperature, needle diameter, the densities of the drop and embedding phases, and aspect ratio.
2. Calibrate the DSA10 by using a water-in-air system. The surface tension should be stable over 10 min and close to 73 mN/m (see Note 7).
3. Peptide solutions (AM1 or AFD4) are filled into a 1 mL syringe with a straight needle of known diameter.
4. Drops (ca. 15 μL) of peptide solutions are formed in air in an 8 mL quartz cuvette.
5. The drop shape is monitored automatically over 10 min. The surface tension readings are made every 5 s over 1,000 s following initial droplet formation.
6. An example of the surface tension results for peptide AM1 is shown in Fig. 3.

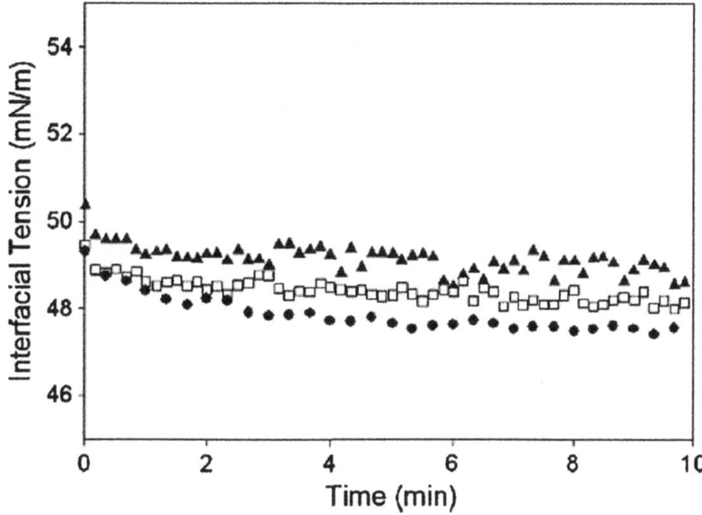

Fig. 3 Air–water surface tension of AM1 versus time for 0.3 mg/mL AM1 solutions. pH 7.4, no added Zn(II) (*square*); pH 7.4, 200 μM Zn(II), 500 μM EDTA (*filled circle*); pH 3.6, 200 μM Zn(II) (*filled triangle*) (12) (Reproduced by permission of The Royal Society of Chemistry (RSC))

3.2.2 Interfacial Tension Measurement

1. Set up the DSA 10; input the known parameter including temperature, needle diameter, the densities of drop phase and embedding phase, and aspect ratio.
2. Calibrate the instrument by using a water-in-oil system in the absence of surfactant. The surface tension should be stable over 10 min and close to the literature value for the chosen water–oil combination.
3. Fill peptide solutions (AM1 or AFD4) into a 1 mL syringe with a straight needle of known diameter.
4. 7.5 mL oil is filled into an 8 mL quartz cuvette.
5. Drops (ca. 15 μL) of peptide solutions are formed in the oil phase in an 8 mL quartz cuvette.
6. The drop shape is monitored automatically over 10 min. Interfacial tension readings are made every 5 s over 1,000 s following initial droplet formation.

3.3 Dilatational Interfacial Viscoelasticity Measurement by Profile Analysis Tensiometer

Dilatational interfacial viscoelasticity is measured with a Profile Analysis Tensiometer PAT1 from Sinterface (Berlin, Germany).

1. Set up the PAT1 system, and flush the tube system with Milli-Q water and make sure the system is clean.
2. Turn off the light and adjust the focus. Light intensity should be between 200 and 255, darkness is between 0 and 50, and drop verticality should be 0.0.

3. Calibrate the instrument with Milli-Q water in oil in the absence of surfactant. Oil phase is filled into a 20 mL quartz cuvette, and Milli-Q water is pumped into a stainless steel capillary. The pendent water droplet is formed at the tip of the capillary.
4. Keep the droplet volume constant for 1,000 s. If the interfacial tension remains constant, the system is clean; proceed to the next step. Otherwise, the system needs to be cleaned by flushing a large amount of water and reagents checked for contamination.
5. The water droplet volume is increased gradually, and adjust the camera calibration parameters (Cz and Cx) to make the interfacial tension constant.
6. The water droplet volume is oscillated with the tensiometer automatic dosing system, and adjust the camera calibration parameters (Cz and Cx) to make the interfacial tension constant.
7. Check if the interfacial tension value is correct, which can be adjusted by nozzle diameter.
8. When all the calibration described above is finished, the measurement can be started.
9. The pendent droplet of peptide solution is formed at the tip of the capillary in an oil phase. Once equilibrium levels of adsorption are approached (that is, the interfacial tension stops decreasing) (see Note 8), measurement of dilatational elasticity and viscosity can be performed.
10. Droplet oscillation is achieved with sinusoidal perturbations at a chosen frequency and amplitude.
11. The dynamic shape changes of an oscillating droplet are recorded, and are then analyzed with a Fourier transform function to extract dilatational elasticity (mN/m) and viscosity (s mN/m).
12. An example of the cyclic interfacial tension for peptide AFD4 droplets versus time measured by PAT1 is shown in Fig. 4. Further data for AM1 and AFD4 are available in the literature (16).

3.4 Mechanical Interfacial Property Measurement by Cambridge Interface Tensiometer

The Cambridge Interfacial Tensiometer (CIT) is used to study the mechanical properties of self-assembled peptide and protein structures at the air–water and oil–water interfaces (17–20). The CIT monitors the transmission of force through an interfacial architecture located between two anchors floating on a test solution at an initial separation of 1,000 μm. Movement of one anchor (attached to a piezoelectric motor) away from the other subjects the interface between the anchors to tensile strain. Force is registered at the second anchor (attached to a sensitive force transducer) if a cohesive

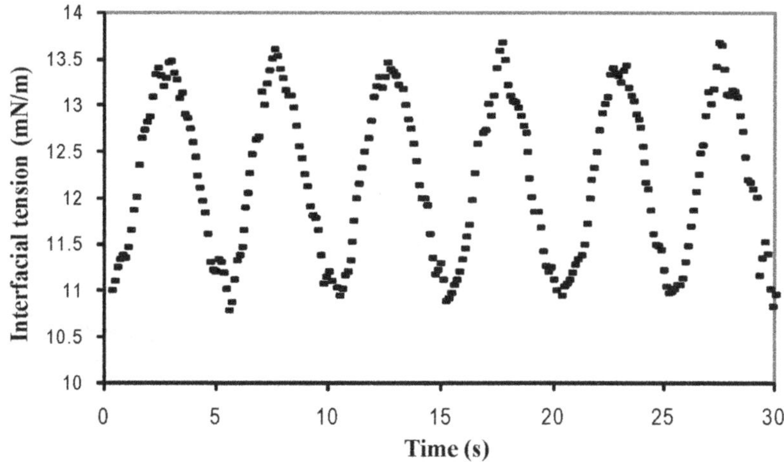

Fig. 4 Cyclic interfacial tension data for an AFD4 peptide droplet versus time

interfacial film is present between the anchors. The instrument is able to determine full interfacial stress–strain curves to high strain in a cyclic fashion, and does not rely on assumptions as to the viscous or elastic behavior of the interface.

1. Start the CIT system including Force Transducer (on), Controller 8200 (on), and Lab view software (open).
2. Clean the Teflon bath and two T-pieces with 5 mM EDTA at pH 11.0 once, 50% ethanol three times, and Milli-Q water ten times (see Note 9).
3. Run a baseline with Milli-Q water.
4. The Teflon bath is filled with 6.5 mL of Milli-Q water by pipette to give an air–water interface located slightly above the edge of the bath.
5. A strain of 300% is applied and reversed at the same motor speed to test the high-strain response. Negative stress will be registered upon compression of the strained film back to its original area if a cohesive network has formed. If the system is clean, there should be no stress response for the pure air–water interface.
6. The Teflon bath is filled with 6.5 mL peptide solutions.
7. After a 60 min aging period, eight tension–compression cycles to 5% strain are applied at a motor speed of 150 μm/s and are used for data averaging.
8. The interfacial elasticity (E) is calculated from the gradient of the linear line of best fit on the plot of the interfacial stress (mN/m) versus strain between 0% and 1% strain.
9. A tension-compression cycle to 300% strain is then applied at the same motor speed to characterize the response of the surfactant film to high strain.

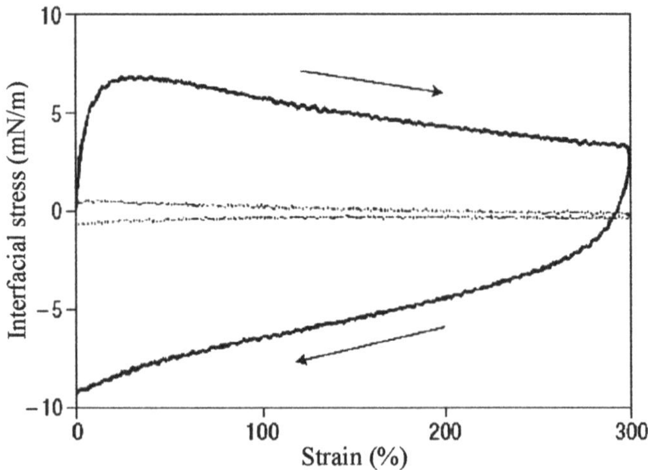

Fig. 5 Mechanical properties of self-assembled AM1 at the air–water interface. AM1 was allowed to adsorb at the interface for 1 h in the absence of added metal ions (*dotted line*) or in the presence of $ZnSO_4$ (*solid line*) (11)

10. Following completion of each experiment the CIT bath is repeatedly rinsed with Milli-Q water, soaked in 5 mM Na^+ EDTA pH 10 for 1 h, and then rinsed again with Milli-Q water.
11. An example of the mechanical properties of interfacial films self-assembled from AM1 measured by CIT is shown in Fig. 5.

3.5 Foam Formation in Column and Switching

1. An aliquot (1 mL) of 0.30 mg/ml AM1 solution in 25 mM HEPES pH 7.4 in the presence of 200 μM $ZnSO_4$ is transferred to a custom foam preparation apparatus (shown in Fig. 1).
2. 7 mL of air in the syringe is bubbled through the liquid via a sintered glass disk at a rate of 20 mL/min which is controlled by the syringe pump.
3. Photographs of the foam formed in the apparatus are taken at timed intervals to monitor foam quality and stability.
4. Foam of approximately 6.5 cm height is formed by passing 7 mL air.
5. Switching of AM1-containing foams is achieved by pipetting an aliquot of H_2SO_4 (12.5 μL, 1.6 N) or EDTA (5 μL, 100 mM, pH 8.0) onto the top of the foam.
6. Reverse switching of AM1 solutions is achieved by adding an aliquot of NaOH (20 μL, 1 M) or $ZnSO_4$ (30 μL, 23.6 mM), depending on whether the switch in the preceding step was performed with acid or EDTA.
7. A sequential switching of foams is shown in Fig. 6.

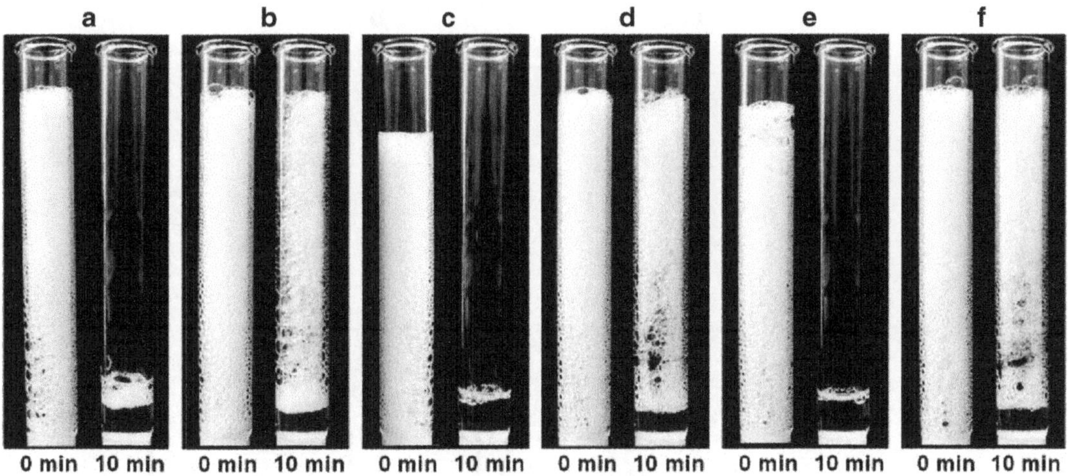

Fig. 6 Sequential switching of foams formed from a single solution of 0.3 mg/ml AM1 through a series (**a**–**f**) of additions of effector solutions. Each panel shows two photographs, the first immediately after the air flow stopped and the second after 10 min aging. Effector solutions were added sequentially to give bulk conditions as follows: (**a**) pH 7.4; (**b**) 200 μM Zn(II), pH 7.4; (**c**) 200 μM Zn(II), pH 3.6; (**d**) 200 μM Zn(II) neutralized to pH 7.4 with NaOH; (**e**) 200 μM Zn(II), 500 μM EDTA, pH 7.4; (**f**) 700 μM Zn(II), 500 μM EDTA, pH 7.4 (12) (Reproduced by permission of The Royal Society of Chemistry (RSC))

3.6 Emulsion Formation by Homogenization and Switch

1. 2.8 mL 60 μM AM1 solution is prepared in 25 mM HEPES in the presence of 250 μM $ZnSO_4$ at pH 7.4.
2. Toluene is used as the oil phase.
3. To facilitate visualization of emulsion switching, Sudan III (50 μM) is added in the oil phase, and methylene blue (10 μM) is added in the aqueous phase.
4. 0.7 mL of toluene is added to 2.8 mL AM1 solution to give an oil fraction of 20% (v/v).
5. The mixture is homogenized for 3 min at 16099 × g in a rotor-stator apparatus (Ystral X10).
6. After homogenization, aliquots (1 mL) of the emulsions are transferred into glass vials and stirred magnetically.
7. To switch the emulsion, an aliquot (8 μL) of acid (1.9 M H_2SO_4) or chelating agent (100 mM EDTA, pH 7.6) is added.
8. Emulsion switching is shown by gross separation of the oil and water phases (shown in Fig. 7).
9. A control vial demonstrates stability of the emulsion during the test period.
10. In control experiments, EDTA is included to a final concentration of 100 μM in the absence of added metal ions, or H_2SO_4 (1.9 M) is added at 1% of the aqueous phase volume, prior to homogenization.

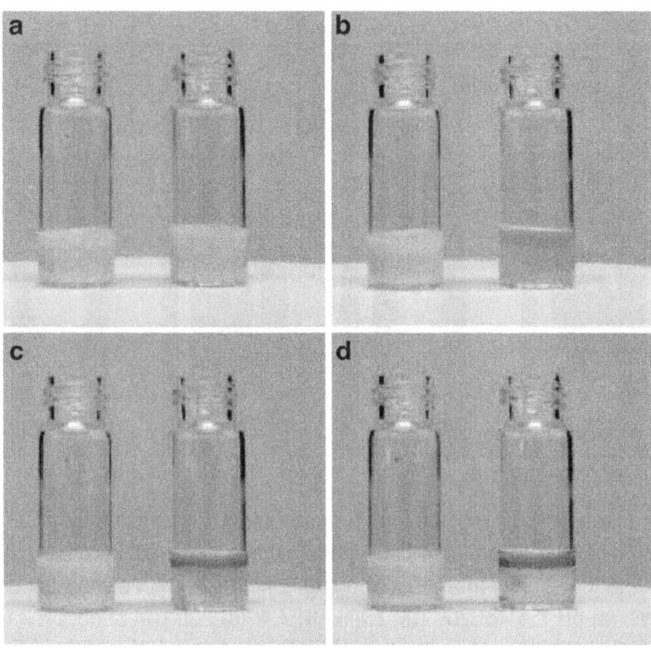

Fig. 7 Coalescence of a Zn(II)-AM1-stabilized toluene-in-water emulsion by addition of acid. An emulsion was prepared and divided into 1 ml aliquots. The initial pH was 7.4. No additions were made to the left-hand vial. An aliquot of H_2SO_4 was added to the right-hand vial, with stirring. The extent of coalescence is shown (**a**), before additions, and (**b**) 10 s; (**c**) 20 s; and (**d**) 10 min after the addition of H_2SO_4 (11)

3.7 Emulsion Formation in Microfluidic Devices and Switch

1. 1 mL gas-tight SGE glass syringe is filled with the continuous phase—100 μM AM1 and 100 μM $ZnSO_4$ in 25 mM HEPES buffer pH 7.0. 0.5 mL glass syringe is filled with the dispersed phase—dodecane; another 100 μL glass syringe is filled with 100 mM EDTA pH 7.4 (see Note 10).
2. A custom T-junction microfluidic chip (Fig. 2), which includes one inlet for the continuous phase, one inlet for the dispersed phase, one inlet for the switching solution, and one outlet, is used to generate and switch the emulsion.
3. The syringes are connected to the T-junction microfluidic chip by connecting tubes, with control by three syringe pumps which are used to control the flow rates of different inlet phases. Flow rates are typically 1 μL/min dodecane, 10 μL/min peptide solution, and 1 μL/min EDTA (see Note 11).
4. A Canon digital camera is mounted on a Nikon microscope for recorded photographs and videos.
5. Firstly, the continuous phase (AM1 + $ZnSO_4$) is pumped into the horizontal channel at a flow rate of 10 μL/min. After the flow stabilizes, the dispersed phase (dodecane) is introduced from the perpendicular channel from T-junction (see Note 12).

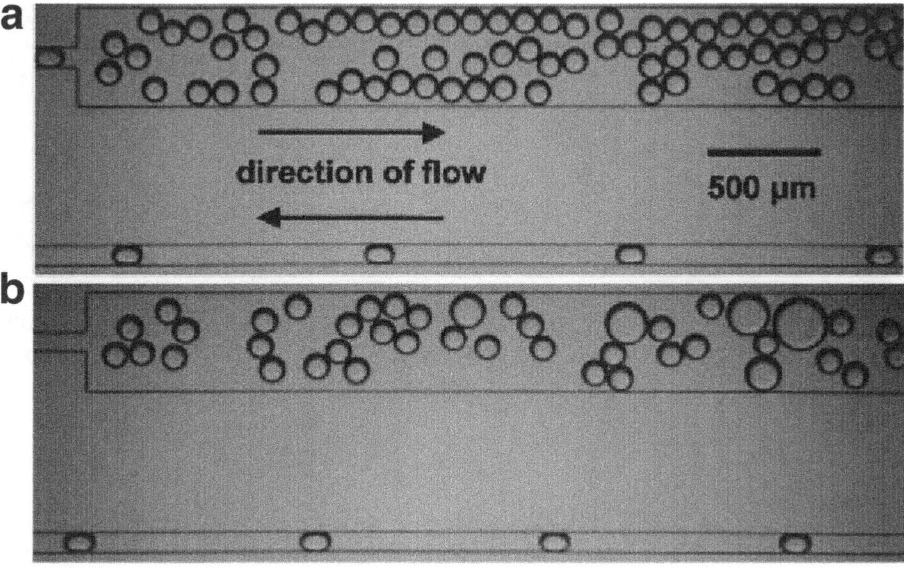

Fig. 8 Control of emulsion coalescence kinetics is achieved under dynamic flow conditions. (**a**) A microfluidic channel containing dodecane droplets, formed at a T-junction in AM1-ZnII solution. (**b**) Upstream addition of excess EDTA to the system shown in (**a**) causes extensive coalescence upon droplet contact (14) (Copyright Wiley-VCH Verlag GmbH & Co. KGaA. Reproduced with permission)

6. Droplet formation is by shearing off dodecane at a T-junction into a cross-flowing aqueous continuous phase. When the droplets travel downstream, they collide. But due to the presence of a cohesive interfacial network on the droplet surface, droplet coalescence is inhibited (shown in Fig. 8).
7. For switching the stable emulsion, 100 mM EDTA solution is introduced from the downstream Y-junction. After a period of time, droplets coalesce rapidly upon contact in the sudden expansion channel (shown in Fig. 8).

3.8 Double Emulsion Formation in Microfluidic Devices (21)

1. Set up the microfluidic device system (as described above) including connecting the syringes with the T-junction chip, setting up the pumps and the camera fitted to the microscope.
2. The aqueous continuous phase is introduced from the horizontal channel, and the dispersed phase is introduced from the perpendicular channel. The droplet is sheared off at the T-junction part. After droplets travel down to the expansion channel, double emulsions or multiple emulsions are formed due to the diffusion of the co-solvent (ethanol) from the dispersed phase to the continuous phase. Therefore, double emulsions or multiple emulsions are observed in the expansion channel.

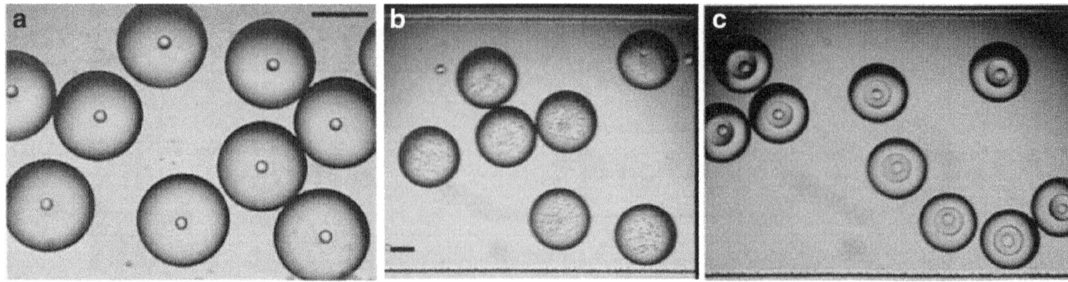

Fig. 9 Multiple emulsions. (**a**) Double emulsions with single inner droplet. (**b**) Double emulsion with multiple inner droplets. (**c**) Triple emulsions (21) (Copyright Wiley-VCH Verlag GmbH & Co. KGaA. Reproduced with permission)

3. For double emulsions with single inner droplet, Miglyol 812–ethanol–water at the volume ratio (1:1:0.04, v/v/v) is used as the dispersed phase, and 100 μM AFD4 with 200 μM $ZnSO_4$ pH 7.0 is used as the continuous phase (shown in Fig. 9a).
4. For double emulsions with multiple inner droplets, the heavy phase of a ternary system sunflower oil–ethanol–water (1:2:0.04, v/v/v) is used as the dispersed phase, and 1.0 mM SDS aqueous solution is used as the continuous phase (shown in Fig. 9b).
5. For multiple emulsions, a ternary Miglyol 812–ethanol–water (1:1:0.04, v/v/v) in the presence of 14–18 mM Span 80 is used as the dispersed phase, and a 1.0 mM SDS aqueous solution is used as the continuous phase (shown in Fig. 9c).
6. The flow rates of the dispersed and continuous phases are 0.01 and 1.0 mL/h, respectively.

4 Notes

1. Unless stated otherwise, all peptide solutions are used within 12 h of preparation.
2. Unless stated otherwise, all glassware used in experiments is acid-cleaned prior to use to ensure the removal of interfacially active contaminants.
3. All personal protective equipment must be correctly fitted and tested; Piranha solution is an extremely corrosive substance.
4. Acid cleaning must be carried out in the fume hood with an appropriate spill tray to ensure any inadvertent spill is contained.

5. Piranha solution must be prepared by slow addition of sulfuric acid to a beaker containing hydrogen peroxide with continuous magnetic stirring slowly and continuously. Peroxide is not to be added to sulfuric acid, to avoid splashing and local heating of the solution. Piranha solution is used in fume hood only. It is recommended that not more than 400 mL solution is made up in a single batch.
6. Piranha solution is only reused for not more than one additional piece of glassware, while still hot.
7. This calibration step has two main purposes. One is to test whether the DSA system is clean. The surface tension must remain constant over a long period of time (e.g., 10 min), otherwise the system needs to be cleaned thoroughly or else reagents need to be tested for inadvertent contamination (e.g., by contact with unclean glassware). The other purpose is to obtain the correct surface tension value by calibrating the well-known surface tension of water in air. The aspect ratio can be adjusted to obtain the correct surface tension. After the calibration is completed, all the parameters should be kept unchanged during subsequent tests.
8. Normally, the measurement starts after aging the droplet for about 1 h which allows the peptide sufficient time to adsorb on the droplet surface.
9. Be careful not to break T-bars which are very delicate and very easy to break. Misalignment of the T-bars, particularly in the *z*-axis perpendicular to the interfacial plane, can also lead to inaccuracies in the force transducer measurement.
10. Air bubbles are undesirable in the glass syringe and will disturb the regularity of droplet formation in the microfluidic chips.
11. It will take several minutes for the flow rate to achieve equilibrium at the first start. Afterwards, after changing any of the flow parameters, at least 100 s of equilibration time is allowed to establish the equilibrium.
12. The continuous phase must be flushed through the microfluidic channel before introducing the dispersed oil phase, to avoid any adhesion of the dispersed phase onto the hydrophilic channel surface.

Acknowledgements

The authors acknowledge funding from the Australian Research Council (DP1093056 and DP1213683) supporting their studies into peptide self-assembly at two-dimensional interfaces. Dr Chun-Xia Zhao acknowledges support from the Australian Research Council in the form of an Australian Postdoctoral Fellow (DP110100394).

References

1. Koopmans RJ, Middelberg APJ (2009) Engineering materials from the bottom up—overview. Adv Chem Eng: Eng Aspects Self-Organ Mater 35:1–10
2. Koopmans RJ, Aggeli A (2010) Nanobiotechnology—quo vadis? Curr Opin Microbiol 13:327–334
3. Leon L, Logrippo P, Tu R (2010) Self-assembly of rationally designed peptides under two-dimensional confinement. Biophys J 99:2888–2895
4. Kwak B, Shin K, Seok S et al (2010) Side chain assisted nanotubular self-assembly of cyclic peptides at the air–water interface. Soft Matter 6: 4701–4709
5. Tanaka M, Ogura K, Abiko S et al (2008) Two-dimensional self-assembly of a designed amphiphilic peptide at air/water interface. Polymer J 40:1191–1194
6. Lepere M, Chevallard C, Hernandez JF et al (2007) Multiscale surface self-assembly of an amyloid-like peptide. Langmuir 23:8150–8155
7. Zhao XB, Pan F, Lu JR (2009) Interfacial assembly of proteins and peptides: recent examples studied by neutron reflection. J R Soc Interface 6:S659–S670
8. Khurana E, DeVane RH, Kohlmeyer A et al (2008) Probing peptide nanotube self-assembly at a liquid–liquid interface with coarse-grained molecular dynamics. Nano Lett 8:3626–3630
9. Middelberg APJ, Radke CJ, Blanch HW (2000) Peptide interfacial adsorption is kinetically limited by the thermodynamic stability of self association. Proc Natl Acad Sci U S A 97:5054–5059
10. Fairman R, Chao HG, Mueller L et al (1995) Characterization of a new 4-chain coiled-coil—Influence of chain-length on stability. Protein Sci 4:1457–1469
11. Dexter AF, Malcolm AS, Middelberg APJ (2006) Reversible active switching of the mechanical properties of a peptide film at a fluid–fluid interface. Nat Mater 5:502–506
12. Malcolm AS, Dexter AF, Middelberg APJ (2006) Foaming properties of a peptide designed to form stimuli-responsive interfacial films. Soft Matter 2:1057–1066
13. Middelberg APJ, He L, Dexter AF et al (2008) The interfacial structure and Young's modulus of peptide films having switchable mechanical properties. J R Soc Interface 5:47–54
14. Malcolm AS, Dexter AF, Katakdhond JA et al (2009) Tuneable control of interfacial rheology and emulsion coalescence. Chemphyschem 10:778–781
15. Dexter AF, Middelberg APJ (2007) Switchable peptide surfactants with designed metal binding capacity. J Phys Chem C 111: 10484–10492
16. Zhao CX, Middelberg APJ (2011) Effects of fluid–fluid interfacial elasticity on droplet formation in microfluidic devices. AIChE J 57: 1669–1677
17. Jones DB, Middelberg APJ (2002) Mechanical properties of interfacially adsorbed peptide networks. Langmuir 18:10357–10362
18. Jones DB, Middelberg APJ (2002) Micromechanical testing of interfacial protein networks demonstrates ensemble behavior characteristic of a nanostructured biomaterial. Langmuir 18:5585–5591
19. Jones DB, Middelberg APJ (2002) Direct determination of the mechanical properties of an interfacially adsorbed protein film. Chem Eng Sci 57:1711–1722
20. Jones DB, Middelberg APJ (2003) Interfacial protein networks and their impact on droplet breakup. AIChE J 49:1533–1541
21. Zhao CX, Middelberg APJ (2009) Microfluidic mass-transfer control for the simple formation of complex multiple emulsions. Angew Chem Int Ed 48:7208–7211

Chapter 11

Designed Self-Assembling Peptides as Templates for the Synthesis of Metal Nanoparticles

Emmanouil Kasotakis and Anna Mitraki

Abstract

Self-assembling peptides are water soluble and form biocompatible nanostructures under mild conditions through non-covalent interactions. They form supramolecular structures such as ribbons, nanotubes, and fibrils. Of particular interest is the possibility of using these peptide fibrils as templates for the growth of inorganic materials, such as metallic nanoparticles. The ability to reliably produce metal-coated fibrils with robust binding of metal nanoparticles is a vital first step towards the exploitation of these fibrils as conducting nanowires with applications in nano-circuitry. One promising strategy consists of the rational introduction of metal-binding amino acids (such as cysteine) at the level of the peptide building block. Upon assembly of the building blocks into fibrils, cysteine residues that remain accessible at the outside of the fibril core could serve as nucleation sites for metals. We will review in this chapter a case study of rationally designed cysteine-containing peptides and basic protocols for their metallization with silver, gold, and platinum nanoparticles.

Key words Peptides, Self-assembly, Cysteine, Metal nanoparticles, Transmission electron microscopy

1 Introduction

Templating of inorganic materials by proteins is a route that is exploited extensively in nature, for example in the fabrication of calcium carbonates in shells, calcium phosphate in bones, and silicate spicules in sponges (1–3). This templating is often mediated by proteins containing sequence repeats with specific amino acids that precisely target the nucleation of the inorganic counterpart—for example negatively charged amino acids for calcium nucleation (4). This strong templating aspect of inorganic structures by biological materials may be used as a fabrication strategy for nanostructured materials destined for novel applications. For example, templating of metals by proteins and peptides is of great technological importance, since the ability to produce metal-coated fibrils is the first step towards their positioning between electrodes and nano-circuit fabrication. Such a proof of concept was demonstrated

Juliet A. Gerrard (ed.), *Protein Nanotechnology: Protocols, Instrumentation, and Applications*, Methods in Molecular Biology, vol. 996, DOI 10.1007/978-1-62703-354-1_11,

by Scheibel and Lindquist by introducing cysteine residues into the sequence of a 250-amino acid self-assembling protein fragment from yeast; the exposed cysteines acted as nucleation sites for gold nanoparticles. Following silver enhancement, formation of conducting nanowires was demonstrated (5). Small building blocks such as short peptides that self-assemble into fibrils can also be used as versatile structural templates for metallization (6, 7). Such peptide-based fibrils are particularly attractive since they can be synthesized under mild, physiological conditions, they are biocompatible, and they can withstand harsh physical and chemical conditions once formed. Furthermore, the possibility of introducing site-specific changes at the sequence level offers the big advantage of tailor-made modifications. Introduction of metal-binding amino acids (such as cysteines) through rational design at the sequence of amyloid peptides is a particularly attractive strategy for templating of metal nanoparticles (8, 9). We will review in this chapter a case study of rationally designed cysteine-containing peptides and basic protocols for their metallization with silver, gold, and platinum nanoparticles.

2 Materials

2.1 Peptides

The following strategy of rational design can be generally followed:

1. Self-assembling building blocks are identified from amyloid-forming proteins, or from beta-structured fibrous proteins made up from repetitive building blocks (10–12).
2. One basic requirement is that the modifications should not affect the capability for self-assembly; hence they should be introduced at positions not engaged in the self-assembling core. A "reductionist approach" is followed with synthesis and study of shorter peptides, subsequences of the original building block (13). This should lead to the identification of a "minimal" building block necessary for assembly. The rest of the amino acids can then be scrutinized for introduction of cysteines, or other metal-binding amino acids such as histidines. Identification of these positions can depend on the nature of each building block and the structural information available. Molecular dynamics simulations can further assist towards identification of residues that are not engaged in the amyloid-forming core and therefore available for substitution (14, 15).

In order to give a practical guide towards the design and fabrication of metal-coated peptide fibrils, we focus on a case study using a parent octapeptide, or original building block, NSGAITIG (Asparagine-Serine-Glycine-Alanine-Isoleucine-Threonine-Isoleucine-Glycine). This was designed using residues 385–392, from the adenovirus fiber protein that corresponds to a short

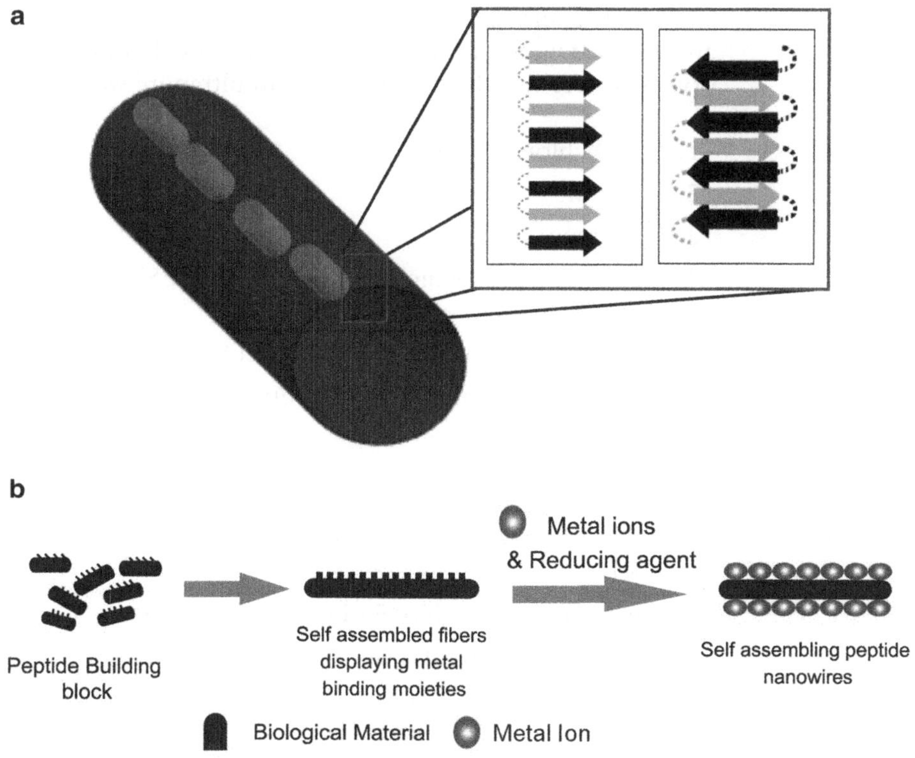

Fig. 1 (**a**) Cartoon of a self-assembled fibril formed from the peptide building block NSGAITIG. The N and S residues do not belong to the amyloid fibril core that consists of the rest of the residues in either parallel or anti-parallel beta-strand conformation. They rather adopt a flexible conformation (*dotted lines*) and remain exposed, protruding from the fibril core. The core is symbolized by the black cylinder and the accessible residues by gray spikes. (**b**) Schematic representation of the strategy followed for the design of self-assembling peptide building blocks that can template the formation of metal nanoparticles. Cysteines are introduced at the positions that remain exposed following the self-assembly into fibrils, and template the nucleation of metal nanoparticles

beta-strand-and-loop region of the natural protein. The NSGAITIG sequence is a part of a 15-residue repeating motif in the fiber shaft (16). Within the NSGAITIG part of the native protein, the N-S-G residues are located in the loop, with the A-I-T-I-G residues in the β-strand. The structure of the peptide sequence within the amyloid fibril state is not known. However, the GAITIG hexapeptide self-assembles into amyloid fibrils, suggesting that these six residues are engaged in the amyloid-forming core (10). Additional insight is obtained from molecular dynamics studies suggesting that the A-I-T-I region of the peptide molecules is engaged in the cross beta core within the fibrils. Therefore, the N-S residues may be exposed at the exterior of the fibril and are chosen as sites for introduction of cysteine residues (14) (Fig. 1a, b).

The following peptides were designed and studied: NCGAITIG, CNGAITIG, and CSGAITIG. The peptides (free N-termini, amidated C-termini) were purchased from Eurogentec (Belgium) and had a degree of purity higher than 95%. Lyophilized peptide powders were dissolved and studied in ultrapure water.

2.2 Microscopy

2.2.1 Formvar Grids (See Note 1)

1. 1,2-Dichloroethane.
2. Formvar.
3. Settlement dish.
4. AGAR Copper or nickel GRIDS 300 MESH (Agar Scientific, or equivalent).
5. Dumont tweezers No. 5a Dumoxel Stainless.
6. Grid Coating Plates (Agar Scientific, or equivalent).
7. Pasteur pipettes.

2.2.2 Transmission Electron Microscopy

1. Square mesh grids: 200 up to 400 MESH, Copper or Nickel, 3.05 mm, Formvar, Carbon, Formvar-Carbon (Agar Scientific, UK).
2. Dumont tweezers No. 5a Dumoxel Stainless.
3. Staining solutions: 1% (w/v) uranyl acetate.
4. JEOL JEM-100C transmission electron microscope operating at 80 kV, or equivalent.
5. Gatan Digital Micrograph software for the analysis of the images (http://www.gatan.com/).

2.3 Metallization of Fibers

1. Gold(III) chloride trihydrate.
2. Silver nitrate.
3. Chloroplatinic acid hydrate.
4. Ascorbic acid.
5. Sodium citrate monobasic anhydrous.
6. Multi-Block® Heater block.

3 Methods

3.1 Evaluation of Peptide Self-Assembly

The modified building blocks should be evaluated for maintaining the ability to self-assemble into fibrils. The samples are examined primarily with TEM analysis using negative staining (Fig. 2) and all other methods for the characterization of the amyloid signature, such as X-ray fiber diffraction, infrared or Raman spectroscopy, CD, and finally thioflavin T or Congo red binding.

3.1.1 Preparation of Formvar Grids

1. A stock solution of Formvar in dichloroethane is needed at least 1 day before the preparation of the coating of grids. The solution concentration should range from 0.7% to 1% (w/v)

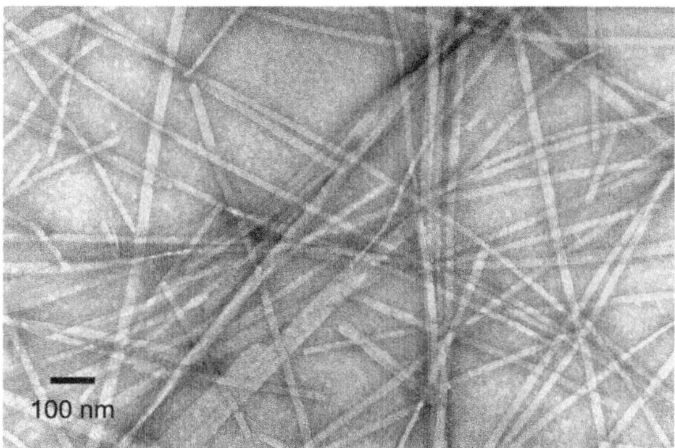

Fig. 2 Fibril formation from a cysteine-containing peptide visualized with negative staining in TEM

and should be stored in the refrigerator. Place the grid coating plate in a settlement dish and cover with pure water.

2. Put the grids gently on the surface of the plate with the shiny side facing up and let the system stabilize for a minute.
3. Use a Pasteur pipette and place one droplet of the Formvar solution above the grids. Open the tap of the settlement dish and collect the water.
4. Take the grid coating plate with the covered grids and let it dry for 1 day before use.
5. The grids can be stored in the refrigerator for some months in order to keep the film fresh.

3.1.2 Preparation of Specimens for TEM Analysis

The peptide solutions are diluted to the desired concentration. 8 μl is placed on a 300 mesh Formvar-coated grid and after 2 min the excess fluid is removed with a filter paper. Finally the samples are negatively stained with 8 μl uranyl acetate 1% for 2 min.

3.2 Metallization of Fibers

3.2.1 Metallization of Fibers with Gold Nanoparticles

1. Prepare a 5 mM stock aqueous solution of $HAuCl_4{\cdot}3H_2O$. Adjust the pH to 5 with some drops of concentrated solution of NaOH.
2. Prepare a fresh aqueous solution of 1% sodium citrate.
3. Boil 20 μl of a 5 mM $HAuCl_4{\cdot}3H_2O$ aqueous solution for 5 min at 100°C and subsequently add to 80 μl of the peptide solution.
4. Incubate for 30 min and subsequently add 8 μl of 1% sodium citrate (reducing agent) for 1 h. This method (referred to as the Turkevich method) yields fairly uniform size gold colloids (17–19).

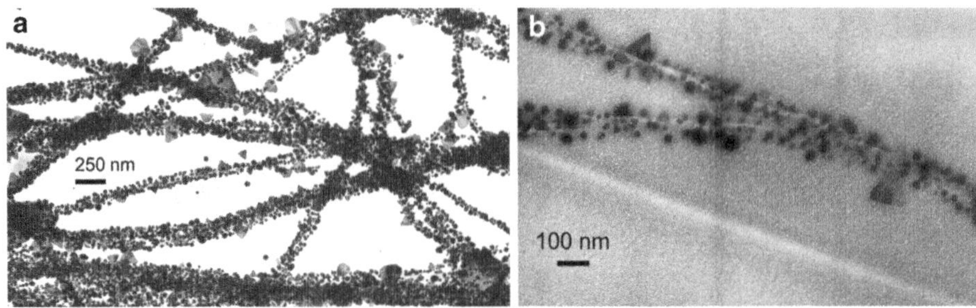

Fig. 3 Transmission electron image of self-assembled fibrils coated with gold nanoparticles without negative staining (**a**) and with negative staining (**b**)

5. After the metallization finishes the sample should have a crimson-dark red color.
6. Place 8 μl of the solution on a Formvar-coated copper grid for TEM analysis.

3.2.2 Metallization of Fibers with Silver Nanoparticles

1. Prepare a 100 mM stock aqueous solution of $AgNO_3$.
2. Prepare a fresh aqueous solution of 1% sodium citrate.
3. Boil 20 μl of a 5 mM $AgNO_3$ aqueous solution for 5 min at 100°C and subsequently add to 80 μl of the peptide solution.
4. Incubate for 30 min.
5. Add 8 μl of 1% sodium citrate (reducing agent) and incubate for 1 h.
6. After the metallization finishes, the sample should have a gray color.
7. Place 8 μl of the solution on a Formvar-coated nickel (see Note 2) grid for TEM analysis (20).

3.2.3 Metallization of Fibers with Platinum Nanoparticles

1. Mix 150 mM of an aqueous solution of ascorbic acid with a 20 mM aqueous solution of $H_2PtCl_6{\cdot}H_2O$ and the peptide solution in a proportion of 1:1:3. After the metallization finishes the sample should have a black color.
2. Place 8 μl of the solution on a Formvar-coated copper grid for TEM analysis (21).

3.3 Characterization of the Metal-Coated Fibrils

1. Evaluate coverage of the self-assembled fibrils and sizes of the metal nanoparticles using TEM with or without negative staining (Fig. 3). Without negative staining the contrast comes only from the metal nanoparticles: additional negative staining with uranyl acetate allows better delineation of the fibril core.
2. Energy Dispersive X-ray Spectroscopy (EDS) analysis should be performed in order to confirm the identity of the elements on the surface of the fibers. Most electron microscopes are nowadays equipped with EDS systems.

4 Notes

1. Formvar grids are also commercially available. However, homemade Formvar grids offer the possibility to adjust the film thickness by using different concentrations of Formvar (thicker films are formed at higher concentrations).
2. The copper grids are oxidized by silver solutions, so the use of nickel grids is necessary. Nickel grids should be handled with antimagnetic tweezers.

5 Acknowledgment

Funding from the European Union (STREP NMP-CT-2006-033256, "BeNatural") is gratefully acknowledged.

References

1. Fratzl P, Gupta HS, Paschalis EP, Roschger P (2004) Structure and mechanical quality of the collagen-mineral nano-composite in bone. J Mater Chem 14:2115–2123
2. Weiner S, Nudelman F, Sone E, Zaslansky P, Addadi L (2006) Mineralized biological materials: a perspective on interfaces and interphases designed over millions of years. Biointerphases 1:P12–P14
3. Aizenberg J, Sundar VC, Yablon AD, Weaver JC, Chen G (2004) Biological glass fibers: correlation between optical and structural properties. Proc Natl Acad Sci USA 101:3358–3363
4. Politi Y, Mahamid J, Goldberg H, Weiner S, Addadi L (2007) Asprich mollusk shell protein: in vitro experiments aimed at elucidating function in $CaCO_3$ crystallization. Crystengcomm 9:1171–1177
5. Scheibel T, Parthasarathy R, Sawicki G, Lin XM, Jaeger H, Lindquist SL (2003) Conducting nanowires built by controlled self-assembly of amyloid fibers and selective metal deposition. Proc Natl Acad Sci USA 100:4527–4532
6. Reches M, Gazit E (2003) Casting metal nanowires within discrete self-assembled peptide nanotubes. Science 300:625–627
7. Lamm MS, Sharma N, Rajagopal K, Beyer FL, Schneider JP, Pochan DJ (2008) Laterally spaced linear nanoparticle arrays templated by laminated beta-sheet fibrils. Adv Mater 20:447–451
8. Kasotakis E, Mossou E, Adler-Abramovich L, Mitchell EP, Forsyth VT, Gazit E, Mitraki A (2009) Design of metal-binding sites onto self-assembled peptide fibrils. Biopolymers 92:164–172
9. Carny O, Shalev DE, Gazit E (2006) Fabrication of coaxial metal nanocables using a self-assembled peptide nanotube scaffold. Nano Lett 6:1594–1597
10. Papanikolopoulou K, Schoehn G, Forge V, Forsyth VT, Riekel C, Hernandez JF, Ruigrok RWH, Mitraki A (2005) Amyloid fibril formation from sequences of a natural beta-structured fibrous protein, the adenovirus fiber. J Biol Chem 280:2481–2490
11. Gazit E (2007) Self-assembled peptide nanostructures: the design of molecular building blocks and their technological utilization. Chem Soc Rev 36:1263–1269
12. Gazit E (2007) Use of biomolecular templates for the fabrication of metal nanowires. FEBS J 274:317–322
13. Gilead S, Gazit E (2005) Self-organization of short peptide fragments: from amyloid fibrils to nanoscale supramolecular assemblies. Supramol Chem 17:87–92
14. Tamamis P, Kasotakis E, Mitraki A, Archontis G (2009) Amyloid-like self-assembly of peptide sequences from the adenovirus fiber shaft: insights from molecular dynamics simulations. J Phys Chem B 113:15639–15647
15. Colombo G, Soto P, Gazit E (2007) Peptide self-assembly at the nanoscale: a challenging target for computational and experimental biotechnology. Trends Biotechnol 25:211–218

16. van Raaij MJ, Mitraki A, Lavigne G, Cusack S (1999) A triple beta-spiral in the adenovirus fibre shaft reveals a new structural motif for a fibrous protein. Nature 401:935–938
17. Kimling J, Maier M, Okenve B, Kotaidis V, Ballot H, Plech A (2006) Turkevich method for gold nanoparticle synthesis revisited. J Phys Chem B 110:15700–15707
18. Daniel MC, Astruc D (2004) Gold nanoparticles: assembly, supramolecular chemistry, quantum-size-related properties, and applications toward biology, catalysis, and nanotechnology. Chem Rev 104:293–346
19. Chiang CL, Hsu MB, Lai LB (2004) Control of nucleation and growth of gold nanoparticles in AOT/Span80/isooctane mixed reverse micelles. J Solid State Chem 177:3891–3895
20. Kamat PV, Flumiani M, Hartland GV (1998) Picosecond dynamics of silver nanoclusters. Photoejection of electrons and fragmentation. J Phys Chem B 102:3123–3128
21. Song YJ, Challa SR, Medforth CJ, Qiu Y, Watt RK, Pena D, Miller JE, van Swol F, Shelnutt JA (2004) Synthesis of peptide-nanotube platinum-nanoparticle composites. Chem Commun 9:1044–1045

Chapter 12

Purification of Molecular Machines and Nanomotors Using Phage-Derived Monoclonal Antibody Fragments

Olga Esteban, Daniel Christ, and Daniela Stock

Abstract

Molecular machines and nanomotors are sophisticated biological assemblies that convert potential energy stored either in transmembrane ion gradients or in ATP into kinetic energy. Studying these highly dynamic biological devices by X-ray crystallography is challenging, as they are difficult to produce, purify, and crystallize. Phage display technology allows us to put a handle on these molecules in the form of highly specific antibody fragments that can also stabilize conformations and allow versatile labelling for electron microscopy, immunohistochemistry, and biophysics experiments.

Here, we describe a widely applicable protocol for selecting high-affinity monoclonal antibody fragments against a complex molecular machine, the A-type ATPase from *T. thermophilus* that allows fast and simple purification of this transmembrane rotary motor from its wild-type source. The approach can be readily extended to other integral membrane proteins and protein complexes as well as to soluble molecular machines and nanomotors.

Key words Phage display, Domain antibodies, Monoclonal antibody fragments, Membrane proteins, ATP synthase, Protein purification, Labelling, Electron microscopy, Crystallization, X-ray crystallography

1 Introduction

Molecular machines and nanomotors are essential components of all biological systems. By definition they use energy (mostly derived from adenosine triphosphate hydrolysis) to perform mechanical work. They also tend to be characterized by multi-subunit complexities on nanometer scales.

The size and dynamic nature of these proteins and protein complexes put challenges on their biochemical, biophysical, and structural characterization. Moreover, many molecular machines are embedded within biological membranes, further complicating the purification process. Consequently, multicomponent membrane-embedded molecular machines are difficult (and often impossible) to produce in heterologous expression systems and typically need to be purified from wild-type sources (such as thermophiles). In most cases these organisms are not amenable to

Juliet A. Gerrard (ed.), *Protein Nanotechnology: Protocols, Instrumentation, and Applications*, Methods in Molecular Biology, vol. 996, DOI 10.1007/978-1-62703-354-1_12, © Springer Science+Business Media New York 2013

genetic modifications, preventing the introduction of affinity tags for purification and labelling.

Many of the above limitations can be overcome through the use of monoclonal antibody fragments directed against subunits of molecular machines. This opens up new avenues to identify, isolate, and label these multi-subunit complexes, representing a valuable tool for many structural and biophysical investigations. Antibody fragments can also stabilize conformations and extend hydrophilic surfaces. This allows the purification of native, folded protein and is likely to increase the chance of crystallization, through the generation of novel and ordered crystal contacts (1–5).

Phage display is a highly suitable technique for antibody selection for this purpose (6). After all, the technique has been successfully used for the isolation of binders against a number of membrane proteins (4, 5, 7) including G-protein-coupled receptors (GPCRs) (8). In addition, this method should be particularly suited for the selection of binders directed against conformational epitopes. As an in vitro method, phage display allows the use of natively folded antigen and provides a great deal of control over selection conditions. This is in marked contrast to the immunization of animals that generally require the use of protein denaturing agents, such as Freud's adjuvant, directing responses against linear epitopes (9).

However, examples of successful selection of antibodies against native membrane-embedded molecular machines via phage display are rare (7, 10). This is mainly due to the difficulty of immobilizing the antigen while keeping it in a native conformation in detergent solution during phage selection (1, 8) and due to the limited availability of suitable phage antibody libraries.

Here we describe the selection of human antibody single domains (dAbs) against a membrane-embedded molecular machine, the A-ATPase from the thermophilic eubacterium *Thermus thermophilus* (10–12). The *T. thermophilus* A-ATPase is a 660 kDa rotary motor composed of nine different subunits that uses the energy of a transmembrane proton gradient to synthesize ATP (Fig. 1). While our approach essentially follows the established phage display selection cycle (13) it contains several variations, which are critical for successful antibody selection against membrane multi-subunit nanomachines. Thus, it is based on (a) an initial selection round against soluble and heterologously expressed individual subunits, followed by (b) additional rounds of selection using the detergent-solubilized and purified molecular machine as an antigen. More specifically, we successfully used this technique for the selection of antibody fragments against the peripheral stalk forming subunits E and G within the A-ATPase from *T. thermophilus.* For this purpose we utilized a second-generation synthetic human domain antibody (V_H) phage display library (kindly provided to D. Stock by Domantis Ltd., Cambridge, UK) and isolated a first pool of binders specific to the A-ATPase peripheral stalk.

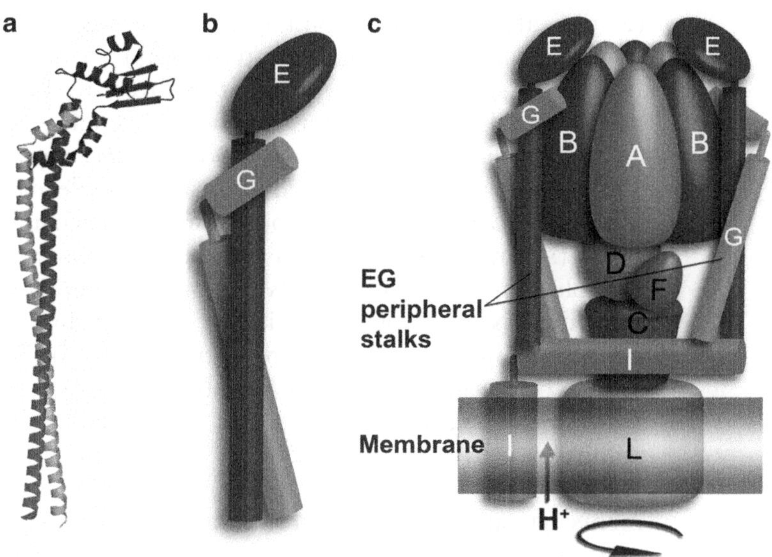

Fig. 1 A-ATPase from *T. thermophilus*. (**a**) X-ray structure of the isolated peripheral stalk complex consisting of subunits E and G. (**b**) Schematic model of the EG complex. (**c**) Schematic model of the intact A-ATPase complex consisting of nine different subunits. Stator subunits are labelled in *white*, rotor subunits in *black*

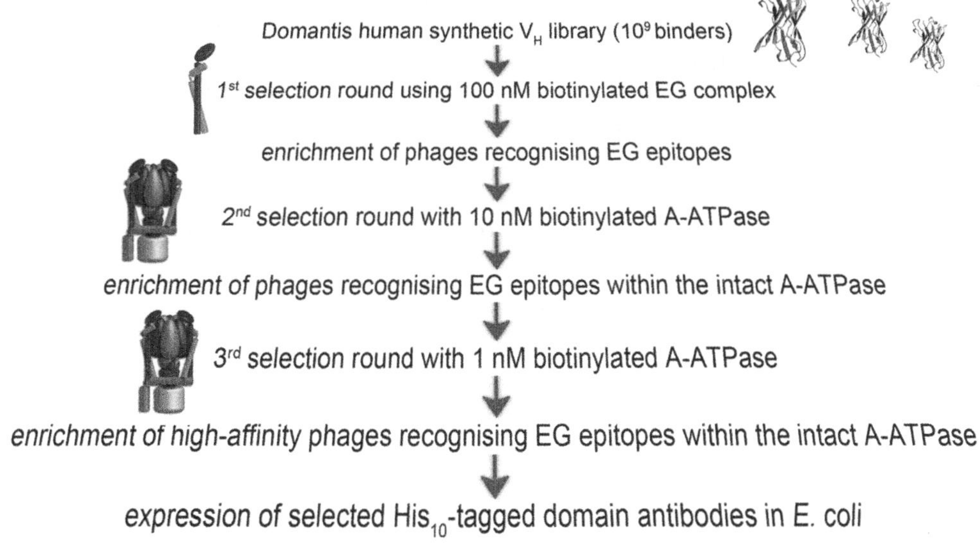

Fig. 2 Flow diagram of phage display cycles for the selection of domain antibodies against the peripheral stalk complex within the intact A-ATPase

In successive rounds we further selected the phage library using the intact A-ATPase as antigen, enriching antibody fragments that bind the EG complex in the context of the native A-ATPase (Fig. 2). This provided us with about one dozen highly specific binders against the peripheral stalk with affinities in the micromolar to nanomolar range. As expected, many of these phage-derived antibody fragments recognized conformational epitopes (as indicated by specific binding to native antigen in ELISA, but not to denatured antigen in Western-blots).

These specific affinity reagents proved to be highly useful research tools that allowed us to develop a single-step protocol for affinity purification of the endogenous *T. thermophilus* A-ATPase (10). This purification procedure is fast (single day) and simple compared to the original protocol using ion exchange chromatography (several days) and yields milligram amounts of pure sample suitable for crystallization trials and other biophysical studies. The procedure is based on expression of decahistidine (His_{10}-tagged) antibody fragments, which can be trapped on nickel affinity columns. They can also be used for labelling of molecular machines for various applications using labels that are coupled to a histidine affinity reagent (such as Ni-NTA) or to an antibody binding reagents (such as protein A). In fact, the use of these antibody fragments has allowed us to specifically stain the EG complex with nano-gold labels to mark the peripheral stalks of the A-ATPase in electron micrographs and provided proof of the existence of two peripheral stalks in the A-ATPase (10). Alternatively, fluorophore and enzyme labels can be covalently attached to the antibody fragments for biophysical and immunochemical analyses (14, 15).

While we outline the details of the procedure in the context of the A-ATPase complex, the approach can be readily extended to other integral membrane proteins and protein complexes as well as to soluble molecular machines and nanomotors. The availability of specific affinity reagents can provide suitable "handles" for purification and imaging of challenging proteins. Finally, the potential of antibody fragments to stabilize dynamic molecular machines by binding to conformational epitopes provides valuable reagents for co-crystallization.

2 Materials

2.1 Preparation of Biotinylated Antigen

1. Antigen: 1 mg/ml purified recombinant EG subunit solution in 20 mM Na-HEPES, pH 8.0, 100 mM NaCl or 1–5 mg/ml purified A-ATPase solution in 20 mM Na-HEPES, pH 8.0, 100 mM sucrose, 10% (v/v) glycerol, 2 mM $MgCl_2$, 100 mM NaCl, 0.05% (w/v) n-dodecyl-β-d-maltoside (DDM; Anatrace, Maumee, OH, USA). Do not use Tris-based buffers as these will interfere with the biotinylation chemistry.
2. Protein buffer: EG buffer—20 mM Tris–HCl, pH 8.0, 100 mM NaCl.
3. A-ATPase buffer: 20 mM Tris–HCl, pH 8.0, 100 mM sucrose, 10% (v/v) glycerol, 2 mM $MgCl_2$, 100 mM NaCl, 0.05% (w/v) DDM.
4. EZ-Link Sulfo-NHS-LC-biotin reagent (Pierce, Rockford, IL, USA).

5. Slide-A-Lyzer® *Dialysis Cassette* (10 kDa MWCO; Pierce).
6. 1 M Tris–HCl, pH 7.5.
7. Ultrapure Milli-Q water.
8. Glycerol.

2.2 Phage Display Selections

1. Antibody phage display library. While the human V_H domain library used in the protocol outlined here is not commercially available (and was kindly provided by Domantis Ltd. to D. Stock) other antibody fragment libraries including scFv (16, 17) and single domains (18, 19) are available.
2. *E. coli* TG1 strain (Agilent, Santa Clara, CA, USA).
3. TYE (tryptone yeast extract) agar plates supplemented with 15 μg/ml tetracycline (library-specific antibiotic).
4. 2× TY medium: 16 g/l bacto-tryptone, 10 g/l yeast extract, 5 g/l NaCl.
5. Streptavidin-coated Dynabeads M-280 and Dynal MPC™ magnet (Invitrogen, Carlsbad, CA, USA).
6. Phosphate Buffered Saline (PBS), pH 7.4.
7. PBST: 0.1% (v/v) Tween-20 in PBS, pH 7.4.
8. Protein buffer: EG buffer—20 mM Tris–HCl pH 8.0, 100 mM NaCl.
9. A-ATPase buffer: 20 mM Tris–HCl, pH 8.0, 100 mM sucrose, 10% (v/v) glycerol, 2 mM $MgCl_2$, 100 mM NaCl, 0.05% (w/v) DDM.
10. 2% (w/v) skimmed milk-protein buffer: dissolve 2 g skimmed milk powder in 100 ml of protein buffer.
11. Trypsin solution: Sigma, St Louis, MO, USA; add 50 μl of 10 mg/ml trypsin stock solution to 450 μl PBS; prepare freshly.
12. Sterile PEG solution: 20% (w/v) polyethylene glycol (PEG 6,000, Sigma), 2.5 M NaCl in deionized water. Filter through 0.22 μm filter (Millipore). Store at room temperature.
13. 500 ml disposable vacuum filter bottles (0.45 μm, Millipore).

2.3 ELISA Screening of Monoclonal Phage Clones

1. Horseradish Peroxidase (HRP)/Anti-M13 antibody conjugate (GE Healthcare, Little Chalfont, UK).
2. MPBS pH 7.4: dissolve 4 g skimmed milk powder in 100 ml of PBS.
3. PBST: 0.1% (v/v) Tween-20 in PBS, pH 7.4.
4. 1× TMB ELISA substrate solution (eBioscience Ltd., Hatfield, UK).
5. 1 M sulfuric acid.

6. Streptavidin-coated plates (Thermo Scientific, Pierce, Rockford, IL, USA).
7. 96-well round-bottom plates.

2.4 Expression and Purification of Soluble Antibody Fragments

1. pET12a-His_{10} vector: this is a modified derivative of the pET12a periplasmic expression vector (Novagen, Gibbstown, NJ, USA) containing a C-terminal decahistidine tag (His_{10}-tag) for increased affinity to Ni-NTA resin.
2. Electrocompetent *E. coli* BL21 Gold (Agilent, Santa Clara, CA, USA).
3. *Sal* I and *Not* I restriction enzymes (New England Biolabs, Ipswich MA, USA).
4. 10× digestion buffer (New England Biolabs).
5. T4 ligase kit (Roche Applied Science, Mannheim, Germany).
6. QIAquick gel extraction kit (Qiagen, Hilden, Germany).
7. QIAprep spin plasmid miniprep (Qiagen).
8. 2× TY/amp/tet/glu media: 2× TY media supplemented with 100 μg/ml ampicillin, 15 μg/ml tetracycline, 4% (v/v) glucose.
9. TYE/amp/tet/glu agar: TYE agar supplemented with 100 μg/ml ampicillin, 15 μg/ml tetracycline, 4% (v/v) glucose.
10. Auto-induction medium: combine 928 ml sterile 2× TY medium, 1 ml 1 M $MgSO_4$, 20 ml 50× 5,052 (0.5% (v/v) glycerol, 0.05% (w/v) glucose, 0.2% (w/v) α-lactose), 50 ml 20× NPS (0.5 M $(NH_4)_2SO_4$, 1 M KH_2PO_4, 1 M Na_2HPO_4), 1 ml ampicillin at 100 mg/ml.
11. Antifoam 204 (Sigma).
12. 10× high salt phosphate buffer (HSPB) (10× PBS pH 7.4 with 5 M NaCl).
13. Streamline rProtein A (GE Healthcare).
14. Gravity flow column.
15. Glycine solution: 0.1 M glycine, 0.15 M NaCl, pH 3.0.
16. 1 M Tris–HCl, pH 7.6.
17. 2.5 L baffled flasks.

2.5 Affinity Purification of A-ATPase Nanomotor Using Polyhistidine-Tagged Antibody Fragments Coupled to Ni-NTA Resin (Fig. 3)

1. 10 g of *Thermus thermophilus* cell pellet.
2. RNAse, DNAse (Roche Applied Science).
3. 0.1% (w/v) phenylmethanesulfonyl fluoride (PMSF) stock solution: dissolve 10 mg of PMSF in 10 ml isopropanol. Toxic and should only be added to buffers if proteolysis is a problem. Unstable in aqueous solutions.
4. EDTA-free protease inhibitor tablets (Roche Applied Science).
5. 1 ml HisTrap columns (GE Healthcare).

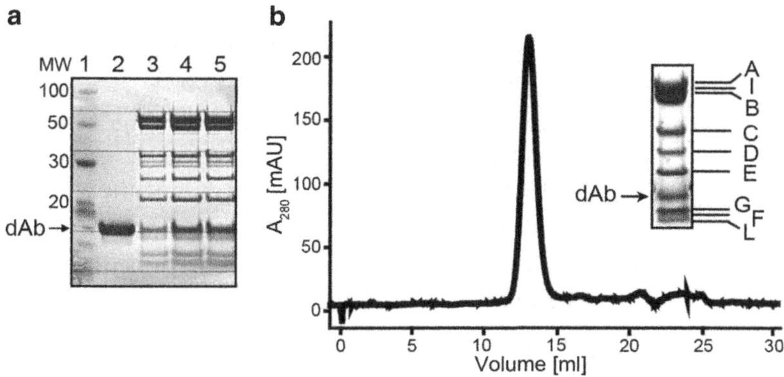

Fig. 3 Domain antibody-mediated purification of intact A-ATPase. (**a**) SDS-PAGE after Ni-NTA purification. *Lane 1*, molecular weight marker; *Lane 2*, purified domain antibody; *Lane 3*-5, Ni-NTA eluate fractions. (**b**) Gel-filtration chromatogram with SDS-PAGE of peak fraction showing bands for the nine different A-ATPase subunits and the domain antibody (Adapted from ref. 10)

6. Buffer A: 20 mM Tris–HCl, pH 8.0, 100 mM sucrose, 10% (v/v) glycerol, 300 mM NaCl, 2 mM $MgCl_2$, 20 mM imidazole, 0.1% (w/v) DDM.
7. Buffer B: 20 mM Tris–HCl, pH 8.0, 100 mM sucrose, 10% (v/v) glycerol, 300 mM NaCl, 2 mM $MgCl_2$, 200 mM imidazole, 0.1% (w/v) DDM; adjust to pH 8.0 with HCl.
8. A-ATPase buffer: 20 mM Tris–HCl, pH 8.0, 100 mM sucrose, 10% (v/v) glycerol, 2 mM $MgCl_2$, 100 mM NaCl, 0.05% (w/v) DDM.
9. Vivaspin 15 centrifugal concentrator (100 kDa MWCO, Vivascience, Hannover, Germany).
10. Ultracentrifuge (e.g., Beckman Coulter Inc., Brea, CA, USA), 45-Ti rotor, and tubes.
11. Gel-filtration column appropriate for molecular weight of sample.
12. Chromatography system.

3 Methods

3.1 Preparation of Biotinylated Antigen

1. Use 0.1–1 mg of antigen at 1–5 mg/ml in the appropriate protein buffer (see Note 1).
2. Immediately before use, prepare a 5 mM solution of EZ-LinkSulfo-NHS-LC-biotin reagent by dissolving 1 mg reagent in 300 μl of ultrapure Milli-Q water (see Note 2).
3. Add the appropriate volume of this solution to the antigen using a fivefold molar excess of NHS-biotin over protein. Mix well and incubate for 1 h at room temperature (see Note 3).
4. Stop reaction by adding 1 M Tris–HCl, pH 7.5 to a final concentration of 20 mM and incubate at room temperature for

10 min. The level of biotinylation can be determined using an EZ Biotin Quantitation Kit (Pierce) or, more accurately, by mass spectrometry.

5. Remove excess, non-reacted biotin reagent by dialysis against protein buffer. A Slide-A-Lyzer® *Dialysis Cassette* (10 kDa MWCO) can be used as a convenient means for dialyzing the biotinylated sample (see Note 4).
6. After dialysis, determine protein concentration in a spectrophotometer at 280 nm.
7. Add an equal volume of glycerol to make a final concentration of 50% glycerol and store at −20°C.

3.2 First Round of Phage Selection: Panning Against Recombinant Subunits

1. Thaw frozen aliquots of antibody phage library on ice (see Note 5).
2. Resuspend the streptavidin-coated Dynabeads by gently shaking the vial and transfer 200 μl of beads to a 1.5 ml tube. Place the tube on a magnet for 2 min and remove the supernatant with a pipette while the tube remains on the magnet (see Note 6).
3. Wash beads once with 1 ml PBST and a second time with PBS. The bead washing procedure is facilitated by using a magnet as described in the previous step.
4. Block beads with 1 ml of 2% skimmed milk-protein buffer and incubate on rocker at room temperature for 1 h (see Note 7).
5. In the meantime, add 50 μl phage library (from Subheading 3.2, step 1) to 1 ml of 2% skimmed milk in protein buffer with 200 μl streptavidin-coated Dynabeads and incubate for 1 h at room temperature on a shaking platform (see Note 8).
6. Deplete phages that bind nonspecifically to beads from phage library by placing the tube on the magnet for 2 min. Transfer depleted supernatant into a 1.5 ml Eppendorf tube (see Note 9).
7. Add biotinylated antigen (EG subunits) to the phage supernatant to a final concentration of 100 nM. Adjust the final skimmed milk concentration to 2%.
8. Mix antigen and phage and incubate on a rotating mixer for 1 h at room temperature.
9. Separate the 200 μl blocked streptavidin-coated Dynabeads (Subheading 3.2, step 4) from 2% skimmed milk-protein buffer using a magnet.
10. Resuspend the blocked beads in phage/biotinylated antigen mix solution and incubate on rotating mixer at room temperature for 10 min to allow bead capture of phage/antigen complexes.
11. Wash beads 10× with 1 ml protein buffer and 2× with 1 ml PBS buffer.

12. Add 500 μl trypsin solution to beads and incubate on rotating mixer for 10 min at room temperature. Place the tube on a magnet and transfer the supernatant with the eluted phages to a new Eppendorf tube. Phage can be stored at 4°C for several days.

3.3 Infection of E coli. and Phage Production for Subsequent Selection Rounds

1. Mix 250 μl of eluted phage from Subheading 3.2, step 12 with 1.75 ml *E. coli* TG1 in log phase (OD_{600} = 0.3–0.6) and incubate for 30 min at 37°C without shaking (as shaking will damage the *E. coli* pili required for phage infection).
2. Centrifuge at 11,600 × *g* for 1 min and resuspend pellet in 100 μl 2× TY. Plate bacterial suspension onto a large (14.5 cm diameter) TYE agar plate supplemented with 15 μg/ml tetracycline. Incubate at 37°C overnight (o/n).
3. Determine the titer of eluted phages by preparing a tenfold dilution series of eluted phage (from 10^{-1} to 10^{-6}) in sterile 2× TY media. Mix 10 μl from each dilution with 90 μl of log phase *E. coli* TG1 and incubate for 30–45 min at 37°C without shaking. Plate 10 μl from each infection onto dried TYE plates supplemented with 15 μg/ml tetracycline. Incubate plates o/n at 37°C. As a control, 10 μl of uninfected *E. coli* TG1 should be plated out.
4. Scrape cells from TYE plates using 2 ml 2× TY per plate and a glass spreader. Mix cells thoroughly, add sterile glycerol to a final concentration of 20%, and store at −80°C. This is the first round of enriched TG1 library. If you plan to repeat the panning within 3 days, keep 50 μl of the enriched TG1 library at 4°C and store the rest at −80°C.
5. Inoculate 50 ml 2× TY plus 15 μg/ml tetracycline with 50 μl of first round enriched TG1 library from previous step (see Note 10). Incubate with shaking o/n at 30°C.
6. Centrifuge culture at 13,000 rpm (20,000 × *g*) for 10 min at 4°C. Alternatively, centrifugation can be performed in Falcon tubes in a bench top centrifuge at 4,000 rpm (2,755 × *g*) for an extended period of time (30 min). Transfer the supernatant (which contains the enriched phage library) to a fresh tube and discard pellet. Filter supernatant through a 0.45 μm filter.
7. Purify the enriched phage library. Add 10 ml PEG solution to 40 ml filtered supernatant. Mix well and incubate on ice for at least 1 h. Spin at 11,000 rpm (19,600 × *g*) for 15 min at 4°C. Alternatively, centrifugation can be performed in Falcon tubes at 4,000 rpm (2,755 × *g*), but the use of higher centrifugation speeds is preferable. Discard supernatant. Re-spin the tube for a few seconds and remove the residual PEG solution.
8. Resuspend pellet in 1 ml PBS buffer. Transfer the resuspended pellet to a sterile 1.5 ml Eppendorf tube and centrifuge at 13,000 rpm (20,000 × *g*) for 10 min at 4°C. Save the

supernatant: this is the enriched (purified) phage library. The phage solution can be filtered using a 0.22 μm filter.

9. Determine phage titers by measuring OD_{260}. Dilute the phage solution tenfold in PBS. Titers can be determined according to the following formula: phage/ml = $OD_{260nm} \times 10 \times 22.14 \times 10^{10}$.
10. Store phage at 4°C until use (see Note 11).

3.4 Second Round of Phage Selection: Panning Against Purified Nanomotor

1. Use biotinylated A-ATPase as antigen at a final concentration of 10 nM. Repeat panning process as described in Subheading 3.2 but use A-ATPase instead of EG subunits. Repeat elution and infection steps as indicated above (see Notes 12 and 13).

3.5 ELISA Screening of Monoclonal Phage Clones

1. After three rounds of selection, individual colonies from the enriched library can be tested for antigen binding by phage ELISA. Dilute an aliquot of the enriched TG1 library after the third panning round and plate out on a TYE plate supplemented with 15 μg/ml tetracycline. Incubate at 37°C o/n.
2. Pick single colonies from the plate using sterile tips and separately inoculate into a 96-well round-bottom plate containing 200 μl 2× TY plus 15 μg/ml tetracycline per well. Pick positive and negative control clones into independent wells from freshly streaked TYE plates supplemented with 15 μg/ml tetracycline. Incubate at 37°C o/n in a humidified atmosphere with shaking at 250 rpm (see Note 14).
3. Centrifuge the 96-well plate at 3,000 rpm (1,500 × *g*) for 10 min at 4°C, transfer supernatant to a new 96-well plate, and store at 4°C (see Note 15).
4. For ELISA screening, add 100 μl of 1-10 nM biotinylated antigen solution (in the appropriate buffer for your antigen) to each well of a streptavidin-coated plate and incubate at room temperature for 2 h or at 4°C o/n.
5. Wash each well three times with PBST and block with MPBS for 1 h at room temperature.
6. Add 100 μl phage supernatant diluted two- to fourfold in MPBS and incubate for 2 h at room temperature. Wash each well five times with PBST.
7. Add 100 μl HRP/anti-M13 antibody conjugate (1:5,000 in MPBS) to each well and incubate for 1 h at room temperature. Wash the wells five times with PBST.
8. Add 100 μl of TMB substrate solution into each well and incubate at room temperature. Wait for blue color to develop.
9. Stop the color reaction in all wells at the same time by adding 50 μl of 1 M sulfuric acid. Blue wells will turn yellow. Read the absorbance at 450 nm (subtract reference at 630 nM).

3.6 Expression and Purification of Soluble Antibody Fragments

1. Isolate insert DNA encoding the selected fragments using a commercially available miniprep kit (e.g., Qiagen) following the manufacturer's instructions. Prepare pET12a-His$_{10}$ vector DNA (see Note 16).
2. Digest insert DNA and pET12a-His$_{10}$ vector with *Sal* I and *Not* I. Separate digested DNA by agarose gel electrophoresis, cut out bands corresponding to digested insert and vector, and gel-purify using a Qiagen QIAquick extraction kit (see Note 17).
3. Combine insert and vector DNA in a 5:1 molar ratio and ligate using a commercially available ligation kit (e.g., Roche T4 ligase).
4. Electroporate 50 μl electrocompetent *E. coli* BL21 Gold with 2 μl ligation mixture. Plate cells on TYE/amp/tet/glu agar plates and incubate at 30°C o/n.
5. Pick 5–10 colonies and grow o/n cultures in 3 ml 2× TY/amp/tet/glu media.
6. Extract plasmid DNA and digest 300–500 ng of DNA with *Sal* I and *Not* I to identify clones containing insert (bands of approximately 350 and 750 bp are expected for single domains and scFv fragments, respectively).
7. Inoculate 5 ml 2× TY/amp/tet/glu media with one of the clones identified in the previous step and incubate o/n at 37°C, shaking at 250 rpm.
8. For large-scale preparation of soluble antibody fragments, add 500 ml of pre-warmed auto-induction medium, 100 μg/ml ampicillin, 15 μg/ml tetracycline, and 15 μl antifoam to a 2.5 L baffled flask and inoculate with the o/n culture. Shake culture at 250 rpm and incubate at 30°C for 36–48 h.
9. Centrifuge at 4,000 rpm (2,755 ×*g*) for 40 min and transfer the supernatant (containing the antibody fragments) to a 500 ml disposable vacuum filter unit (0.45 μm pore size). Filter supernatant and keep on ice until use.
10. Add 50 ml 10× HSPB to the filtered supernatant (500 ml) and mix well. Add 7.5 ml resuspended Streamline rProtein A resin and mix on a roller for 3 h at room temperature.
11. Collect supernatant (save for later analysis of capture efficiency) and load resin with bound sample into a column. Then let the resin settle by gravity.
12. Wash the column with 75 ml 1× HSPB and elute bound antibody fragments with 37.5 ml of glycine solution. Immediately neutralize to pH 8.0 with 11 ml of 1 M Tris–HCl pH 7.6. Optional: Repeat elution.
13. Dialyze against PBS buffer and determine protein concentration in a UV spectrophotometer at 280 nm.

14. Store purified antibody fragments at 4°C if used within days, otherwise snap freeze in liquid nitrogen and store at −20°C (see Note 18).

3.7 Affinity Purification of A-ATPase Nanomotor Using Polyhistidine-Tagged Antibody Fragments Coupled to Ni-NTA Resin

1. Resuspend 10 g of *T. thermophilus* cell pellet in 100 ml of buffer A in a glass beaker. Add 100 μl RNAse (10 mg/ml), 10 μl DNAse (10 mg/ml), 3 tablets of protease inhibitors, 0.001% PMSF (final concentration), and 1 g DDM (1% (w/v) final concentration for initial solubilization step). Place the beaker on ice and disrupt cells by sonication (see Note 19).
2. Transfer the cell suspension into Ti-45 tubes and centrifuge at 35,000 rpm (142,000 × *g*) in a Ti-45 rotor for 30 min at 4°C to remove cell debris. The supernatant contains the solubilized A-ATPase. Keep on ice.
3. Pass 1–5 mg of His-tagged antibody fragments from previous Subheading 3.6, step 14 over a 1 ml HisTrap column connected to a chromatography system. Wash with buffer A until baseline is reached.
4. Apply supernatant obtained in Subheading 3.7, step 2 to the column and wash with 50 ml of buffer A or until baseline is reached.
5. Elute the antibody fragment-A-ATPase complex and excess antibody fragments from the column using 5 ml buffer B.
6. Concentrate the antibody fragment-A-ATPase complex solution to 0.5 ml using a Vivaspin 15 centrifugal concentrator (100 kDa MWCO).
7. Apply to gel-filtration column connected to chromatography system and equilibrated in A-ATPase buffer.
8. Run an aliquot of the gel-filtration peak fractions on SDS-PAGE and pool antibody fragment-A-ATPase containing fractions.

4 Notes

1. Different antigen concentrations can be used but the coupling ratio should be increased if the antigen is more diluted. The biotinylation reaction has to be carried out in a buffer that does not contain primary amines (e.g., not Tris–HCl) and is free of reducing agents (such as DTT). Use gel-filtration or dialysis to adjust buffer conditions if necessary.
2. To ensure optimal display of the protein on the surface of the streptavidin, the use of a flexible linker is recommended between the target protein and the biotin label (as present in the EZ-Link Sulfo-NHS-LC-biotin reagent).

3. Proteins should contain as few biotins as possible with an average of 1–3 for recombinant subunits and 1–8 for larger multi-subunit complexes. Excess biotinylation can alter epitopes and induce aggregation.
4. Excess biotinylation reagent can be removed by using a desalting column (such as the Zeba Spin desalting column from Pierce).
5. Always use filtered pipette tips when manipulating phage to avoid cross-contamination. Handle the libraries carefully as phage contamination can be difficult to eradicate.
6. As an alternative to using a magnet, any type of streptavidin-coated beads can be collected by centrifugation at moderate *g*-forces.
7. The first round of panning is performed using recombinant subunits as antigens and can in principle be carried out in a buffer of choice. However, it is recommended to use the storage buffer of the multi-subunit nanomotor to allow phage selection under identical buffer conditions throughout the selection process.
8. Phage numbers should exceed the complexity of the library by at least 100-fold to allow sufficient coverage of diversity.
9. The beads will retain nonspecific phage binders whereas the supernatant will contain remaining blocked phages. Depletion of bead-bound phages during this step will avoid unspecific phage binders to be amplified in subsequent rounds.
10. Cell numbers should exceed the size of the first round enriched TG1 library by at least 100-fold to allow sufficient coverage of diversity.
11. The enriched phage library can be stored at 4°C for up to 2 weeks. For long-term storage, add sterile glycerol to a final concentration of 20% and store at –80°C.
12. When using membrane proteins an appropriate amount and type of detergent has to be included in all buffer solutions to maintain solubility and a native conformation. This is a key determinant for the successful generation of conformation-specific antibody fragments.
13. To select for high-affinity binders, antigen concentration should be lowered throughout the selection process. In our experience, a first round of selection using recombinant subunits at 100 nM, followed by two rounds of selection against 10 nM and 1 nM of purified A-ATPase, respectively, were sufficient to select binders down to nanomolar affinities.
14. An adhesive gas-permeable sealing membrane can be used to cover the surface of the plate to avoid contamination during culture in 96-well microtiter plates.

15. Glycerol stocks of the original 96-well overnight cultures should be prepared by adding glycerol to the plate (20% final concentration) and storing it at −80°C.
16. We modified the *E. coli* T7 expression vector pET12a to enable cloning of antibody genes in frame with a region encoding a C-terminal His_{10}-tag. Alternatively, other vectors for periplasmic expression as well as other epitope tags can be used. Note that periplamsic expression is not compatible with Ni-affinity purification as the culture medium interferes with the His-tag capture. Antibody fragments are best purified using protein A resin.
17. The use of these restriction enzymes is specific to the pET12a-His_{10} and the single domain library as outlined here and may have to be adjusted for different libraries and expression vectors. This can be readily achieved by PCR amplification of antibody fragment inserts and introduction of new restriction sites in the PCR primers.
18. Note that protein yield may vary widely between different clones. Antibody fragments typically express at levels between 0.1 and 5 mg/L of culture medium.
19. The protocol can be scaled up to larger volumes using a continuous flow cell disrupter.

Acknowledgements

The authors would like to thank Domantis Ltd. for providing the phage display library and Alastair Stewart for critical reading of the manuscript and help with the figures. This work was funded by the Medical Research Council U.K. and the Australian Research Council (ARC DP110101387).

References

1. Hunte C, Michel H (2002) Crystallisation of membrane proteins mediated by antibody fragments. Curr Opin Struct Biol 12:503–508
2. Lam AY, Pardon E, Korotkov KV, Hol WG, Steyaert J (2009) Nanobody-aided structure determination of the EpsI:EpsJ pseudopilin heterodimer from Vibrio vulnificus. J Struct Biol 166:8–15
3. Lee JE, Fusco ML, Abelson DM et al (2009) Techniques and tactics used in determining the structure of the trimeric ebolavirus glycoprotein. Acta Crystallogr D Biol Crystallogr 65:1162–1180
4. Conrath K, Pereira AS, Martins CE et al (2009) Camelid nanobodies raised against an integral membrane enzyme, nitric oxide reductase. Protein Sci 18:619–628
5. Rothlisberger D, Pos KM, Pluckthun A (2004) An antibody library for stabilizing and crystallizing membrane proteins—selecting binders to the citrate carrier CitS. FEBS Lett 564:340–348
6. Winter G, Griffiths AD, Hawkins RE, Hoogenboom HR (1994) Making antibodies by phage display technology. Annu Rev Immunol 12:433–455
7. Rubinstein JL, Holt LJ, Walker JE, Tomlinson IM (2003) Use of phage display and high-density screening for the isolation of an antibody against the 51-kDa subunit of complex I. Anal Biochem 314:294–300
8. Mirzabekov T, Kontos H, Farzan M, Marasco W, Sodroski J (2000) Paramagnetic proteoliposomes containing a pure, native, and

oriented seven-transmembrane segment protein, CCR5. Nat Biotechnol 18:649–654

9. Paus D, Winter G (2006) Mapping epitopes and antigenicity by site-directed masking. Proc Natl Acad Sci U S A 103:9172–9177
10. Esteban O, Bernal RA, Donohoe M et al (2008) Stoichiometry and localization of the stator subunits E and G in *Thermus thermophilus* H^+-ATPase/synthase. J Biol Chem 283: 2595–2603
11. von Ballmoos C, Wiedenmann A, Dimroth P (2009) Essentials for ATP synthesis by F1F0 ATP synthases. Annu Rev Biochem 78: 649–672
12. Yokoyama K, Imamura H (2005) Rotation, structure, and classification of prokaryotic V-ATPase. J Bioenerg Biomembr 37:405–410
13. Lee CM, Iorno N, Sierro F, Christ D (2007) Selection of human antibody fragments by phage display. Nat Protoc 2:3001–3008
14. Patel AR, Kanazawa KK, Frank CW (2009) Antibody binding to a tethered vesicle assembly using QCM-D. Anal Chem 81:6021–6029
15. Jin T, Tiwari DK, Tanaka S et al (2010) Antibody-ProteinA conjugated quantum dots for multiplexed imaging of surface receptors in living cells. Mol Biosyst 6:2325–2331
16. Ascione A, Flego M, Zamboni S et al (2005) Application of a synthetic phage antibody library (ETH-2) for the isolation of single chain fragment variable (scFv) human antibodies to the pathogenic isoform of the hamster prion protein (HaPrPsc). Hybridoma (Larchmt) 24:127–132
17. de Wildt RM, Mundy CR, Gorick BD, Tomlinson IM (2000) Antibody arrays for high-throughput screening of antibody-antigen interactions. Nat Biotechnol 18:989–994
18. Christ D, Famm K, Winter G (2006) Tapping diversity lost in transformations—in vitro amplification of ligation reactions. Nucleic Acids Res 34:e108
19. Christ D, Famm K, Winter G (2007) Repertoires of aggregation-resistant human antibody domains. Protein Eng Des Sel 20:413–416

Chapter 13

Determination of Enzyme Thermal Parameters for Rational Enzyme Engineering and Environmental/Evolutionary Studies

Charles K. Lee, Colin R. Monk, and Roy M. Daniel

Abstract

Of the two independent processes by which enzymes lose activity with increasing temperature, *irreversible thermal inactivation* and *rapid reversible equilibration with an inactive form*, the latter is only describable by the Equilibrium Model. Any investigation of the effect of temperature upon enzymes, a mandatory step in rational enzyme engineering and study of enzyme temperature adaptation, thus requires determining the enzymes' thermodynamic parameters as defined by the Equilibrium Model. The necessary data for this procedure can be collected by carrying out multiple isothermal enzyme assays at 3–5°C intervals over a suitable temperature range. If the collected data meet requirements for V_{max} determination (i.e., if the enzyme kinetics are "ideal"), then the enzyme's Equilibrium Model parameters (ΔH_{eq}, T_{eq}, $\Delta G^{\ddagger}_{cat}$, and $\Delta G^{\ddagger}_{inact}$) can be determined using a freely available iterative model-fitting software package designed for this purpose.

Although "ideal" enzyme reactions are required for determination of all four Equilibrium Model parameters, ΔH_{eq}, T_{eq}, and $\Delta G^{\ddagger}_{cat}$ can be determined from initial (zero-time) rates for most nonideal enzyme reactions, with substrate saturation being the only requirement.

Key words Enzyme activity, Enzyme stability, Thermal denaturation, Temperature adaptation, Enzyme reactor, Enzyme kinetics, Enzyme thermodynamics, Equilibrium model

1 Introduction

For the last forty years, thermophilic enzymes have become increasingly available and widely utilized (1–3). In addition to thermal stability and, consequently, sustained catalytic activity at high temperatures, they tend to exhibit general resistance to denaturation by organic solvents and detergents, and lowered susceptibility to proteolysis (4–6). These attributes allow them to remain functional under a wide range of conditions, and they consequently have widespread industrial and research applications. However, the ability of thermophilic enzymes to maintain their activity at high temperatures has largely been assessed by determining their thermal

Juliet A. Gerrard (ed.), *Protein Nanotechnology: Protocols, Instrumentation, and Applications*, Methods in Molecular Biology, vol. 996, DOI 10.1007/978-1-62703-354-1_13, © Springer Science+Business Media New York 2013

stability, generally by measuring $\Delta G^{\ddagger}_{inact}$, the Gibbs' free energy of activation for the irreversible thermal inactivation of an enzyme. Attempts to engineer enzymes that possess thermophilic qualities have also been focused on preventing irreversible thermal inactivation, an approach that has seen quite limited success in relation to the amount of effort. It has now become clear that there is an additional mechanism, unrelated to that described by $\Delta G^{\ddagger}_{inact}$, through which enzymes lose activity at high temperatures. Irreversible thermal inactivation alone is therefore a poor predictor of the ability of an enzyme to function at high temperatures and in particular is inadequate to allow rational enzyme engineering or evolutionary study of enzyme temperature adaptation.

2 The Equilibrium Model

The way enzymes respond to temperature is fundamental to many areas of biology. Until recently, the effect of temperature on enzyme activity has been understood in terms of raised temperature increasing activity and simultaneously causing activity to be lost by denaturation (7–10). However, it is now clear that these two opposing effects are insufficient to explain the effect of temperature on enzymes over time (11, 12), which thus cannot be predicted from the enzymes' $\Delta G^{\ddagger}_{cat}$ and $\Delta G^{\ddagger}_{inact}$ values alone (11–14). The Equilibrium Model provides a quantitative explanation of enzyme thermal behavior, by introducing an intermediate and inactive (but not denatured) form that is in rapid equilibrium with the active form (13–15):

$$E_{act} \overset{K_{eq}}{\rightleftharpoons} E_{inact} \xrightarrow{K_{inact}} X$$

where E_{act} is the active form of the enzyme, which is in equilibrium with the inactive form, E_{inact}; K_{eq} is the equilibrium constant describing the ratio of E_{inact}/E_{act}; K_{inact} is the rate constant for the E_{inact} to X reaction; X is the irreversibly denatured form of the enzyme. For a more detailed description of the model, see ref. 15.

In short, the key discovery of the Equilibrium Model is that there are two unrelated ways through which enzymes lose activity as the temperature is raised: irreversible thermal inactivation, and a rapid reversible equilibration between the active and inactive forms of enzyme. The latter (described by T_{eq} and ΔH_{eq}) can only be described by fitting enzyme activity data collected across a temperature range and over time to the Equilibrium Model, and deriving a set of thermodynamic parameters that best describes the enzyme's behavior in relation with temperature.

So far, all the enzymes for which Equilibrium Model parameters have been determined fitted the Model. The enzymes cover most reaction classes and could all be measured directly and continuously, ensuring rapid and accurate collection of V_{max} assay data.

It is apparent from the range of those enzymes' quaternary structures, from monomeric to hexameric, including a citrate synthase whose active site is at a subunit interface (15–22), that conformity with the Equilibrium Model is independent of tertiary and quaternary structures as well the reaction mechanism. Evidence so far thus suggests that the Equilibrium Model is universally applicable to enzymes (15, 22).

Using the Equilibrium Model, the variation of enzyme activity with temperature can be expressed by

$$V_{max} = \frac{k_{cat} E_0 e^{-\frac{k_{inact} K_{eq} t}{1+K_{eq}}}}{1+K_{eq}}$$

where

$$K_{eq} = e^{\frac{\Delta H_{eq}}{R}\left(\frac{1}{T_{eq}} - \frac{1}{T}\right)}$$

$$k_{cat} = \frac{k_B T}{h} e^{-\frac{\Delta G^{*}_{cat}}{RT}}$$

and

$$k_{inact} = \frac{k_B T}{h} e^{-\frac{\Delta G^{*}_{inact}}{RT}}$$

where k_{cat} enzyme catalytic rate constant; t assay duration; $[E_0]$ enzyme concentration; ΔH_{eq} the change in enthalpy associated with the E_{act}/E_{inact} equilibrium; T_{eq} the temperature midpoint of the E_{act}/E_{inact} equilibrium; k_B Boltzmann's constant; R Gas constant; T temperature; h Planck's constant; $\Delta G^{\ddagger}_{cat}$ activation energy of the catalyzed reaction; $\Delta G^{\ddagger}_{inact}$ activation energy of the thermal inactivation process.

3 The Equilibrium Model and Protein Engineering

The new parameters associated with the Equilibrium Model provide tools for understanding and quantifying the temperature dependence of enzyme activity, and the adaptation of enzymes and organisms both to temperatures and to ranges of temperature. T_{eq}, the temperature of the midpoint of the equilibrium between the active and reversibly inactive forms of the enzyme, is an evolved property of enzymes related to the organism's growth temperature, being better correlated with the environmental temperature of the enzyme than its stability (18). ΔH_{eq}, the enthalpic change associated with the equilibrium, governs the temperature range over which the equilibrium occurs and thus the ability of the enzyme to function at different temperatures and temperature ranges (18).

The Equilibrium Model quantitatively explains the effect of temperature on all enzymes for which V_{max} can be measured over a

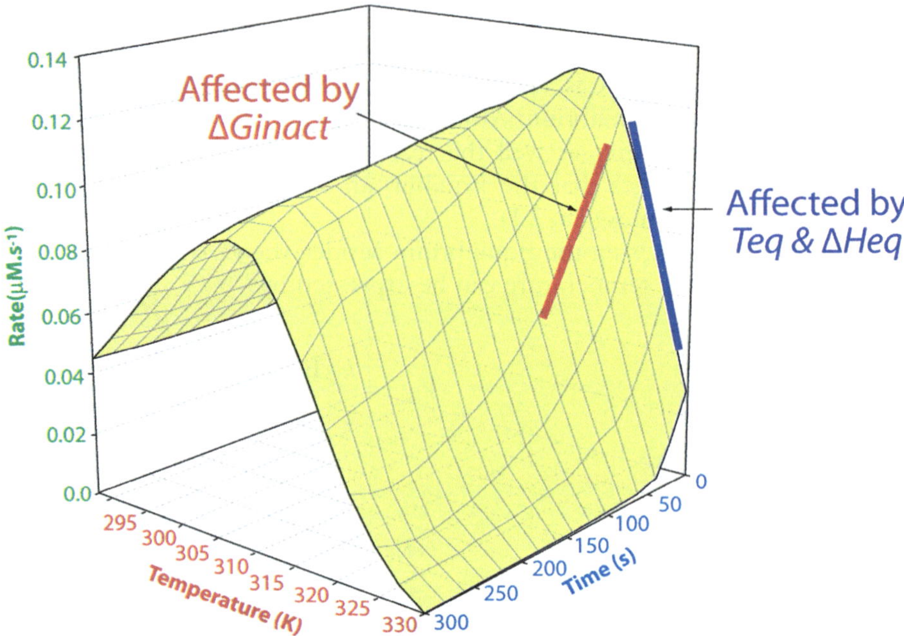

Fig. 1 A plot showing the major effects of the Equilibrium Model parameters on the temperature-dependent activity of an enzyme. The value of $\Delta G^{\ddagger}_{inact}$ determines the time-dependent loss of activity at any temperature due to irreversible thermal inactivation. When measuring initial (zero-time) rates, the rapid reduction in activity at high temperatures arises from the conversion of E_{act} to E_{inact} and is governed by T_{eq} and ΔH_{eq}. At any point along the time axis, the activity of the enzyme is thus determined by all three thermodynamic parameters

range of temperatures (11, 13–18, 21, 23) and has predicted and explained the counterintuitive behavior of enzyme reactors at some temperatures (23, 24). In particular, it explains why enzyme engineering has to date been so relatively unsuccessful. As indicated in Fig. 1, while $\Delta G^{\ddagger}_{inact}$ (traditionally the focus of protein engineering) determines the rate at which activity is lost with time, the model shows that an increase in stability (i.e., in $\Delta G^{\ddagger}_{inact}$) will not necessarily lead to increased activity at high temperatures unless T_{eq} (exclusive to the Equilibrium Model) is also shifted to a higher temperature and alters the rapidly reversible equilibrium between the active and reversibly inactivated forms of the enzyme. If the method chosen to detect increases in $\Delta G^{\ddagger}_{inact}$ is to determine increased activity at a higher temperature after a fixed period of time, then any increase may go undetected because T_{eq} and ΔH_{eq} together limit activity. Furthermore, any actual increase in $\Delta G^{\ddagger}_{inact}$ may have little apparent effect if observable activity is restricted by T_{eq} and ΔH_{eq}. In other words, to effectively improve enzyme activity at high temperatures, both irreversible inactivation and a shift in the E_{act}/E_{inact} equilibrium, the two independent mechanisms through which enzymes lose activity as temperature increases, must be addressed (Fig. 1). However, an increase in T_{eq} could in principle improve enzyme activity at elevated temperatures in the absence of any change in $\Delta G^{\ddagger}_{inact}$.

4 Mathematical Basis of the Equilibrium Model

The Equilibrium Model has four data inputs: enzyme concentration, temperature, concentration of product, and time. From the last two, an estimate of the rate of reaction (in M s^{-1}) can be obtained. The quantitative expression of the dependence of rate on temperature and time is given by Eq. 1:

$$V_{\max} = \frac{k_B T e^{-\left(\frac{\Delta G^{\ddagger}_{\text{cat}}}{RT}\right)} E_0 e^{\left(-\frac{k_B T e^{-\left(\frac{\Delta G^{\ddagger}_{\text{inact}}}{RT}\right)} e^{\left(\frac{\Delta H_{\text{eq}}\left(\frac{1}{T_{\text{eq}}}-\frac{1}{T}\right)}{R}\right)} t}{h\left(1+e^{\left(\frac{\Delta H_{\text{eq}}\left(\frac{1}{T_{\text{eq}}}-\frac{1}{T}\right)}{R}\right)}\right)}\right)}}{h\left(1+e^{\left(\frac{\Delta H_{\text{eq}}\left(\frac{1}{T_{\text{eq}}}-\frac{1}{T}\right)}{R}\right)}\right)} \tag{1}$$

Experimentally, however, rates are rarely measured directly; rather, product concentration is determined at regular intervals, either by continuous or discontinuous assay, producing a series of progress curves. The quantitative expression relating product concentration, time, and temperature for the Equilibrium Model can be obtained by integrating Eq. 1, giving Eq. 2.

$$[P] = -\frac{e^{\left(\frac{\Delta G^{\ddagger}_{\text{cat}}}{RT}\right)} E_0 e^{\left(-\frac{k_B T e^{\left(\frac{\Delta G^{\ddagger}_{\text{inact}}}{RT}\right)} e^{\left(\frac{\Delta H_{\text{eq}}\left(\frac{1}{T}-\frac{1}{T_{\text{eq}}}\right)}{R}\right)} t}{h\left(1+e^{\left(\frac{\Delta H_{\text{eq}}\left(\frac{1}{T}-\frac{1}{T_{\text{eq}}}\right)}{R}\right)}\right)}\right)}}{e^{\left(-\frac{\Delta G^{\ddagger}_{\text{inact}}}{RT}\right)} e^{\left(-\frac{\Delta H_{\text{eq}}\left(\frac{1}{T}-\frac{1}{T_{\text{eq}}}\right)}{R}\right)}} + \frac{e^{\left(-\frac{\Delta G^{\ddagger}_{\text{cat}}}{RT}\right)} E_0}{e^{\left(-\frac{\Delta G^{\ddagger}_{\text{inact}}}{RT}\right)} e^{\left(-\frac{\Delta H_{\text{eq}}\left(\frac{1}{T}-\frac{1}{T_{\text{eq}}}\right)}{R}\right)}} \tag{2}$$

We find that data processed both as enzyme rates using Eq. 1 and as product concentration changes using Eq. 2 give essentially the same results. However, since Eq. 2 involves a more direct measurement, it is the preferred approach.

The Equilibrium Model in itself does not explain the molecular basis of these effects, or the physical nature of E_{inact}. However, it enables a comprehensive description of the effects of temperature on enzyme activity, revealing a new structurally localized and apparently universal mechanism for enzyme activity loss with increasing temperature, independent of and additional to irreversible thermal inactivation (15).

5 Enzyme Assays and Temperature-Related Issues

Determination of Equilibrium Model thermal parameters requires accurate measurement of the change in concentration of substrate or product with time, i.e., of enzyme activity (as V_{max}) across a range of temperatures using a fixed amount of enzyme. Enzyme assays must be carried out under V_{max} reaction conditions. If reaction rates are not "ideal" (e.g., if substrate is not saturating for the entire reaction duration, or if substrate and/or product is inhibiting), it may be possible to derive some of the parameters using initial rates (see below). The linearity of assay data must be ensured by considering the detection limits of the instrument (e.g., spectrophotometer). It is difficult to obtain the required accuracy with discontinuous assays (20), and we have found continuous spectrophotometric assays to be the most straightforward method. Here we provide a brief overview of factors that may compromise the valid comparison of enzyme activity over a wide range of temperatures and thus the accurate determination of Equilibrium Model parameters. Further discussion of these factors is available elsewhere (20, 25, 26).

5.1 Temperature Control

Enzyme assays should be initiated by the rapid addition (and mixing) of small amounts (μL) of enzyme into the temperature-equilibrated reaction mixture, so that the addition of ice-cold enzyme solution has limited effect on the temperature of the final reaction mixture.

Equilibrating the content of a cuvette in a temperature-controlled spectrophotometer at 37°C is a relatively quick procedure, although a plastic cuvette may still require over 5 min (27). At higher temperatures, not only is the slower temperature equilibration a more serious drawback for plastic cuvettes, they may also deform. Above 60°C, the temperature equilibration of any cuvette (including quartz) is slow, and the heat loss from liquid-jacketed cuvette holders during circulation will lead to a significant offset between the (higher) water-bath temperature and the temperature in the cuvette. The water-bath temperature will therefore need to be adjusted based on readings from a temperature sensor in the cuvette itself for reliable results. Above 80°C, electrically heated cuvette holders will be required. In all cases, the only reliable estimate of the reaction mixture temperature is a direct measurement inside the cuvette using a calibrated and sensitive temperature sensor. Temperature gradients within cuvettes can be significant, and if stirred cuvettes are not used, temperature should be measured at both top and bottom of the cuvette. Other potential problems associated with temperature are evaporation from the cuvette at high temperatures (use a cover on the cuvette), and condensation on the surface of the cuvette at low temperatures (blow dry air into the cuvette surroundings).

5.2 Buffers

There are many factors to consider when choosing a buffer (28, 29), the most significant being the temperature dependence of buffer pH ($\Delta pK_a/\Delta t$), which means many buffers will need to be reformulated at each temperature to ensure consistency in pH. Effective buffering around neutral pH with only moderate temperature dependence can be achieved with few (and mostly simple) buffers; phosphate buffer is useful in this respect. At high temperatures, the stability of buffer components (esp. complex buffers such as HEPES, MOPS, etc.) may need to be considered.

5.3 Assay Component Stability

Substrate and/or product and/or cofactor stability (depending on how the reaction rate is measured) need to be checked if assays are to be carried out at high temperatures. For example, NAD(P) is quite unstable at high temperature (30), and glutamine degrades significantly even at 80°C (31). An additional complication is that many compounds have temperature-dependent extinction coefficients; NADH and potassium ferricyanide, for example, have significantly lower extinction coefficients at 80°C than at 20°C (32); the absorbance and λ_{max} of p-nitrophenol vary with temperature, leading to substantial errors in k_{cat} values measured by continuous release of p-nitrophenol (33).

5.4 Variation of K_M with Temperature

A variety of enzymes are known to have different K_M values at different temperatures. In almost all cases, K_M rises with temperature (12, 26, 34), possibly because of changes associated with the shift from the E_{act} to the E_{inact} form of the enzyme (22). These findings emphasize the need to carry out K_M determinations at the highest temperature at which the enzyme will be assayed to ensure substrate saturation, so that V_{max} is maintained over the whole temperature range.

6 Data Collection and Parameter Generation

6.1 Instrumentation

For continuous spectroscopic assays, we have used a Thermo spectronic Helios γ spectrophotometer, equipped with a Thermo spectronic single-cell Peltier-effect cuvette holder. It is connected to a computer running Vision32 (Version 1.25, Unicam Ltd.) software, including the Vision Enhanced Rate Program capable of recording absorbance changes over time intervals down to 0.125 s.

6.2 Data Collection and Processing

For each enzyme, reaction progress curves at a variety of temperatures are collected with absorbance readings recorded at convenient intervals (e.g., one second), with three replicate progress curves at each temperature. If the slopes for these triplicates deviate by more than 10%, more replicates should be performed. Nonenzymatic control reactions are measured at each temperature unless no significant background rate exists at the highest assay temperature.

Prior to collecting the final data, a preliminary characterization needs to be carried out over a temperature range from 30°C below the enzyme's physiological operating temperature to 30°C above, at intervals of 10°C, to determine the assay conditions to be used in the main experiment. These conditions include the temperature range to be covered and the enzyme concentration which will produce, at V_{max} and over the duration of the assay, a significant linear change in substrate concentration without approaching the instrument's linear measurement limit. The stability of the reactants should also be determined at the highest assay temperature, and the extinction coefficient of the compound being measured should be assessed over the temperature range, so that accurate conversion of absorbance values to concentrations can be made.

The accumulated data sets are initially processed by converting absorbance values to molar concentrations, and assay temperatures from °C to K. Data point times are expressed in seconds. All data sets need to contain the same number of data points over the same duration. The data sets are assembled into an Excel spreadsheet, which will be accessed by the analysis software (see below). This spreadsheet will also require the molar enzyme concentration used, and a set of initial estimates of the thermal parameters. A template spreadsheet is included in the software package.

Experimental data can be fitted to the Equilibrium Model using the methods described at http://www.hdl.handle.net/10289/3791. This software package includes a Matlab implementation of the Equilibrium Model (expressed as Eq. 2), which performs a recursive, nonlinear minimization of least squares fitting of experimental data to generate Equilibrium Model parameters. The minimization routine utilizes Powell's algorithm to find a local minimum, which can be but is not necessarily the global minimum (therefore it is necessary to repeat the calculations using different initial estimates until the final parameters stabilize), of the sum of squared deviations between the experimental data and the model calculations, and calculates the fitting error.

Although the software package automatically generates the fitting error, this is small compared with the experimental error, which can be estimated in several ways. One method that we have found satisfactory is to separately process each of the three replicate data sets (i.e., containing only a single assay at each temperature). By taking the 5–95% confidence intervals generated by the software package and selecting the individual values of the respective lowest and highest 5% and 95% confidence values from the three data sets to give the widest possible confidence interval, realistic estimates of the combined fitting and experimental errors can be obtained.

When required (see below), the initial (zero-time) rate of reaction for each assay triplicate can be determined by using the "Linear Search" function in the Vision32 Rate Program. An Excel spreadsheet containing sets of assay temperatures and initial rate values can be used as input data by a zero-time function in the software to generate a zero-time fit.

6.3 Robustness of the Fitted Constants

If the enzyme preparation is not pure, or the enzyme concentration inaccurate, then errors in estimated enzyme amount in assays are likely. Few methods for determining protein concentration give answers that are accurate in absolute terms, since, in addition to sensitivity limitations and interferences, most methods are based on a comparison with a standard of uncertain equivalence to the enzyme being quantified. The determination of enzyme concentration is therefore a potential source of error. However, even tenfold variations in the enzyme concentration result in only small differences in values for $\Delta G^{\ddagger}_{inact}$, ΔH_{eq}, and T_{eq}, although $\Delta G^{\ddagger}_{cat}$ may be affected (20).

6.4 Data Sampling Requirements

Some enzyme assays are difficult to carry out continuously. A comparison of continuous and discontinuous data collection indicated that in principle discontinuous enzyme assays can be used for the determination of T_{eq}, with a key requirement being data accuracy rather than the number of data points (20). However, the Equilibrium Model equation contains exponential of exponentials, and small data errors can lead to poor results. It may be difficult to generate sufficiently accurate data using "stopped" reactions, and continuous assays are to be preferred.

The minimum number of points per progress curve required to give accurate values for the Equilibrium Model parameters depends upon the length of the assay and the curvature of the progress curve, but an absolute minimum of 10 points should be taken at each temperature, and a larger number of data points gives more accurate results (20). Much of the work on the Equilibrium Model to date has been carried out with assay durations of 2–5 min.

A key requirement for data to be used with the Equilibrium Model is a clear temperature optimum at zero time (T_{opt}), and at least two temperature points above T_{eq} must be included. Furthermore, T_{eq} (a mathematical parameter not directly observable from data) may be well above the apparent T_{opt}, so more temperature points at the high end will improve the robustness of the resulting parameters.

All the foregoing discussion is based on an ab initio presumption that the temperature dependence of enzyme activity is described by the Equilibrium Model. So far, all enzymes for which accurate V_{max} data has been gathered follow the model. However, data that do not show clear evidence of a zero-time temperature optimum might be fitted equally well to the simpler Classical model. It must therefore be stressed that if only one or two points above the apparent temperature optimum (T_{opt}) are determined, the measured initial rates at those temperatures must be sufficiently lower than that at T_{opt} for the assumption of the Equilibrium Model to be justified. A minimum of two rate measurements above T_{opt} showing a clear trend of falling rates should be obtained to apply the Equilibrium Model with confidence.

6.5 Enzymes Operating Under "Nonideal" Conditions: The Use of Initial Rates

To use data from progress curves collected over extended periods of time, valid fitting to the Equilibrium Model requires that any decrease in activity observed is due solely to thermal factors and not some other process. In other words, the enzyme reaction kinetics should be "ideal": no substrate or product inhibition, a reaction essentially irreversible over the course of the assay, and the enzyme operating at V_{max} for the entire assay.

However, many enzyme reactions are necessarily assayed under nonideal conditions. For example, the reaction may be sufficiently reversible that the back reaction contributes to the observed rate during the assay, and/or the products/substrates of the reaction may be inhibitors of the enzyme. Application of the Equilibrium Model to these "nonideal" enzyme reactions can usually be achieved by restricting assays to the initial rate of reaction. Setting $t=0$ in Eq. 1 gives Eq. 3. Using Eq. 3, it is possible to fit the experimental data for "zero time" (i.e., initial rates) to the Equilibrium Model to determine $\Delta G^{\ddagger}_{cat}$, ΔH_{eq}, and T_{eq}, although the time-dependent thermal parameter, $\Delta G^{\ddagger}_{inact}$, cannot be determined.

At $t=0$,

$$V_{max} = \frac{k_B T \cdot e^{\left(-\frac{\Delta G^{\ddagger}_{cat}}{RT}\right)} \cdot E_0}{h\left(1+e^{\left(\frac{\Delta H_{eq}\left(\frac{1}{T_{eq}}-\frac{1}{T}\right)}{R}\right)}\right)} \quad (3)$$

Another situation where initial rates may have to be used is when substrate saturation cannot be sustained, and substrate depletion causes a decrease in reaction rate (the enzyme still needs to be saturated with substrate at the start of the assay). It should be noted that in general terms, use of initial rates will give less accurate results.

6.6 Checking Validity of the Calculated Parameters

All the foregoing discussion is based on a presumption that the enzyme in question follows behavior described by the Equilibrium Model. So far, all enzymes for which accurate V_{max} data can be gathered obeyed the model (15, 22). For newly characterized enzymes, their conformity to the Equilibrium Model can be verified by 1. comparing the temperature-dependent behavior of the enzyme predicted from its Equilibrium Model parameters with the experimental data and 2. examining the magnitude of fitting and experimental errors. Full details related to parameter validation are available at http://www.hdl.handle.net/10289/3791.

Acknowledgments

We thank Michelle Peterson for all the early experimental work she did to validate the Equilibrium Model, and Martin Seefeld and Andreas Pickl for technical assistance. We also thank the Royal Society of New Zealand Marsden Fund for financial support [grant number UOW0501].

References

1. Brock TD (1978) Thermophilic microorganisms and life at high temperatures. Springer, New York, NY
2. Doig AR (1974) Stability of enzymes from thermophilic micro organisms. In: Pye EK, Wingard LB (eds) Enzyme engineering. Plenum Press, New York, NY, pp 17–21
3. Sonnleitner B, Fiechter A (1983) Advantages of using thermophiles in biotechnological processes: expectations and reality. Trends Biotechnol 1:74–80
4. Amelunxen RE, Murdock AL (1978) Mechanisms of thermophily. CRC Crit Rev Microbiol 6:343–393
5. Daniel RM, Cowan DA, Morgan HW, Curran MP (1982) A correlation between protein thermostability and resistance to proteolysis. Biochem J 207:641–644
6. Owusu RK, Cowan DA (1989) Correlation between microbial protein thermostability and resistance to denaturation in aqueous: organic solvent two-phase systems. EnzymeMicrob Technol 11:568–574
7. Copeland RA (2000) Enzymes: a practical introduction to structure, mechanism, and data analysis, 2nd edn. Wiley, Hoboken, NJ
8. Dixon M, Webb EC (1979) Enzymes. Longman, London
9. Garrett RH, Grisham CM (2009) Biochemistry, 4th edn. Brook/Cole (Cengage), Stamford, CT.
10. Wiseman A (1983) Principles of biotechnology, 1st edn. Surrey University Press, Glasgow
11. Daniel RM, Danson MJ, Eisenthal R (2001) The temperature optima of enzymes: a new perspective on an old phenomenon. Trends Biochem Sci 26:223–225
12. Thomas TM, Scopes RK (1998) The effects of temperature on the kinetics and stability of mesophilic and thermophilic 3-phosphoglycerate kinases. Biochem J 330(Pt 3):1087–1095
13. Peterson ME, Eisenthal R, Danson MJ, Spence A, Daniel RM (2004) A new intrinsic thermal parameter for enzymes reveals true temperature optima. J Biol Chem 279:20717–20722
14. Peterson ME, Eisenthal R, Danson MJ, Spence A, Daniel RM (2005) Erratum: a new intrinsic thermal parameter for enzymes reveals true temperature optima. J Biol Chem 280:41784
15. Daniel RM, Danson MJ (2010) A new understanding of how temperature affects the catalytic activity of enzymes. Trends Biochem Sci 35:584–591

16. Daniel RM, Danson MJ, Eisenthal R, Lee CK, Peterson ME (2008) The effect of temperature on enzyme activity: new insights and their implications. Extremophiles 12:51–59
17. Daniel RM, Danson MJ, Hough DW, Lee CK, Peterson ME, Cowan DA (2008) Enzyme stability and activity at high temperatures. In: Siddiqui KS, Thomas TM (eds) Protein adaptation in extremophiles. Nova Publishers, New York
18. Lee CK, Daniel RM, Shepherd C, Saul DJ, Cary SC, Danson MJ, Eisenthal R, Peterson ME (2007) Eurythermalism and the temperature dependence of enzyme activity. FASEB J 21:1934–1941
19. Moore V (2008) PhD Thesis: A computational and experimental study of the thermal stability of citrate synthase. University of Bath, Bath, UK.
20. Peterson ME, Daniel RM, Danson MJ, Eisenthal R (2007) The dependence of enzyme activity on temperature: determination and validation of parameters. Biochem J 402: 331–337
21. Daniel RM, Danson MJ, Eisenthal R, Lee CK, Peterson ME (2007) New parameters controlling the effect of temperature on enzyme activity. Biochem Soc Trans 35:1543–1546
22. Daniel RM, Peterson ME, Danson MJ, Price NC, Kelly SM, Monk CR, Weinberg CS, Oudshoorn ML, Lee CK (2010) The molecular basis of the effect of temperature on enzyme activity. Biochem J 425:353–360
23. Eisenthal R, Peterson ME, Daniel RM, Danson MJ (2006) The thermal behaviour of enzymes: implications for biotechnology. Trends Biotechnol 24:289–292
24. Oudshoorn ML (2008) MSc Thesis: Tests of predictions made by the Equilibrium Model for the effect of temperature on enzyme activity. University of Waikato, Hamilton, New Zealand http://hdl.handle.net/10289/2418
25. Lee CK (2007) PhD Thesis: Eurythermalism of a deep-sea symbiosis system from an enzymological aspect. University of Waikato, Hamilton, New Zealand, http://hdl.handle.net/10289/2588
26. Daniel RM, Danson MJ (2001) Assaying activity and assessing thermostability of hyperthermophilic enzymes. Method Enzymol 334:283–293
27. John RA (2002) Photometric assays. In: Eisenthal R, Danson MJ (eds) Enzyme assays: a practical approach. Oxford University Press, Oxford, UK, p 59
28. Beynon RJ, Easterby JS (2003) Buffer solutions: the basics. Taylor & Francis, Oxford, UK
29. Price NC, Stevens L (2002) Techniques for enzyme extraction. In: Eisenthal R, Danson MJ (eds) Enzyme assays: a practical approach. Oxford University Press, Oxford, UK, p 221
30. Daniel RM, Danson MJ (1995) Did primitive microorganisms use nonhem iron proteins in place of NAD/P? J Mol Evol 40:559–563
31. Ratcliffe H, Drozd J, Bull A (1978) The utilization of l-glutamine and the products of its thermal decomposition by *Klebsiella pneumoniae* and *Rhizobium leguminosarum*. FEMS Microbiol Lett 3:65–69
32. Walsh KA, Daniel RM, Morgan HW (1983) A soluble NADH dehydrogenase (NADH: ferricyanide oxidoreductase) from *Thermus aquaticus* strain T351. Biochem J 209:427–433
33. Fourage L, Helbert M, Nicolet P, Colas B (1999) Temperature dependence of the ultraviolet-visible spectra of ionized and un-ionized forms of nitrophenol: consequence for the determination of enzymatic activities using nitrophenyl derivatives—a warning. Anal Biochem 270:184–185
34. Hudson RC, Ruttersmith LD, Daniel RM (1993) Glutamate dehydrogenase from the extremely thermophilic archaebacterial isolate AN1. Biochim Biophys Acta 1202:244–250

Part III

Tools of the Trade

Chapter 14

Rational-Based Protein Engineering: Tips and Tools

Meghna Sobti and Bridget C. Mabbutt

Abstract

The rational engineering of proteins is driven by contemporary needs for new and altered biomolecular forms. Utilizing manipulative procedures of molecular biology, it is relatively straightforward to alter protein structure and function to create mutated or fused sequences. We here give an overview of procedures and strategies for site-directed mutagenesis, construction of fusion proteins, and insertion of tags. The design of new protein constructs as well as their over-expression as recombinant products is considered. We also summarize approaches for the engineering of protein complexes by co-expression, a valuable route to generate bioactive multicomponent systems.

Key words Codon bias, Affinity tag, Site-directed PCR, Overlap PCR, Polyproteins, Linker peptide

1 Introduction

Engineering of new protein forms is of paramount importance for emerging products and applications in biotechnology and pharmaceuticals (1). The manipulative power of genetic engineering to alter or exploit the complex functional properties of proteins is realized through the relatively straightforward procedures of recombinant DNA. "Rational-based" engineering of proteins engages an intuitive approach, based on detailed knowledge of structure and function to generate desired changes to the molecular system. At the outset, the solubility and viability of any recombinant product likely requires engineering of a sequence to correctly incorporate appropriate protein domain boundaries (2). As well, internal deletions or mutations in a sequence may ensure a stable and folded protein, or functionally tune a wild-type form. The position, chemistry, and size of any extraneous tag (whether required for affinity purification or detection purposes) can also impact on the solubility of a final expressed product (3, 4).

Proteins are today perceived as key interaction systems which engage partner biomolecules (proteins, nucleic acids, metabolites)

Juliet A. Gerrard (ed.), *Protein Nanotechnology: Protocols, Instrumentation, and Applications*, Methods in Molecular Biology, vol. 996, DOI 10.1007/978-1-62703-354-1_14, © Springer Science+Business Media New York 2013

within stable or transient complexes (5). Thus, rather than being studied as monomeric entities, analysis of both the architecture and function of intact protein assemblies serves to enhance the emerging concept of the cell as a collection of multi-subunit protein machines (6). The preparation of large quantities of protein complexes has traditionally been accomplished by recombinant expression of distinct component proteins followed by in vitro reconstitution, or purification of the endogenous protein assemblies. Protein engineering allows us to deliberately alter these components and combinations so as to better suit the processes of recombinant production and/or structure determination. Over the last decade, co-expression of multiple proteins within the one host cell has proved successful for generating mixed protein complexes (7, 8). This allows reconstitution of complexes should protein components not fold properly when expressed in isolation, yet form a viable complex when simultaneously produced in a cellular environment (9).

Heterologous protein production can today be efficiently achieved utilizing a selection of prokaryotic or eukaryotic (yeast, insect, mammalian) cellular systems developed for expression of a wide range of protein products (10, 11). By modulating experimental factors such as temperature or media components, tRNA levels, or the context and copy number of the gene itself, it is relatively straightforward to optimize production of specific products within a given expression host. In those cases where a desired gene product is toxic to the host cell, cell-free expression systems designed for protein synthesis have recently offered an alternative route to production (12). Regardless of host system, expression of a specific protein product is often a complex balance of parameters, and specific protocols might be selected for (a) molecular biology procedures (vector design, rapid cloning capability, gene design), (b) gene over-expression (gene dosage, promoter strength, mRNA stability, translation initiation and termination, codon usage), or (c) generation of biomass (fermentation factors).

We survey here experimental steps and laboratory procedures appropriate for generation of engineered proteins and protein complexes. The expression host most often employed for heterologous production of recombinant proteins remains the bacterium *Escherichia coli* (13, 14). Detrimental factors associated with the use of this host, particularly when handling genes of eukaryotic origin, include its lack of capacity for posttranslational modification and formation of incorrectly folded product. However, its advantages include a well-understood genetics, a potential for rapid biomass accumulation and inexpensive carbon source requirements. Its common use has driven the development of a large array of compatible tools for biotechnology, especially a variety of plasmids, recombinant fusion partners, and mutant strains.

2 Methods

2.1 Codon Usage

Synonymous codons are used with varying frequencies across different organisms. Codons under-represented in *E. coli* (e.g., those for Arg: AGA, AGG, CGA; Ile: AUA; Leu: CUA) represent <8% of their corresponding codon partners (15). This codon bias can therefore impede translation of a desired heterologous target due to the demand for tRNAs rare in the host population, resulting in premature termination or amino acid mis-incorporation (16). Codon bias problems become most pronounced when transcripts contain clustered rare codons, particularly when located near the N-terminus of a coding sequence (17).

1. Check the sequence of the target gene for codon usage with reference to the expression host prior to attempting protein expression. Several online tools (e.g., GCUA (18), EMBOSS CAI (19)) analyze and compare codon adaptive index of two or more organisms. Other tools are available, both from the community (e.g., RaCC, maintained at UCLA) and commercial vendors, allow codon bias analysis of an input sequence with reference to common expression hosts.
2. If the gene is to be expressed in an organism utilizing different codon usage, two alternative strategies are available:
   (a) Alter the target sequence (e.g., by site-directed mutagenesis, Subheading 2.2.1) to incorporate codons best matching the tRNA pool of the host. This approach, while highly effective (20), may prove time-consuming in high-throughput applications. The online server OPTIMIZER (21) allows assessment of alien gene sequences, as well as optimization of levels of gene expression.
   (b) Co-transformation of the host with a multiple-copy plasmid harboring the appropriate gene encoding the tRNA for the problematic codon (22). The tRNA gene can either be inserted into the expression vector itself or placed on a second compatible plasmid. Commercial systems (plasmids and expression hosts) are available.

2.2 Generating Gene Variants

It is often desirable to engineer variation into a target protein: perhaps to study a specific effect on structure or function; to improve protein expression, solubility, or crystallization; or to incorporate a biophysical probe. These might involve mutation, deletion, or insertion of one or several amino acids in the native protein sequence. The following techniques are appropriate for creation of designed protein variants, and contrast with the approaches of directed evolution (which employ methods such as error-prone PCR (23), see Chapter 15).

2.2.1 PCR-Based Site-Directed Mutagenesis

Site-directed mutagenesis allows engineering of a desired alteration directly within a gene. The most commonly used method is oligonucleotide-mediated mutagenesis (24), in which an oligonucleotide incorporating the desired mutation is annealed to one strand of target DNA template for PCR to generate copies of new, mutated sequence. Further variations of this PCR-based method include megaprimer-based PCR (25), inverse PCR (26), and overlap extension based PCR (27). The method requires some experimental care, as follows:

1. Design and synthesize a pair of primers (one for each strand), both corresponding to the desired mutation. To ensure efficient hybridization to target DNA, the following should apply: (a) Locate the desired mutation centrally in the primer, with 10–15 perfectly matched nucleotides either side; (b) total primer length should be 25–45 bp, with a high melting temperature (T_m), usually ≥75°C; (c) the sequence should not generate secondary structures that inhibit annealing to the target; and (d) primers should have a minimum GC content of 40%, and terminate in G or C bases.
2. The web-based program PrimerX (hosted at bioinformatics.org) is useful for automatic mutagenic PCR primer design. Desired mutations can be specified as either protein or DNA sequences. The program generates mutagenic primers based on all possible DNA sequences that encode the desired change, and also accounts for codon degeneracy.
3. For a single-site mutation, deletion, or insertion, a 50 μl PCR ideally contains 2–10 ng of template, 100–200 ng of each primer, 200 μM dNTPs, and a high-fidelity DNA polymerase. A primer concentration far in excess of template is used to ensure template-primer interactions dominate. A high-fidelity polymerase is crucial to minimize unwanted mutation as the amplification reaction proceeds around the entire plasmid. Thermostable polymerases (e.g., Pfu, Vent or a Pfu/Taq mixture) are commercially available and are recommended.
4. Typically, PCR cycles are initiated at 95°C for 1–5 min to denature plasmid and generate single-stranded regions. This is followed by 12–18 amplification cycles consisting of (a) denaturation at 95°C for 1 min, (b) primer annealing at (T_m - 5) °C for 1 min, and (c) extension of the primer/s by polymerases at 68–72°C for x min (x= 1 min/500–1,000 bp plasmid length, according to enzyme used). Finally, there is an extension step for 10 min at 72°C. The PCR mix is often treated with DpnI to digest any methylated parent plasmid. The resulting circular, un-methylated nicked vector is then ready for ligation procedures (28).
5. Multiple primers, each with one or more mutations, substitutions, or deletions, can often be used in a single reaction.

6. A secondary mutation is also often introduced into the vector to allow screening of mutants. Commonly used secondary mutations include repair/disruption of antibiotic resistance (29), introduction/elimination of a unique restriction site (30), and activation/deactivation of a gene detectable by bioassay, deoxyuridine, (31) or phosphorthioate (32) introduction.
7. Commercial kits (such as the Stratagene's QuickChange™ product) are commonly used to accomplish single or simultaneous multiple mutations in a single tube.

2.2.2 Gene Truncations and Deletions

Truncated variants of a protein are valuable in mapping specific regions of functionality. Alternatively, truncated variants may display improved yield of soluble product, as the composition of N- and C-termini can influence protein behavior (2, 33). Thus, multiple constructs of a protein of interest are ideally prepared and screened during development of a new production protocol. Variants trialed should include affinity tag locations at both termini as well as diverse gene lengths (34). Molecular features that negatively correlate with protein solubility, and hence may need to be altered or removed, include transmembrane regions, contiguous lengths of hydrophobicity or low-complexity, disulfide bonds, and long unstructured termini or loops.

A bioinformatics evaluation of the protein of interest should thus be performed prior to construct design and cloning, incorporating both sequence and structural knowledge.

1. Use an alignment tool (e.g., NCBI search tool, PSI-BLAST (35)) to align the target gene/protein sequence to known sequences. Known protein sequences have already been aligned and grouped into homologous families within various databases (e.g., Pfam (36), Uniprot (37), KEGG (38)).
2. Identify most closely related proteins (ideally, ~80% sequence identity) and align sequences to map conserved regions (39). These conserved regions commonly constitute core structural domains and should be retained within expression constructs to ensure fold integrity.
3. For many protein families, representative tertiary structures may have been determined, with coordinates available in the Protein Data Bank (40). For proteins belonging to such families, and of sufficient homology, the sequence can be modeled or threaded on the known tertiary structure using prediction tools (e.g., SWISS-MODEL (41)). This process will predict core structural elements, establishing sequence boundaries for engineering discrete domains of interest.
4. For sequences with no known protein relatives (orphan proteins), disordered regions can be predicted using online tools (e.g., GlobPlot (42)) and deleted from the sequence prior to cloning.

An experimental alternative to identifying core structural segments is to carry out limited proteolysis on protein preparations, followed by mass spectrometry or chromatographic analysis (43). Alternatively, deuterium–hydrogen exchange coupled to mass spectrometry (DXMS) will identify fast-exchanging amides that originate in unstructured fragments (44).

5. The above procedures and predictions will define domain boundaries and intervening disordered portions of the target protein. This should guide the design of a panel of successive N- and C-terminal truncations for the complete protein, or its component domains.

2.3 Construction of Polyproteins

If two or more genes are fused to one another via encoded polypeptide-linkers, a single gene can be cloned into a vector for expression under one promoter. The method has been successfully used to create recombinant polyproteins in which components are linked head to tail, e.g., a version of GroEL, containing seven subunits covalently linked by short Ala/Gly-rich segments (45). As a part of construction of polyproteins, appropriate peptide linkers need to be incorporated between protein pairs to allow folding of subunits in an unstrained manner, so as to maintain native protein interfaces. Alterations in linker regions have been found to impact on the stability, oligomeric state, proteolytic resistance, and solubility of polyproteins (46, 47). The order of protein components within a fused polyprotein may strongly impact expression of the recombinant product, and different combinations may need to be trialed.

2.3.1 Design of Linker Peptides

1. The length of the linker must be adequate to span the minimum distance between the C-terminus of one protein and the N-terminus of the second in an orientation compatible with correct folding. It must also allow adequate flexibility between the component domains. Correct lengths can be judged using relevant protein structures or molecular models. In cases where tertiary information is not available, a panel of expressed constructs with varying linker lengths is best trialed, with linker length optimized for maximum stability and solubility (46).
2. Linker composition is usually based on glycine, so imparting full conformational freedom of polypeptide dihedral angles. Too much linker flexibility will, however, destabilize the polyprotein entropically. Thus, poly-Gly linkers are best interspersed with simple side chains (such as Ser, Ala). In the case of a single-chain version of *Arc* repressor, the composition of the linker (rather than its specific sequence) determined overall stability of the polyprotein (46).
3. Natural sequences can be incorporated into a polyprotein linker, rather than artificial peptide lengths. If the N- and

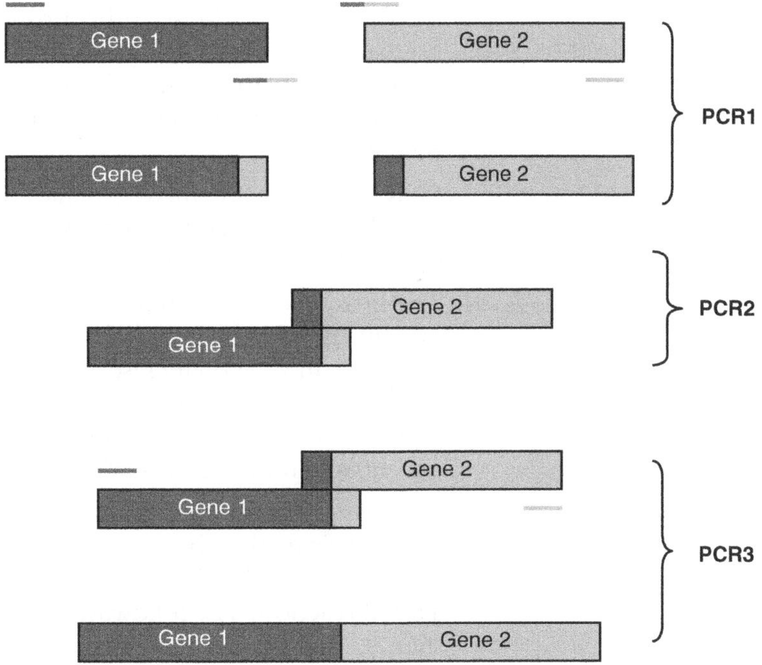

Fig. 1 Method of overlap PCR. The first step (PCR1) generates appropriate sequence overlap regions for genes to be eventually fused. The overlap region should minimally encode five amino acids and have $T_m \geq 50°C$. In step PCR2, the genes are annealed via these overlapping regions. A third step, PCR3, allows extension of annealed genes

C-terminal regions of the component proteins are flexible, these provide unstructured segments accessible beyond the component protein folds. This linker composition has been utilized to engineer simplified circular polyproteins incorporating RNA-binding proteins (48).

2.3.2 Method for Linking Genes

For a small construct, it may be possible to directly synthesize the full-length gene appropriate to the desired polyprotein sequence. Alternatively, manipulation by overlap PCR allows pairs of genes to be recombined into a single ORF at precise junctions, irrespective of nucleotide sequences at the recombination site (49). The method, outlined in Fig. 1, does not require restriction endonucleases or ligase:

1. Genes for recombination are generated in separate PCRs. Overlap primers are designed (a) to create single genes of each polycistron containing regions (15–18 bp) overlapping the required neighboring gene and (b) to have a theoretical T_m of the overlap region ≥ 50°C.
2. PCRs typically contain genomic DNA (0.5 μg), polymerase (1.25 U), manufacturer's reaction buffer, dNTPs (200 μM

each), and the respective primer pairs (1 μM). PCR conditions used are 1 cycle at 94°C for 3 min; 30 cycles of 94°C (30 s), 50–65°C, 5°C lower than the T_m (30 s), and 72°C (1 min/kb gene length); and 1 cycle at 72°C for 5 min. PCR products can then be purified.

3. PCR products are mixed, denatured and re-annealed. The strands containing matching sequences overlap, and act as primers for one another.
4. The first gene fusion step is carried out in a 50 μl reaction: 50 or 100 ng of each gene of a pair, dNTPs (200 μM), 1 mM $MgCl_2$ (1 mM), polymerase (1.25 U), and reaction buffer. No primers are added at this stage. Thermocycler conditions suggested 1 cycle at 94°C (3 min); 15 cycles of 94°C (30 s), 55°C (30–60 s), and 72°C (30–60 s); and 1 cycle at 72°C (5 min).
5. Following purification (commercial clean-up kit), 2.5–5 μl of material from the above reaction is combined with 5′ and 3′ primers (0.5 μM of each) for cloning into the target vector in a 50 μl reaction. The following components are included as for step 4: dNTPs, $MgCl_2$, polymerase, and reaction buffer. Thermocycler conditions as follows: 1 cycle at 94°C (3 min); 5 cycles of 94°C (30 s), 45°C (30 s), 72°C (30–60 s); 25 cycles of 94°C (30 s), 55°C (30 s), 72°C (30–60 s); and 1 cycle at 72°C (5 min). This step allows extension of annealed genes from the second PCR step to become linked.
6. Linked genes can now be cloned into a suitable expression vector using conventional cloning techniques.

2.4 Fusion Proteins Incorporating Tags

Fusion proteins can be organized to incorporate a partner or "tag" linked to the target protein for a variety of detection, purification, or stability purposes (Table 1) (50–53). Tags that display high affinity to a chemical ligand or antibody allow simple batch purification of a recombinant product. The choice of a tag depends on the protein application, the cost of the specific purification matrix, and the scalability of the process. Immobilized supports for capturing fusion proteins with affinity tags are commercially available, as are vector expression systems for encoding a desired tag (see Table 1a). These expression systems have provisions for the eventual removal of the tag from the recombinant product by chemical (54) or, more commonly, enzymatic means (55).

Specific protein fusions can alternatively be engineered to assist the yield and detection of a folded fusion partner (56). Green fluorescent protein (GFP), which emits colored fluorescence, can be fused to indicate protein products correctly folded in either cytosolic or membrane environments (57). Fusion proteins incorporating maltose-binding protein (MBP) or glutathione S-transferase (GST) may increase overall expression levels, solubility, or stability of a specific target (Table 1b). N-terminal fusion with MBP has

Table 1
Commonly used fusion tags

Tag	Location[a]	# Residues	Size (kDa)	Role[b]	Affinity matrix	Elution	Advantages	Disadvantages
A. Tags for protein expression, purification, and detection								
poly-His	N,C,I	2–10; us. 6	0.84	P	Immobilized divalent metal (e.g., Ni^{2+}-NTA, Co^{2+} IDA)	Imidazole, Low pH	Small size, Little effect on structure and activity of target, Adsorption in denaturing conditions	Co-elution of His-rich host proteins
FLAG	N,C	8–24	1–3	P	Antibody	Glycine/pH 2, EDTA, FLAG peptide	Small size, High specificity	Costly matrix, Unstable antibody matrix
S-Tag	N,C,I	15	1.75	P, R	S-protein	Guanidine thiocyanate, Citrate/pH 2, $MgCl_2$	Small size, High specificity, Straightforward protein assay	Harsh elution conditions, Costly matrix
T7-Tag	N,I	11–16	1	P	Antibody	pH 2.2	Small size, High specificity, Enhances expression	Harsh elution conditions, Unstable antibody matrix
Streptag II	N,C	8	1.2	P	Strep-Tactin Sepharose	Desthiobiotin	Small size, High specificity, Mild elution conditions	Costly matrix
Streptavidin-binding peptide	N,C	38	4	P	Streptavidin	Biotin	High specificity, Mild elution conditions	Costly matrix
Calmodulin-binding peptide	N,C	26	3	P	Calmodulin-derivative resin	EDTA with 1 M NaCl	Low metabolic burden, High specificity	Cross-reactivity for eukaryotic proteins, Costly matrix
Cellulose-binding domains	N,C,I	27–189	3–20	P, T	Carbohydrate-based resin	Urea, Guanidine, Ethylene glycol, High pH, Cellobiose	Inert matrix, High specificity	Harsh elution conditions

(continued)

Table 1
(continued)

Tag	Location[a]	# Residues	Size (kDa)	Role[b]	Affinity matrix	Elution	Advantages	Disadvantages
B. Tags for recombinant protein stability and solubility								
Glutathione S-transferase	N	211	26	P,S	Glutathione-derivatized matrix	Glutathione	Enhances translation initiation and solubility, Mild elution conditions, Inexpensive affinity resin	Large size, Tends to dimerize
Maltose-binding protein	N,C	396	40	P,T,S	Amylose resin	Maltose	Enhances translation initiation and solubility, Mild elution conditions, Tag detected by immunoassay	Large size, Incompatible with denaturing agents
Thioredoxin	N,C	109	12	S	–	–	Enhances translation initiation and solubility, Assists disulphide bond formation	Large size, Not an affinity tag
NusA	N.I	495	55	S	–	–	Increases solubility	Large size, Not an affinity tag
Mistic	N	330	13	S	–	–	Beneficial for membrane protein expression	Not an affinity tag
DsbA	N	208	23	T,S	–	–	Beneficial for periplasmic targeting, Assists disulfide bond formation	Not an affinity tag
Ubiquitin/SUMO	N	128/101	15/12	S	–	–	Enhances translation initiation and solubility	Not an affinity tag

[a]Sequence location is indicated as N-terminal (N), C-terminal (C), or internal (I)
[b]Role is indicated as purification (P), reporter (R), targeting (T), or enhanced solubility (S)

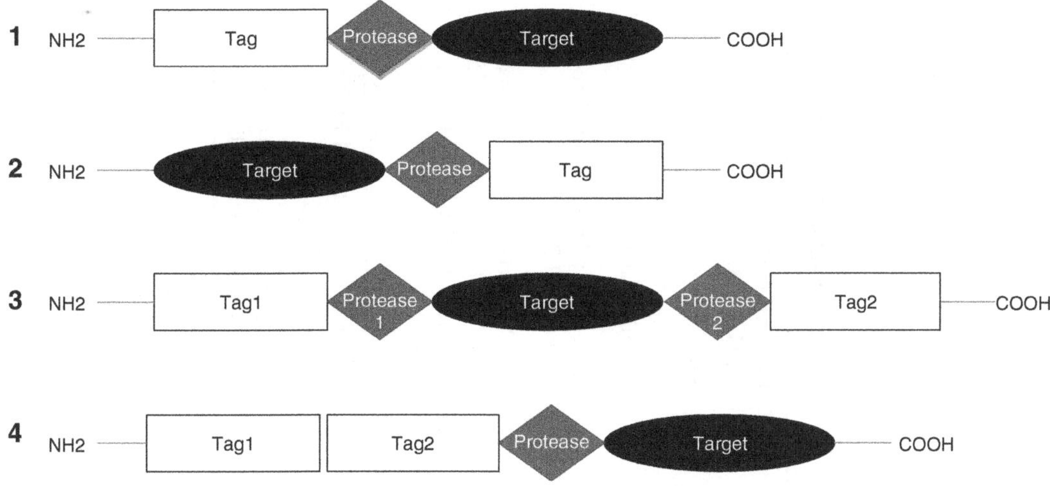

Fig. 2 Various arrangements of tags within recombinant fusion proteins. The tags can be selected for affinity, solubility, or detection. Dual or multiple tagging can be used to gain more than one tag function. In each case, a protease cleavage site is located between tag and protein target

been particularly beneficial for expression of recombinant membrane proteins in *E. coli*, due to insertion of the hydrophobic polypeptide into the cytoplasmic membrane of the host (58). A Mistic protein tag has also been recently successful in increasing bacterial expression of membrane proteins (59). Protein fusions may need to be used in conjunction with a separate affinity tag to expedite purification.

2.4.1 Placement of Tags

1. The polyhistidine-tag (His-tag), which binds to divalent transition metals, is by far the most commonly used affinity tag for high-throughput protein purification (60). The immobilized affinity matrix (IMAC) utilized has a high binding capacity, is relatively inexpensive and can be regenerated repeatedly. While protein elution conditions are usually mild, the affinity media also works well under denaturing conditions, enhancing its versatility.
2. The effect on tertiary structure and biological activity of the fusion protein product must be considered when determining tag location and composition. A fusion tag can be placed at N- or C-terminal to the target sequence, or in a region with appropriate surface exposure to allow binding and recognition (Fig. 2).
3. A linker region is often incorporated to sterically separate tag and target protein sequences. A protease cleavage site, if included, will allow eventual removal of the tag from the target (Table 2).
4. Multiple/combinatorial-tagging is now often used to derive maximum benefit from each tag (61). Mixed tags can be placed at one or both ends of the target sequence (Fig. 2). This might allow a second stage of affinity chromatography, or enhance the solubility or detection of the target protein.

Table 2
Site-specific proteases suitable for fusion tag removal

Protease	Cleavage site	M_R (kDa)	Removal feature or method	Optimal reaction conditions
Enterokinase	Asp–Asp–Asp–Asp–Lys/Xaa	150	Anti-enterokinase-agarose conjugate	RT; Tris or MES buffers (50 mM), pH 7.0–8.0
Factor Xa protease	Ile–Glu(Asp)–Gly–Arg/Xaa	48	Anti-factor Xa-agarose conjugate	22°C–25°C; Tris–HCl buffer (50 mM, with 150 mM NaCl, 1 mM $CaCl_2$), pH 7.5
Thrombin[a]	Xaa4–Xaa3–Pro–Arg/(Lys)–Xaa1′–Xaa2′ [b]	37	*p*-amino agarose, Benzamidine sepharose, SEC	22°C–25°C; Phosphate-buffered saline, pH 7.4
TEV protease	Glu–Asn–Leu–Tyr–Phe–Gln/Gly(Ser)	27	Affinity tagged, SEC	4°C RT (recom.), 34°C (optimal); Tris–HCl buffer (50 mM, with 0.5 mM EDTA,1 mM DTT), pH 8.0
PreScission protease	Leu–Glu–Val–Leu–Phe–Gln/Gly–Pro	46	Affinity (GST) tagged	4°C; Tris–HCl buffer (50 mM, with 150 mM NaCl, 1 mM EDTA, 1 mM DTT), pH 7.0
SUMO protease (Ulp1)	Structural feature	26	Affinity (poly-His) tagged	4–30°C (recom.), 30°C (optimal); phosphate-buffered saline, pH 7.4

RT room temperature, *DTT* dithiothreitol, *Xaa* any amino acid, except Pro or Arg
[a]Introduces two new amino acids (Gly, Ser) at N-terminus of cleavage product
[b]Xaa4 and Xaa3 are hydrophobic amino acids; Xaa1′ and Xaa2′ are nonacidic

5. To allow isolation of highly pure protein complexes from within cells, tandem affinity purification (TAP) tagging incorporates multiple tags (62). A gene of interest is fused in frame so as to incorporate tandem affinity tags separated by an endoproteinase cleavage site. The method is highly successful for monitoring protein interactions, particularly in conjunction with mass spectrometry (63).

2.4.2 Removal of Tags

Proteases commonly used for cleavage of tags are tabulated here, with their optimal reaction conditions (Table 2). Pilot experiments are best carried out on a specific tagged protein to establish optimal protease concentration, temperature, and incubation time for complete tag removal (55, 64, 65):

1. "On-column" cleavage involves proteolysis of the tag while the recombinant protein remains adsorbed to the affinity matrix. This method is highly recommended, as contaminants (including protease) are washed out prior to elution of the pure target protein.
2. Alternatively, the fusion tag is removed after elution of target from the affinity matrix ("in-solution" cleavage). Following the proteolysis step, the enzyme must then be removed (e.g., by affinity chromatography).
3. Cleavage of a tag without a protease is possible by incorporation of inteins (self-splicing protein elements) (66). Several commercial systems are available which encode an intein modified to catalyze an inducible cleavage at one terminus, fused between target protein and affinity tag. This allows the fusion product to be captured and purified on an affinity matrix. Once the intein splicing is initiated, the target protein (now free of intein tag) can be eluted from the adsorbent in a single step.

2.5 Multi-parallel Protein Expression

Successful protein engineering often requires evaluation of multiple versions of constructs. Many mutated, truncated, or orthologous versions of a protein might be screened to identify an optimal variant; a panel of fusion partners might be investigated for enhanced expression or purification capacity. For these purposes a high-throughput matrix cloning system, in which clones of gene variants are established across a set of vectors for parallel cloning, is best utilized (67, 68).

1. Select a backbone plasmid for the construction of parallel cloning vectors. Depending on end-use application, any commercial vector with a suitable origin of replication, promoter, and antibiotic resistance elements can be used.
2. Design a set of vectors that share common features and encode sequences for tags (often a combination of affinity and solubility tag) at one or both ends of the target gene. An encoded protease recognition site (see Table 2) following each tag region will assist its ultimate removal.

3. Ensure all vectors in the panel have common restriction sites in their multiple cloning sites, to allow a single set of primers for cloning. A large number of cloning steps is likely required, particularly in the case of co-expression applications (Subheading 2.6), and thus the use of a single set of restriction enzymes for cloning is recommended. It may be necessary to mutate the target sequence to remove sequences sensitive to the restriction enzymes selected for cloning procedures.
4. Generate parallel cloning vectors by replacing the multiple cloning site of the backbone vector with the newly designed multiple cloning sites (which now contain a combination of tags and a common restriction site) using conventional molecular biology techniques (28).
5. These vectors can now be prepared for parallel cloning of multiple inserts using restriction digestion and ligation (69), ligation-independent cloning (70), or relevant commercial systems (e.g., TOPO™ or GATEWAY™). Recently, these cloning techniques have been adapted to 96-well format in several laboratories (71).

2.6 Combination of Gene Products by Co-expression

Co-expression of two or more components of a multi-protein complex, or of a protein with a chaperone, may alleviate problems of solubility or functional loss, or greatly increase yield. The strategy has also been successfully used to co-express genes with a chaperonin (72) to promote proper folding or with rare tRNA to circumvent codon bias (73). A choice of approaches is available for co-expression; all compatible with one another should they be needed to be used in combination (74). The order of gene placement within a co-expression vector, boundaries of the domains incorporated, as well as internal deletions may all strongly affect expression. Imbalances in protein product ratios may occur as a consequence of differential transcription, translation, and RNA or protein stability; these will affect the choice of production route.

2.6.1 Many Vectors: Many Genes

The multiple vector approach utilizes two or more expression vectors, each encoding a single gene (ORF) under its own promoter (75, 76). The method is particularly flexible and robust, and allows separate screening of various constructs, tags, or promoters in parallel.

1. Select two or more expression vectors, each containing a distinct antibiotic selection marker, different compatible replicons, but a similar copy number (to maintain stoichiometry and yield). Vectors with same origin of replication can be used, provided they carry different selection markers (7).
2. Each of the genes is first cloned separately into the vectors by conventional cloning methods (28).

3. All vectors are then co-transformed into the expression host by chemical transformation or electroporation. As co-transformation decreases transformation efficiencies, both highly competent cells and elevated plasmid concentrations are required.
4. Transformed cells are then selected on agar plates with the appropriate combination of antibiotics expressed by all plasmids within into the host. Normal levels of antibiotics should be used for selection. The antibiotic concentration may, however, need to be reduced during expression phase to prevent slow cell growth.

2.6.2 Single Vector: Many Genes

In this technique, different ORFs, possibly under the control of a separate promoter, are cloned into the one plasmid. This strategy generates multiple mRNA transcripts, each encoding a separate protein. A vector is said to be polycistronic if more than one insertion site is available under the one promoter.

1. Select a vector with more than one multiple cloning site (MCS), each designed for a distinct set of restriction enzymes and antibiotic selection. The promoter can be selected according to the gene to be expressed; each gene placed at a different MCS will result in differentially tagged recombinant products.
2. The first target gene is cloned into MCS1 using conventional cloning methods. This first gene should lack any sequences sensitive to restriction enzymes required for the second gene insertion.
3. The second gene is then cloned into MCS2.
4. The technique is particularly exemplified by Novagen's Duet™ system, in which pET-based vectors allow for co-expression of up to 8 genes in a single *E. coli* cell (77, 78). These vectors contain two expression units, preceded by a T7 promoter/*lac* operator and a ribosome-binding site. The nature and positioning of the cloning sites in these vectors generates two protein products: one N-terminally fused with a His-tag and/or the other with a C-terminal S-Tag.
5. As an alternative to the method outlined above, cloning of a single polyprotein construct (Subheading 2.3) into a monocistronic system could also be considered.

References

1. Grunberg R, Serrano L (2010) Strategies for protein synthetic biology. Nucleic Acids Res 38:2663–2675
2. Graslund S, Sagemark J, Berglund H, Dahlgren LG, Flores A, Hammarstrom M, Johansson I, Kotenyova T, Nilsson M, Nordlund P, Weigelt J (2008) The use of systematic N- and C-terminal deletions to promote production and structural studies of recombinant proteins. Protein Expr Purif 58:210–221
3. Esposito D, Chatterjee DK (2006) Enhancement of soluble protein expression through the use of fusion tags. Curr Opin Biotechnol 17:353–358

4. Woestenenk EA, Hammarstrom M, van den Berg S, Hard T, Berglund H (2004) His-tag effect on solubility of human proteins produced in *Escherichia coli*: a comparison between four expression vectors. J Struct Funct Genomics 5:217–229
5. Gavin AC, Superti-Furga G (2003) Protein complexes and proteome organization from yeast to man. Curr Opin Chem Biol 7:21–27
6. Ahnert SE, Teichmann SA (2008) Networks for all. Genome Biol 9:324
7. Tolia NH, Joshua-Tor L (2006) Strategies for protein co-expression in *Escherichia coli*. Nat Methods 3:55–64
8. Romier C, Ben Jelloul M, Albeck S, Buchwald G, Busso D, Celie PH, Christodoulou E, De Marco V, van Gerwen S, Knipscheer P, Lebbink JH, Notenboom V, Poterszman A, Rochel N, Cohen SX, Unger T, Sussman JL, Moras D, Sixma TK, Perrakis A (2006) Co-expression of protein complexes in prokaryotic and eukaryotic hosts: experimental procedures, database tracking and case studies. Acta Crystallogr D Biol Crystallogr 62:1232–1242
9. Perrakis A, Romier C (2008) Assembly of protein complexes by co-expression in prokaryotic and eukaryotic hosts: an overview. Methods Mol Biol 426:247–256
10. Palomares LA, Estrada-Mondaca S, Ramírez OT (2004) Production of Recombinant Proteins: challenges and solutions. Methods Mol Biol 267:15–52
11. Aricescu AR, Assenberg R, Bill RM, Busso D, Chang VT, Davis SJ, Dubrovsky A, Gustafsson L, Hedfalk K, Heinemann U, Jones IM, Ksiazek D, Lang C, Maskos K, Messerschmidt A, Macieira S, Peleg Y, Perrakis A, Poterszman A, Schneider G, Sixma TK, Sussman JL, Sutton G, Tarboureich N, Zeev-Ben-Mordehai T, Jones EY (2006) Eukaryotic expression: developments for structural proteomics. Acta Crystallogr D Biol Crystallogr 62:1114–1124
12. Endo Y, Sawasaki T (2006) Cell-free expression systems for eukaryotic protein production. Curr Opin Biotechnol 17:373–380
13. Studier FW (2005) Protein production by auto-induction in high density shaking cultures. Protein Expr Purif 41:207–234
14. Sivashanmugam A, Murray V, Cui C, Zhang Y, Wang J, Li Q (2009) Practical protocols for production of very high yields of recombinant proteins using *Escherichia coli*. Protein Sci 18:936–948
15. Welch M, Villalobos A, Gustafsson C, Minshull J (2009) You're one in a googol: optimizing genes for protein expression. J R Soc Interface 6(Suppl 4):S467–S476
16. Kane JF (1995) Effects of rare codon clusters on high-level expression of heterologous proteins in *Escherichia coli*. Curr Opin Biotechnol 6:494–500
17. Chen GF, Inouye M (1990) Suppression of the negative effect of minor arginine codons on gene expression; preferential usage of minor codons within the first 25 codons of the *Escherichia coli* genes. Nucleic Acids Res 18: 1465–1473
18. Fuhrmann M, Hausherr A, Ferbitz L, Schodl T, Heitzer M, Hegemann P (2004) Monitoring dynamic expression of nuclear genes in *Chlamydomonas reinhardtii* by using a synthetic luciferase reporter gene. Plant Mol Biol 55:869–881
19. Sharp PM, Li WH (1987) The codon Adaptation Index–a measure of directional synonymous codon usage bias, and its potential applications. Nucleic Acids Res 15:1281–1295
20. Kane JF, Violand BN, Curran DF, Staten NR, Duffin KL, Bogosian G (1992) Novel in-frame two codon translational hop during synthesis of bovine placental lactogen in a recombinant strain of *Escherichia coli*. Nucleic Acids Res 20:6707–6712
21. Puigbo P, Guzman E, Romeu A, Garcia-Vallve S (2007) OPTIMIZER: a web server for optimizing the codon usage of DNA sequences. Nucleic Acids Res 35:126–131
22. Dieci G, Bottarelli L, Ballabeni A, Ottonello S (2000) tRNA-assisted overproduction of eukaryotic ribosomal proteins. Protein Expr Purif 18:346–354
23. Yuan L, Kurek I, English J, Keenan R (2005) Laboratory-directed protein evolution. Microbiol Mol Biol Rev 69:373–392
24. Shimada A (1996) PCR-Based Site-Directed Mutagenesis Vol. 57, Methods Mol Med 57: 157–165
25. Xu Z, Colosimo A, Gruenert DC (2003) Site-directed mutagenesis using the megaprimer method. Methods Mol Biol 235:203–207
26. Dominy CN, Andrews DW (2003) Site-directed mutagenesis by inverse PCR. Methods Mol Biol 235:209–223
27. Lee J, Shin MK, Ryu DK, Kim S, Ryu WS (2010) Insertion and deletion mutagenesis by overlap extension PCR. Methods Mol Biol 634:137–146
28. Sambrook J (2000) Molecular cloning: A laboratory manual. CSH press, America
29. Lewis MK, Thompson DV (1990) Efficient site directed in vitro mutagenesis using ampicillin selection. Nucleic Acids Res 18:3439–3443
30. Deng WP, Nickoloff JA (1992) Site-directed mutagenesis of virtually any plasmid by eliminating a unique site. Anal Biochem 200:81–88
31. Kunkel TA (1985) Rapid and efficient site-specific mutagenesis without phenotypic selection. Proc Natl Acad Sci USA 82:488–492
32. Taylor JW, Ott J, Eckstein F (1985) The rapid generation of oligonucleotide-directed mutations at high frequency using phosphorothioate-modified DNA. Nucleic Acids Res 13: 8765–8785

33. Klock HE, Koesema EJ, Knuth MW, Lesley SA (2008) Combining the polymerase incomplete primer extension method for cloning and mutagenesis with microscreening to accelerate structural genomics efforts. Proteins 71:982–994
34. Graslund S, Nordlund P, Weigelt J, Hallberg BM, Bray J, Gileadi O, Knapp S, Oppermann U, Arrowsmith C, Hui R, Ming J, dhe-Paganon S, Park HW, Savchenko A, Yee A, Edwards A, Vincentelli R, Cambillau C, Kim R, Kim SH, Rao Z, Shi Y, Terwilliger TC, Kim CY, Hung LW, Waldo GS, Peleg Y, Albeck S, Unger T, Dym O, Prilusky J, Sussman JL, Stevens RC, Lesley SA, Wilson IA, Joachimiak A, Collart F, Dementieva I, Donnelly MI, Eschenfeldt WH, Kim Y, Stols L, Wu R, Zhou M, Burley SK, Emtage JS, Sauder JM, Thompson D, Bain K, Luz J, Gheyi T, Zhang F, Atwell S, Almo SC, Bonanno JB, Fiser A, Swaminathan S, Studier FW, Chance MR, Sali A, Acton TB, Xiao R, Zhao L, Ma LC, Hunt JF, Tong L, Cunningham K, Inouye M, Anderson S, Janjua H, Shastry R, Ho CK, Wang D, Wang H, Jiang M, Montelione GT, Stuart DI, Owens RJ, Daenke S, Schutz A, Heinemann U, Yokoyama S, Bussow K, Gunsalus KC (2008) Protein production and purification. Nat Methods 5:135–146
35. Altschul SF, Madden TL, Schaffer AA, Zhang J, Zhang Z, Miller W, Lipman DJ (1997) Gapped BLAST and PSI-BLAST: a new generation of protein database search programs. Nucleic Acids Res 25:3389–3402
36. Finn RD, Tate J, Mistry J, Coggill PC, Sammut SJ, Hotz HR, Ceric G, Forslund K, Eddy SR, Sonnhammer EL, Bateman A (2008) The Pfam protein families database. Nucleic Acids Res 36:D281–D288
37. Apweiler R, Bairoch A, Wu CH, Barker WC, Boeckmann B, Ferro S, Gasteiger E, Huang H, Lopez R, Magrane M, Martin MJ, Natale DA, O'Donovan C, Redaschi N, Yeh LS (2004) UniProt: the Universal Protein knowledgebase. Nucleic Acids Res 32:D115–D119
38. Kanehisa M, Goto S, Kawashima S, Nakaya A (2002) The KEGG databases at GenomeNet. Nucleic Acids Res 30:42–46
39. Edgar RC, Batzoglou S (2006) Multiple sequence alignment. Curr Opin Struct Biol 16:368–373
40. Berman HM, Westbrook J, Feng Z, Gilliland G, Bhat TN, Weissig H, Shindyalov IN, Bourne PE (2000) The Protein Data Bank. Nucleic Acids Res 28:235–242
41. Kiefer F, Arnold K, Kunzli M, Bordoli L, Schwede T (2009) The SWISS-MODEL Repository and associated resources. Nucleic Acids Res 37:D387–D392
42. Linding R, Russell RB, Neduva V, Gibson TJ (2003) GlobPlot: Exploring protein sequences for globularity and disorder. Nucleic Acids Res 31:3701–3708
43. Gao X, Bain K, Bonanno JB, Buchanan M, Henderson D, Lorime D, Marsh C, Reynes JA, Sauder JM, Schwinn K, Thai C, Burley SK (2005) High-throughput limited proteolysis/mass spectrometry for protein domain elucidation. J Struct Funct Genomics 6:129–134
44. Hamuro Y, Coales SJ, Southern MR, Nemeth-Cawley JF, Stranz DD, Griffin PR (2003) Rapid analysis of protein structure and dynamics by hydrogen/deuterium exchange mass spectrometry. J Biomol Tech 14:171–182
45. Farr GW, Furtak K, Rowland MB, Ranson NA, Saibil HR, Kirchhausen T, Horwich AL (2000) Multivalent binding of nonnative substrate proteins by the chaperonin GroEL. Cell 100:561–573
46. Robinson CR, Sauer RT (1998) Optimizing the stability of single-chain proteins by linker length and composition mutagenesis. Proc Natl Acad Sci USA 95:5929–5934
47. Volkel T, Korn T, Bach M, Muller R, Kontermann RE (2001) Optimized linker sequences for the expression of monomeric and dimeric bispecific single-chain diabodies. Protein Eng 14:815–823
48. Sobti M, Cubeddu L, Haynes PA, Mabbutt BC (2010) Engineered rings of mixed yeast Lsm proteins show differential interactions with translation factors and U-rich RNA. Biochemistry 49:2335–2345
49. Wurch T, Lestienne F, Pauwels PJ (1998) A modified overlap extension PCR method to create chimeric genes in the absence of restriction enzymes. Biotechnol Tech 12:653–657
50. Hearn MT, Acosta D (2001) Applications of novel affinity cassette methods: use of peptide fusion handles for the purification of recombinant proteins. J Mol Recognit 14:323–369
51. Terpe K (2003) Overview of tag protein fusions: from molecular and biochemical fundamentals to commercial systems. Appl Microbiol Biotechnol 60:523–533
52. Uhlen M, Nilsson B, Guss B, Lindberg M, Gatenbeck S, Philipson L (1983) Gene fusion vectors based on the gene for staphylococcal protein A. Gene 23:369–378
53. Walls D, Loughran ST (2011) Tagging recombinant proteins to enhance solubility and aid purification. Methods Mol Biol 681:151–175
54. Rais-Beghdadi C, Roggero MA, Fasel N, Reymond CD (1998) Purification of recombinant proteins by chemical removal of the affinity tag. Appl Biochem Biotechnol 74:95–103
55. Walker PA, Leong LE, Ng PW, Tan SH, Waller S, Murphy D, Porter AG (1994) Efficient and rapid affinity purification of proteins using recombinant fusion proteases. Biotechnology (N Y) 12:601–605
56. Hammarstrom M, Hellgren N, van Den Berg S, Berglund H, Hard T (2002) Rapid screening for improved solubility of small human proteins

produced as fusion proteins in *Escherichia coli*. Protein Sci 11:313–321
57. Drew DE, von Heijne G, Nordlund P, de Gier JW (2001) Green fluorescent protein as an indicator to monitor membrane protein overexpression in *Escherichia coli*. FEBS Lett 507:220–224
58. Korepanova A, Moore JD, Nguyen HB, Hua Y, Cross TA, Gao F (2007) Expression of membrane proteins from *Mycobacterium tuberculosis* in *Escherichia coli* as fusions with maltose binding protein. Protein Expr Purif 53:24–30
59. Roosild TP, Greenwald J, Vega M, Castronovo S, Riek R, Choe S (2005) NMR structure of Mistic, a membrane-integrating protein for membrane protein expression. Science 307:1317–1321
60. Loughran ST, Walls D (2011) Purification of Poly-Histidine-Tagged Proteins. Methods Mol Biol 681:311–335
61. Nilsson J, Larsson M, Stahl S, Nygren PA, Uhlen M (1996) Multiple affinity domains for the detection, purification and immobilization of recombinant proteins. J Mol Recognit 9:585–594
62. Prinz B, Schultchen J, Rydzewski R, Holz C, Boettner M, Stahl U, Lang C (2004) Establishing a versatile fermentation and purification procedure for human proteins expressed in the yeasts *Saccharomyces cerevisiae* and *Pichia pastoris* for structural genomics. J Struct Funct Genomics 5:29–44
63. Gingras AC, Gstaiger M, Raught B, Aebersold R (2007) Analysis of protein complexes using mass spectrometry. Nat Rev Mol Cell Biol 8:645–654
64. Arnau J, Lauritzen C, Petersen GE, Pedersen J (2006) Current strategies for the use of affinity tags and tag removal for the purification of recombinant proteins. Protein Expr Purif 48:1–13
65. Charlton A, Zachariou M (2011) Tag removal by site-specific cleavage of recombinant fusion proteins. Methods Mol Biol 681:349–367
66. Chong S, Mersha FB, Comb DG, Scott ME, Landry D, Vence LM, Perler FB, Benner J, Kucera RB, Hirvonen CA, Pelletier JJ, Paulus H, Xu MQ (1997) Single-column purification of free recombinant proteins using a self-cleavable affinity tag derived from a protein splicing element. Gene 192:271–281
67. Wang HM, Shih YP, Hu SM, Lo WT, Lin HM, Ding SS, Liao HC, Liang PH (2009) Parallel gene cloning and protein production in multiple expression systems. Biotechnol Prog 25: 1582–1586
68. Eschenfeldt WH, Lucy S, Millard CS, Joachimiak A, Mark ID (2009) A family of LIC vectors for high-throughput cloning and purification of proteins. Methods Mol Biol 498:105–115
69. Sievert V, Ergin A, Bussow K (2008) High throughput cloning with restriction enzymes. Methods Mol Biol 426:163–173
70. Aslanidis C, de Jong PJ (1990) Ligation-independent cloning of PCR products (LIC-PCR). Nucleic Acids Res 18:6069–6074
71. Abdullah JM, Joachimiak A, Collart FR (2009) "System 48" high-throughput cloning and protein expression analysis. Methods Mol Biol 498:117–127
72. Widersten M (1998) Heterologous expression in *Escherichia coli* of soluble active-site random mutants of haloalkane dehalogenase from *Xanthobacter autotrophicus* GJ10 by co-expression of molecular chaperonins GroEL/ES. Protein Expr Purif 13:389–395
73. Wakagi T, Oshima T, Imamura H, Matsuzawa H (1998) Cloning of the gene for inorganic pyrophosphatase from a thermoacidophilic archaeon, *Sulfolobus* sp. strain 7, and overproduction of the enzyme by co-expression of tRNA for arginine rare codon. Biosci Biotechnol Biochem 62:2408–2414
74. Kerrigan JJ, Xie Q, Ames RS, Lu Q (2011) Production of protein complexes via co-expression. Protein Expr Purif 75:1–14
75. Fribourg S, Romier C, Werten S, Gangloff YG, Poterszman A, Moras D (2001) Dissecting the interaction network of multi-protein complexes by pairwise co-expression of subunits in *E. coli*. J Mol Biol 306:363–373
76. Johnston K, Clements A, Venkataramani RN, Trievel RC, Marmorstein R (2000) Co-expression of proteins in bacteria using T7-based expression plasmids: expression of heteromeric cell-cycle and transcriptional regulatory complexes. Protein Expr Purif 20:435–443
77. Held D, Yaeger K, Novy R (2003) New co-expression vectors for expanded compatibilities in *E. coli*. Innovations 18:4–6
78. Novy R, Yaeger K, Held D, Mierendorf R (2002) Co-expression of multiple target proteins in *E. coli*. Innovations 15:2–6

Chapter 15

Construction and Analysis of Randomized Protein-Encoding Libraries Using Error-Prone PCR

Paulina Hanson-Manful and Wayne M. Patrick

Abstract

In contrast to site-directed mutagenesis and rational design, directed evolution harnesses Darwinian principles to identify proteins with new or improved properties. The critical first steps in a directed evolution experiment are as follows: (a) to introduce random diversity into the gene of interest and (b) to capture that diversity by cloning the resulting population of molecules into a suitable expression vector, en bloc. Error-prone PCR (epPCR) is a common method for introducing random mutations into a gene. In this chapter, we describe detailed protocols for epPCR and for the construction of large, maximally diverse libraries of cloned variants. We also describe the utility of an online program, PEDEL-AA, for analyzing the compositions of epPCR libraries. The methods described here were used to construct several libraries in our laboratory. A side-by-side comparison of the results is used to show that, ultimately, epPCR is a highly stochastic process.

Key words Directed evolution, Random mutagenesis, Error-prone PCR, GeneMorph II, Mutazyme II DNA polymerase, Library, Mutation spectrum, Mutational bias, PEDEL-AA

1 Introduction

In the past two decades, directed evolution has emerged as a powerful method for altering the properties of proteins. It involves mimicking the process of Darwinian evolution on a single gene, on a laboratory timescale. In the first step of a directed evolution experiment, mutations are introduced at random into copies of the target gene, resulting in a large and diverse library of variants (typically 10^3–10^9 clones). The members of this library are subjected to a suitably high-throughput screen or genetic selection, in order to identify rare variants with improvements in the desired property. Multiple rounds of mutagenesis and screening/selection enable the accumulation of beneficial mutations that may have been impossible to predict, a priori. Directed evolution has been adopted widely for tailoring industrially relevant biocatalysts with improvements in properties such as substrate specificity, enantioselectivity, and

Juliet A. Gerrard (ed.), *Protein Nanotechnology: Protocols, Instrumentation, and Applications*, Methods in Molecular Biology, vol. 996, DOI 10.1007/978-1-62703-354-1_15, © Springer Science+Business Media New York 2013

thermostability (1, 2). It has also been used to address fundamental questions about protein structure, function, and evolution (3).

Many methods have been developed for introducing molecular diversity into parent sequences (4, 5). One of these methods, error-prone polymerase chain reaction (epPCR), remains a particularly common means of generating random mutations at any position in the target gene. Conceptually, epPCR is simple: the target gene is amplified exponentially, under conditions in which the fidelity of the polymerase is reduced. Cloning the randomly mutagenized PCR product into an appropriate expression vector yields a library of variants that can be used in downstream screening or selection. While outside the scope of this chapter, it is worth noting that these downstream steps are also critically important for the success of any directed evolution experiment (6).

The original—and still the cheapest—way to carry out an epPCR is to reduce the fidelity of *Taq* DNA polymerase, by adding Mn^{2+} ions and unbalanced ratios of dNTPs (7, 8). Detailed protocols for using *Taq* polymerase to construct epPCR libraries have been described previously (9, 10). However, *Taq*-generated epPCR libraries suffer from biases in the types of mutations that are observed; in particular, mutations at A:T base pairs are massively overrepresented (11).

We have argued that an unbiased and maximally diverse library has the highest probability of containing variants with the desired function (12). The GeneMorph II Random Mutagenesis Kit from Agilent Technologies (http://tinyurl.com/3mb6x66) is designed specifically for the construction of unbiased epPCR libraries. In the protocols below, we describe the use of the GeneMorph II kit, as well as the subsequent steps that are required to construct a library. We illustrate the protocols with data from a library we have recently constructed. We also describe the analyses that we routinely perform to assess the compositions of our libraries. Our target for randomization was the *Escherichia coli cynT* gene, which was identified in a previous experiment because of its contribution to antibiotic resistance when it was over-expressed (13). We conclude the chapter by comparing the *cynT* epPCR library with two other libraries and with the results that are predicted by the manufacturer of the GeneMorph kit (Agilent). This analysis highlights the stochastic nature of epPCR.

2 Materials

2.1 Error-Prone PCR

1. Plasmid containing the gene that is to be amplified by epPCR.
2. Spectrophotometer and cuvettes for measuring DNA concentration, e.g., an Eppendorf Biophotometer and UVettes.
3. Oligonucleotide primers for the epPCR amplification (see Note 1).

4. GeneMorph II Random Mutagenesis Kit (Agilent). The kit contains Mutazyme II DNA polymerase (2.5 U/μL), 10× Mutazyme II reaction buffer, and a dNTP mix (10 mM each dNTP).
5. Thermocycler with a heated lid.
6. Agarose gels, stained with ethidium bromide at 0.5 μg/mL.
7. DNA ladder with bands that contain known amounts of DNA.
8. Apparatus for agarose gel electrophoresis.
9. QiaQuick PCR Purification Kit (Qiagen). Equivalent kits from other manufacturers are also suitable.

2.2 Vector and Insert Preparation

1. Protein expression vector, into which the epPCR product will be cloned (see Note 2).
2. QiaPrep Spin Miniprep Kit (Qiagen). Equivalent kits from other manufacturers are also suitable.
3. Restriction enzyme(s) that facilitate directional, sticky-ended cloning (e.g., SfiI from New England Biolabs).
4. Restriction enzyme that cuts within the expression vector's stuffer fragment (see Note 3).
5. Restriction enzyme DpnI (New England Biolabs).
6. Agarose gels stained with 1× SYBR Safe DNA gel stain (Invitrogen).
7. Safe Imager 2.0 Blue-Light Transilluminator (Invitrogen).
8. Clean razor blades for excising bands from gels.
9. MinElute Gel Extraction Kit (Qiagen).

2.3 Preparation of a Test Library

1. T4 DNA ligase and ligation buffer. We obtain comparable results with the T4 DNA ligases from New England Biolabs, Fermentas, and Enzymatics Inc.
2. Aliquots (50 μL) of electrocompetent *E. coli* cells (see Note 4).
3. Gene Pulser electroporation cuvettes (BioRad, 0.2 cm electrode gap).
4. Gene Pulser electroporation unit with Pulse Controller (BioRad).
5. Sterile SOC medium: 20 g/L tryptone; 5 g/L yeast extract; 10 mM NaCl; 2.5 mM KCl; 20 mM glucose.
6. LB-agar plates containing the correct antibiotic for selecting plasmid-containing cells.

2.4 Analysis of Library Composition

1. Thermocycler with a heated lid (e.g., an MJ Mini from BioRad).
2. Primers for amplifying cloned inserts from the epPCR library (see Note 1).

3. Reagents for a standard PCR screen. While there are many alternate (and equally good) suppliers, we routinely use 5× Green GoTaq Reaction Buffer (Promega), i-Taq DNA polymerase (iNtRON Biotechnology), and the dNTP mix that is supplied with the polymerase (which contains 2.5 mM of each dNTP).
4. Ethidium bromide-stained agarose gels and electrophoresis apparatus.
5. QiaQuick PCR Purification Kit (Qiagen), or equivalent.

2.5 Construction and Storage of the Full-Sized Library

1. T4 DNA ligase and ligation buffer, as listed in Subheading 2.3.
2. QiaQuick PCR Purification Kit (Qiagen), or equivalent.
3. Standard-sized LB-agar plates (circular, 85–90 mm diameter) containing the correct antibiotic for selecting plasmid-containing cells.
4. Two square bioassay dishes (245 mm × 245 mm) from Corning or Nunc. Each bioassay dish holds 200 mL of LB agar, supplemented with the appropriate antibiotic for maintaining the library vector.
5. A fresh batch of electrocompetent *E. coli* cells (see Note 4).
6. Electroporation cuvettes and apparatus, as described in Subheading 2.3.
7. SOC medium, as described in Subheading 2.3.
8. Supercoiled pUC19 control plasmid (10 pg/μL; Invitrogen).
9. Sterile 50 mL tube (Falcon or similar).
10. LB medium supplemented with the appropriate antibiotic (~20 mL, total).
11. Refrigerated centrifuge with a rotor that takes the 50 mL tube listed above (item 9).
12. Spectrophotometer and cuvettes for measuring cell density (OD_{600}), e.g., an Eppendorf Biophotometer and UVettes.
13. Sterile glycerol (50% v/v).
14. Cryogenic vials, suitable for storage at −80°C.

3 Methods

3.1 Error-Prone PCR

1. Prepare purified plasmid DNA, containing the gene that is to be mutagenized. Measure its concentration spectrophotometrically.
2. The amount of template used in the epPCR affects the mutation rate (see Note 5). Calculate the amount of plasmid DNA that is required for the desired mutation rate (see Notes 6 and 7).

3. Prepare the epPCR reagents in a thin-walled, 0.2 mL tube:

x μL	Plasmid DNA template (see Note 6)
y μL	Water to a total volume of 50 μL
5 μL	10× Mutazyme II reaction buffer
1 μL	dNTP mix (gives 200 μM of each dNTP, final concentration)
2 μl	Forward primer (from 10 μM stock solution)
2 μL	Reverse primer (from 10 μM stock solution)
1 μL	Mutazyme II DNA polymerase (2.5 U)

4. Mix the sample and place the tube in the thermocycler.
5. Run the epPCR program:

Step 1:	1 min	95°C
Step 2:	20 s	94°C
Step 3:	20 s	Annealing temperature for primers (see Note 1)
Step 4:	1 min	72°C (for a ~1 kb gene; see Note 8)
Step 5:	Repeat steps 2–4 for an additional 29 cycles	
Step 6:	2 min	72°C
Step 7:	Hold	4°C (for product storage, if necessary)

6. Run 2 μL of the product on an ethidium bromide-stained agarose gel, alongside a DNA ladder. Determine the total yield of the epPCR product by comparing the intensity of the epPCR sample with the intensity of the bands in the ladder. The total yield of epPCR product is required to calculate the PCR efficiency (see Note 9).
7. Purify the remainder of the epPCR product using the PCR Purification Kit. Elute the purified DNA from the spin column in 30 μL elution buffer (EB).
8. Estimate the concentration of the purified sample by running 1 μL on an agarose gel, alongside a suitable DNA ladder.

3.2 Vector and Insert Preparation

1. Prepare 3–4 μg of the plasmid that will be used for cloning and expression of the epPCR library (see Note 2). In the example discussed below, we used vector pCA24N (14), which was purified from a saturated overnight culture using the QiaPrep Spin Miniprep Kit (Qiagen).

2. Digest the vector (~3 μg) and epPCR product (~1 μg) to completion, with restriction enzyme(s) that introduce sticky ends and facilitate directional cloning. We typically digest pCA24N and the epPCR product with 20 U of SfiI, in total reaction volumes of 30 μL. Under these conditions, incubating for 5 h at the enzyme's optimal temperature (50°C) is generally sufficient for complete digestion.
3. For additional improvements in the quality of the final library, add fresh restriction enzymes, as follows:
   (a) Vector preparation—add 10 U of an enzyme that cuts within the stuffer fragment (see Note 3).
   (b) Insert preparation—add 10 U of DpnI, to eliminate any of the methylated, unmutated template that may have carried over from the epPCR.

   Incubate the reactions for a further 2 h at 37°C, then heat inactivate the enzymes (where possible), according to the manufacturer's guidelines.
4. Run the two reactions on separate agarose gels, with agarose concentrations that are appropriate for the fragments being resolved (e.g., 0.8% agarose for the vector and 1.2% agarose for the insert).
5. Excise the bands that correspond to the digested vector and insert. We strongly recommend the use of a blue-light transilluminator and a compatible stain (SYBR Safe), rather than ethidium bromide and a UV transilluminator, for this step. See Note 10.
6. Purify the vector and insert DNA from the excised gel bands. We use the MinElute Gel Extraction Kit (Qiagen) and elute the DNA from each spin column in 12 μL EB.
7. Determine the concentrations of the purified vector and insert DNA by running 2 μL aliquots of each on an ethidium bromide-stained agarose gel, as described above (Subheading 3.1, step 6).
8. Store the purified DNA at −20°C, as necessary.

3.3 Preparation of a Test Library (See Note 11)

1. Prepare two ligation reactions: one with vector DNA only and one with the vector and a threefold molar excess of the insert DNA. Each reaction should contain the following: 1× ligation buffer; 50 ng of vector DNA (see Subheading 3.2, step 6); T4 DNA ligase (1 U); plus or minus the insert DNA; and water to a final volume of 10 μL. Add the T4 DNA ligase last and mix gently.
2. Incubate the ligation reactions at 16°C for 16 h.
3. Use a 1 μL aliquot of each ligation reaction to transform 50 μL aliquots of *E. coli*, by electroporation. See Note 12.
4. Immediately after electroporation, add 500 μL of SOC medium to the cuvette and transfer the cells to a sterilized, capped test

tube or a 15 mL tube (Falcon or similar). Allow the cells to recover by incubating them at 37°C, with shaking, for 1 h.

5. Store the remaining 9 μL of each ligation reaction (leftover from step 3, above) at −20°C.
6. Spread aliquots (10 and 50 μL) of the two recovery cultures on LB-agar plates. Incubate the plates at 37°C for 12–16 h.
7. Count the number of colonies on each plate. Use the results from the "vector only" plates to calculate the fraction of the library (as represented on the "vector + insert" plates) that contains recircularized vector. This background must be minimized, to avoid wasting time on futile library screens. If the "vector only" background is >1% of the total library, we recommend preparing a fresh batch of the vector (Subheading 3.2, above), and lengthening the incubation time with each restriction enzyme.
8. The number of colonies on the "vector + insert" plates also allows the size of the final, scaled-up library to be estimated. The final library is likely to be ~10^3 times larger than the total number of colonies on the "vector + insert" test plates (see Note 13).

3.4 Analysis of Library Composition (See Note 14)

1. Use 2 μL pipette tips (or sterile toothpicks) to pick 10–20 colonies at random from the "vector + insert" test plates (Subheading 3.3, step 7).
2. Transfer each colony into a thin-walled, 0.2 mL tube containing 5 μL of sterile water.
3. Lyse the cells by incubating the tubes at 95°C for 5 min, in a thermocycler.
4. Amplify the randomly mutagenized gene inserts from each colony by PCR. We have listed our routine protocol, for guidance. However, many variations are possible; the goal here is merely to generate enough of the amplified product for DNA sequencing. We typically set up 25 μL PCRs in thin-walled 0.2 mL tubes, as follows:

14.75 μL	Water
5 μL	5× Green GoTaq buffer
2 μL	dNTP mix (gives 200 μM of each dNTP, final concentration)
1 μl	Forward primer (from 10 μM stock solution)
1 μL	Reverse primer (from 10 μM stock solution)
0.25 μL	*Taq* DNA polymerase (1.25 U)
1 μL	Cell lysate (from step 2, above)

Table 1
Matrix of point mutations identified in 16 *cynT* variants

		Mutation To			
		T	C	A	G
Mutation From	T	–	15	16	6
	C	10	–	8	0
	A	18	3	–	13
	G	10	4	18	–

4. Mix each sample and place the tubes in the thermocycler.
5. Run an appropriate PCR program, such as the one listed in Subheading 3.1, step 5.
6. Run a 2 μL aliquot of each PCR product on an agarose gel, to confirm successful amplification.
7. Purify the remainder of each PCR product using the QiaQuick PCR Purification Kit (or equivalent). Elute the purified DNA from each spin column in 30 μL EB.
8. Sequence each PCR product. Use the forward and/or reverse primers from the PCR as the sequencing primer(s), as necessary.
9. Align the sequence of each PCR product with the known sequence of the unmutated parental gene. Computer programs such as MacVector are useful for this analysis.
10. Tabulate all of the point mutations in the sequenced samples. Also note any insertions or deletions that may have arisen during the epPCR. The point mutations should be grouped by type. For example, we randomized the *E. coli cynT* gene and sequenced 16 clones from the resulting test library. The 121 point mutations that we identified in the 16 variants are summarized in Table 1. There were also two deletions and one insertion in the data set; in total, the sequencing revealed 124 mutations.
11. Use the tabulated data to calculate the overall mutation rate and to assess biases in the mutation spectrum of the epPCR library. There are three key indicators of bias (see Note 15): (a) the ratio of transition (Ts) to transversion (Tv) mutations; (b) the ratio of AT→GC transitions to GC→AT transitions; and (c) the frequency of mutations at A:T base pairs, to mutations at G:C base pairs. The mutation rate and bias measures for our *cynT* epPCR library are shown in Table 2.

Table 2
Mutational spectrum of the *cynT* epPCR library

Type(s) of mutations	Frequency	Proportion of total
Transitions		
A→G, T→C	28	22.6%
G→A, C→T	28	22.6%
Transversions		
A→T, T→A	34	27.4%
A→C, T→G	9	7.3%
G→C, C→G	4	3.2%
G→T, C→A	18	14.5%
Insertions and deletions		
Insertions	1	0.8%
Deletions	2	1.6%
Summary of bias		
Transitions/transversions	0.86	NA[a]
AT→GC/GC→AT	1	NA[a]
A→N, T→N	71	57.3%
G→N, C→N	50	40.3%
Mutation rate		
Mutations per kb	11.8	NA[a]
Mutations per *cynT* gene[b]	7.8	NA[a]

[a]NA: not applicable
[b]The cloned *cynT* insert was 657 bp

12. The library analysis program PEDEL-AA (15), available online at http://guinevere.otago.ac.nz/stats.html, should now be used to predict the utility of the final epPCR library. PEDEL-AA has an easy-to-use web interface and takes the following parameters as its inputs:
    (a) The sequence of the gene that was randomized
    (b) The estimated size of the scaled-up library (Subheading 3.3, step 8)
    (c) The nucleotide mutation matrix (Subheading 3.4, step 10; see Table 1 for an example)
    (d) The mean number of mutations per gene in the library (Subheading 3.4, step 11; see Table 2 for an example)

Table 3
PEDEL-AA outputs for the *cynT* epPCR library

Property	Estimate
Total library size	1.4×10^7
Number of variants with no insertions, deletions, or stop codons	9.0×10^6
Mean number of amino acid substitutions per variant	5.5
Unmutated (wild-type) sequences (% of total library)	3.0%
Number of distinct, full-length proteins in the library[a]	7.4×10^6

[a]Calculated using the PCR efficiency parameter. PEDEL-AA also calculates a less accurate estimate the number of distinct, full-length proteins in the library by using the simplifying assumption of Poisson statistics

(e) The number of cycles in the epPCR (Subheading 3.1, step 3)

(f) The PCR efficiency parameter for the epPCR (see Note 9)

(g) The mean number of insertions per gene in the library (Subheading 3.4, step 10)

(h) The mean number of deletions per gene in the library (Subheading 3.4, step 10)

The program outputs a variety of statistics about the protein variants that are encoded by the epPCR library. A selection of these statistics, calculated for our *cynT* library, is shown in Table 3. Together, the data in Tables 2 and 3 allow an informed decision to be made about whether to scale up the library (or whether to start over, with different epPCR and ligation conditions).

3.5 Construction and Storage of the Full-Sized Library

1. Prepare "vector only" and "vector + insert" ligation reactions that are tenfold larger than those described in Subheading 3.3, step 1 (see Note 16). Each reaction should contain the following: 1× ligation buffer; 500 ng of vector DNA; T4 DNA ligase (10 U); plus or minus insert DNA (3-fold molar excess over vector); and water to a final volume of 100 μL.
2. Incubate the ligation reactions at 16°C for 16 h.
3. Add the remaining 9 μL of each test ligation (Subheading 3.3, step 5) to the scaled-up "vector only" and "vector + insert" ligation reactions.
4. Purify the products from each ligation reaction using the QiaQuick PCR Purification Kit (or equivalent). Elute the purified DNA from each spin column in 42 μL EB.
5. Prepare the LB-agar plates on which the transformed cells of the library will be spread. For each library, we typically use two 245 mm × 245 mm square bioassay dishes (see Note 17).

6. Prepare a fresh batch of electrocompetent *E. coli* cells (see Note 4). In our hands, the transformation efficiencies of freshly prepared cells are four to fivefold higher than cells that have undergone a freeze/thaw cycle.
7. Add 3 μL aliquots of the "vector + insert" ligation to 14 × 50 μL aliquots of electrocompetent *E. coli* cells. Transform each aliquot, and recover the transformed cells, as described previously (Subheading 3.3, steps 3 and 4).
8. Transform a single 50 μL aliquot of cells with 3 μL of the "vector only" ligation.
9. Transform one more 50 μL aliquot of the electrocompetent *E. coli* with an appropriate plasmid for determining the transformation efficiency of the cells. We routinely use 10 pg of supercoiled pUC19.
10. Pool all of the cells that were transformed with the "vector + insert" ligation, in a sterile 15 mL tube. The total volume should be 7.7 mL (14 electroporations; 550 μL per recovery culture).
11. Mix the cells briefly, by inverting the tube 2–3 times.
12. Spread 1, 5, and 25 μL aliquots on regular LB-agar plates (diluting as necessary to obtain a spreadable volume).
13. Spread the remainder of the library on the two large plates (see step 5, above); ~3.85 mL per plate.
14. Spread 1, 5, and 25 μL aliquots of the "vector only" control on regular LB-agar plates.
15. Spread suitable aliquots (typically 2 and 10 μL) of the cells transformed with the pUC19 control on LB-agar plates that contain ampicillin (100 μg/mL).
16. Incubate all of the dilution and control plates at 37°C for 16 h. Incubate the two large library plates at 30°C, to avoid the formation of a confluent lawn.
17. Count the number of colonies on each plate, except for the large library plates (which should be covered in dense lawns of small colonies).
18. Use the pUC19 control to calculate the transformation efficiency of your electrocompetent cells. The easiest way to increase the size of an epPCR library is to improve the transformation efficiency of the cells. A good batch of *E. coli* cells should yield >10^9 colonies per microgram of pUC19 used in the transformation.
19. The regular LB-agar plates with aliquots of the library and the "vector only" control should be used to estimate the final size of the library, and to verify that the "vector only" background is <1% of the total library (see Subheading 3.3, step 7).

20. The large library plates should each be covered in thousands (or millions) of small colonies. To recover the library from one of the plates, pipette 4 mL of LB medium (supplemented with the appropriate antibiotic) into the center of the plate.
21. Use a glass spreader to scrape cells off the surface of the plate; pool them in one corner of the plate.
22. Remove the resuspended cells with a P1000 pipettor, and transfer them to a sterile 50 mL tube.
23. Pipette a second 4 mL aliquot of LB onto the same plate, and repeat the scraping step.
24. Use another 2 × 4 mL aliquots of LB to recover the cells from the second library plate.
25. All of the recovered cells should be pooled in the same 50 mL tube.
26. Break up cell clumps and mix well, by pipetting up and down repeatedly with a P1000 pipettor.
27. Pellet the cells by centrifugation at 3,000 × *g*, 4°C, for 15 min. Use a P1000 pipettor to remove the supernatant.
28. Resuspend the cell pellet in 1 mL of LB medium, plus antibiotic. The pellet is likely to be large; resuspension is likely to involve vigorous pipetting and/or gentle vortexing.
29. Knowing the cell density is likely to be useful for planning downstream screening/selection experiments. Mix 1 μL of the cell suspension with 999 μL of sterile water, and measure the OD_{600} (against a water blank). The final OD_{600} of the resuspended library is usually >100, corresponding to a cell density of $>2.5 \times 10^{10}$ cells per milliliter (see Note 18).
30. Split the library into 100 μL aliquots and transfer each aliquot to a cryogenic vial.
31. Add 50 μL of sterile glycerol (50% v/v) to each aliquot and mix well by pipetting.
32. Store the aliquots at −80°C until you are ready to proceed with screening/selection to identify improved variants in the library.

3.6 Summarizing Stochasticity in epPCR Library Construction

1. Tables 1 and 2 show that the *cynT* epPCR library, constructed with the GeneMorph II kit, is not free of mutational bias. For example, G→C and C→G mutations occur much less frequently than A→T and T→A mutations (Table 2)—indicating that the Mutazyme II polymerase retains some of the bias of *Taq* polymerase. For reference, we have included a side-by-side comparison of the *cynT* data with two other epPCR libraries that were constructed using the same protocol, together with the guidelines from the manufacturer (Table 4). All three randomized genes (*cynT*, *ydfW*, and *yeaD*) are latent contributors to antibiotic resistance (13).

Table 4
Comparison of three epPCR libraries with Agilent's product guidelines

	cynT	*ydfW*	*yeaD*	[Agilent][a]
epPCR details				
Gene length	657 bp	147 bp	882 bp	NA[b]
Plasmid in the epPCR	90 ng	90 ng	90 ng	NA[b]
Amount of target DNA[c]	15 ng	7 ng	18 ng	0.1–1,000 ng
Library bias indicators				
Transitions/transversions	0.86	0.56	0.56	0.90
AT→GC/GC→AT	1	1.50	1.17	0.60
A→N, T→N	57.3%	71.4%	61.1%	50.7%
G→N, C→N	40.3%	28.6%	38.9%	43.8%
Mutation rates				
Mutations per kb	11.8	9.5	4.1	0–16
Mutations per gene	7.8	1.4	3.6	NA[b]

[a]Source: Tables 1 and 2 of the GeneMorph II Random Mutagenesis Kit manual, available for download from http://tinyurl.com/3mb6x66
[b]NA: not applicable
[c]Calculated as described in Note 6

2. The data in Table 15.4 highlight the stochasticity that is inherent in epPCR. No two libraries are identical. While the GeneMorph II kit introduces less mutational bias than other epPCR methods, variation in mutational spectra is still to be expected. Mutation rates are also variable, and the recommendations given here (see Note 6) should be considered a rough guide only.
3. Bearing these facts in mind, an epPCR practitioner should be prepared to construct several libraries, combining the analyses that we have described here with considerable trial and error!

4 Notes

1. Primers should be noncomplementary and should have melting temperatures that are within 5°C of each other. Melting temperatures can be estimated accurately using the OligoAnalyzer tool from Integrated DNA Technologies: http://www.idtdna.com/analyzer/Applications/OligoAnalyzer/. The optimal annealing temperature to use in a PCR is typically 3–5°C cooler than the lowest primer melting

temperature. We routinely resuspend lyophilized primers in TE buffer (10 mM Tris, 1 mM EDTA, pH 8.0), to a concentration of 100 μM. Working stocks (10 μM) are made by tenfold dilution of these master stocks, using sterile water.

2. A high-quality vector preparation is critical for constructing a large library. We find it useful to use a plasmid with a stuffer fragment in the cloning cassette. Excision of this stuffer fragment allows the progress of the restriction digestion to be monitored. It also ensures that the doubly digested vector (with stuffer fragment removed) can be resolved from undigested and singly digested material on an agarose gel.

3. Using a restriction enzyme that cuts within the stuffer fragment minimizes the number of "vector only" clones in the final library. We routinely use BglII (New England Biolabs) for this purpose.

4. The choice of *E. coli* strain will depend on the downstream selection or screen that is being employed. In general, the final size of the epPCR library is directly proportional to the transformation efficiency of the host strain. Therefore, strains with high transformation efficiencies (such as *E. coli* DH5α-E) are preferable. We prepare electrocompetent cells according to the method of Hanahan (16).

5. The overall mutation frequency depends on the error rate of the polymerase and also the number of times that each template is duplicated in the reaction. If the initial amount of template is high, it will undergo few duplications in the epPCR. On the other hand, a low amount of template will result in a greater number of duplications, and more mutations will be introduced. This is discussed further in the GeneMorph II Random Mutagenesis Kit manual, available for download from http://tinyurl.com/3mb6x66.

6. Agilent recommends 500–1,000 ng of template DNA for a low mutation rate (0–4.5 mutations/kb); 100–500 ng of template for a medium mutation rate (4.5–9 mutations/kb), and 0.1–100 ng of template for a high mutation rate (9–16 mutations/kb). Note that the amount of template is not the amount of purified plasmid DNA. Instead, it is the amount of target DNA to be amplified. For example, we used 90 ng of plasmid pCA24N-*cynT* as the starting point for one of our epPCR libraries. In total, this plasmid was 5,180 bp in size. However, the amplified product (i.e., the *cynT* gene, plus flanking sequences) was 857 bp. Therefore, the amount of template DNA in the reaction was given by

$$\left(\frac{857\ \text{bp}}{5{,}180\ \text{bp}}\right) \times 90\ \text{ng} = 15\text{ng}$$

7. We typically aim for a medium-to-high mutation rate, because this generates libraries that contain minimal numbers of "wasted" variants (i.e., unmutated copies of the template, or multiple copies of variants with any one point mutation). This strategy is discussed in more detail elsewhere (12, 15).
8. We routinely use extension times that are calculated at a rate of 1 min/kb. For example, a 30 s extension time is used for a 500 bp product, and a 90 s extension time is used for a 1,500 bp product.
9. Calculating the PCR efficiency parameter allows robust statistical analysis of library composition (*see* Subheading 3.4, step 12). When the total product yield and the amount of starting template are known, the number of doublings in the PCR, *d*, can be calculated as follows:

$$d = \frac{\log(\text{Product} / \text{Template})}{\log 2}$$

The PCR efficiency (i.e., the probability that any particular sequence is duplicated in any one cycle of the PCR, *eff*) is then given by

$$eff = 2^{(d/n)} - 1$$

where n is the number of PCR cycles ($n = 30$ in our protocol). An online tool for calculating *eff*, given d and n, can be found at http://guinevere.otago.ac.nz/cgi-bin/aef/PCReff.pl.
10. It is well known that UV transillumination of ethidium bromide-stained DNA can induce damage, resulting in lower cloning and transformation efficiencies (17). Even short exposures to UV (<60 s) can have dramatic and deleterious effects. Constructing a large epPCR library (>10^6 variants) requires the highest possible quality of DNA. Therefore, we use SYBR Safe stain and a blue-light transilluminator for preparation of our library vector and epPCR insert. In our hands, this results in libraries that are 5–10 times larger than equivalent libraries prepared with ethidium bromide-stained DNA.
11. Before scaling up to a full-sized library, we find it useful and expedient to construct a test library. This allows the epPCR mutation spectrum to be determined. It also ensures that the ligation protocol is optimized for constructing a full-sized library with low "vector only" background and the maximum number of insert-containing clones.
12. Aliquots of cells should be thawed on ice. DNA is added to each 50 μl aliquot of cells and chilled on ice in a sterile Gene Pulser cuvette. Samples are electroporated at 2.5 kV, 200 Ω, and 25 μF in a Gene Pulser unit with Pulse Controller.

13. This is a rough estimate, based on the following: (a) Tenfold scale-up of the ligation reaction, (b) transforming 10–20 aliquots of electrocompetent cells, and (c) spreading all 550 μL of each recovery culture (instead of 10–50 μL aliquots).

14. Clones from the test library (Subheading 3.3) should be sequenced in order to analyze the mutation rate and the spectrum of mutations that arose in the epPCR. As discussed in Notes 5 and 6, some control over the mutation rate is possible. However, in our experience, there is considerable experiment-to-experiment variation in the outcomes of the epPCR process (see Subheading 3.6). Therefore, we recommend conducting the analyses described in Subheading 3.4, to avoid wasting time and resources on a scaled-up library that contains little molecular diversity.

15. A library with an unbiased spectrum of mutations will be maximally diverse; that is, it will have the lowest probability of duplicated variants. Therefore, it is more likely to contain at least one improved variant (12). One indicator of bias is the ratio of transitions (i.e., purine-to-purine and pyrimidine-to-pyrimidine mutations) to transversions (purine-to-pyrimidine and pyrimidine-to-purine mutations). There are four possible transitions and eight possible transversions (listed in Table 15.2). Therefore, a completely unbiased error-prone polymerase would generate libraries with transition/transversion (Ts/Tv) ratios of 0.5. Provided that the GC content of the gene is ~50%, the ratio of AT→GC transitions to GC→AT transitions (i.e., AT→GC/GC→AT) in an unbiased epPCR library should also be 1. Similarly, the number of mutations at A:T base pairs (A→N, T→N) should also be the same as the number of mutations at G:C base pairs (G→N, C→N). The effects of mutational bias on overall library composition can be assessed by altering the input parameters for PEDEL-AA analysis (Subheading 3.4, step 12).

16. When constructing the full-sized library, the focus should be on scaling everything up by as much as possible. All of the remaining epPCR insert (Subheading 3.2, step 7) should be used in a scaled-up ligation, and as many aliquots of electrocompetent *E. coli* as possible should be transformed with the ligated products. The protocol that we describe is a typical example from our laboratory.

17. The large volume and high surface area of the square bioassay dishes make them prone to "sweating" when they are incubated at 30–37°C (particularly if the LB agar is too hot when the plates are poured). They may need to be pre-warmed at 37°C for 4–6 h and/or dried in a laminar flow hood or a class II biosafety cabinet (10–15 min), before they are dry enough to use.

18. For *E. coli* strain DH5α-E, we find that $OD_{600} = 1$ corresponds to ~2.5×10^8 cells/mL.

Acknowledgment

The authors gratefully acknowledge financial support for this work from the New Zealand Marsden Fund.

References

1. Turner NJ (2009) Directed evolution drives the next generation of biocatalysts. Nat Chem Biol 5:567–573
2. Jäckel C, Hilvert D (2010) Biocatalysts by evolution. Curr Opin Biotechnol 21:753–759
3. Peisajovich SG, Tawfik DS (2007) Protein engineers turned evolutionists. Nat Methods 4:991–994
4. Lutz S, Patrick WM (2004) Novel methods for directed evolution of enzymes: quality, not quantity. Curr Opin Biotechnol 15:291–297
5. Otten LG, Quax WJ (2005) Directed evolution: selecting today's biocatalysts. Biomol Eng 22:1–9
6. Arnold FH, Georgiou G (eds) (2003) Directed enzyme evolution: screening and selection methods. *Methods in Molecular Biology*, vol 230, Humana Press, Totowa, New Jersey.
7. Leung DW, Chen E, Goeddel DV (1989) A method for random mutagenesis of a defined DNA segment using a modified polymerase chain reaction. Technique 1:11–15
8. Cadwell RC, Joyce GF (1992) Randomization of genes by PCR mutagenesis. PCR Methods Appl 2:28–33
9. Cirino PC, Mayer KM, Umeno D (2003) Generating mutant libraries using error-prone PCR. Methods Mol Biol 231:3–9
10. McCullum EO, Williams BA, Zhang J, Chaput JC (2010) Random mutagenesis by error-prone PCR. Methods Mol Biol 634:103–109
11. Shafikhani S, Siegel RA, Ferrari E, Schellenberger V (1997) Generation of large libraries of random mutants in *Bacillus subtilis* by PCR-based plasmid multimerization. Biotechniques 23:304–310
12. Patrick WM, Firth AE, Blackburn JM (2003) User-friendly algorithms for estimating completeness and diversity in randomized protein-encoding libraries. Protein Eng 16:451–457
13. Soo VWC, Hanson-Manful P, Patrick WM (2011) Artificial gene amplification reveals an abundance of promiscuous resistance determinants in *Escherichia coli*. Proc Natl Acad Sci USA 108:1484–1489
14. Kitagawa M, Ara T, Arifuzzaman M, Ioka-Nakamichi T, Inamoto E, Toyonaga H, Mori H (2005) Complete set of ORF clones of *Escherichia coli* ASKA library (a complete set of *E. coli* K-12 ORF archive): unique resources for biological research. DNA Res 12:291–299
15. Firth AE, Patrick WM (2008) GLUE-IT and PEDEL-AA: new programmes for analyzing protein diversity in randomized libraries. Nucleic Acids Res 36:W281–W285
16. Hanahan D, Jessee J, Bloom FR (1991) Plasmid transformation of *Escherichia coli* and other bacteria. Methods Enzymol 204:63–113
17. Hartman PS (1991) Transillumination can profoundly reduce transformation frequencies. Biotechniques 11:747–748

Chapter 16

Droplets as Reaction Compartments for Protein Nanotechnology

Sean R.A. Devenish, Miriam Kaltenbach, Martin Fischlechner, and Florian Hollfelder

Abstract

Extreme miniaturization of biological and chemical reactions in pico- to nanoliter microdroplets is emerging as an experimental paradigm that enables more experiments to be carried out with much lower sample consumption, paving the way for high-throughput experiments. This review provides the protein scientist with an experimental framework for (a) formation of polydisperse droplets by emulsification or, alternatively, of monodisperse droplets using microfluidic devices; (b) construction of experimental rigs and microfluidic chips for this purpose; and (c) handling and analysis of droplets.

Key words In vitro compartmentalization, Directed evolution, Protein engineering, Microfluidics, Water-in-oil emulsion

1 Introduction

Compartmentalization is an important and widespread phenomenon that allows distinct elements to be contained in a single enveloping chamber separated from the surrounding milieu. In Nature compartmentalization in cells has allowed the emergence of complex life forms by providing a mechanism for retention of important small molecules: the cell wall contains these along with the protein machinery for their production and the DNA that codes for the proteins, thus maintaining a link between genotype and phenotype. The principle of compartmentalization can be transferred to the laboratory environment: here compartmentalization is useful for maintaining a correspondence between different chemical species (Fig. 1a). Macroscale containers—test tubes or multi-well plates—are conventional laboratory reaction compartments, but a massive scale-down from the microliter to the nano- or picoliter scale is possible through the use of water-in-oil emulsion microdroplets.

Juliet A. Gerrard (ed.), *Protein Nanotechnology: Protocols, Instrumentation, and Applications*, Methods in Molecular Biology, vol. 996, DOI 10.1007/978-1-62703-354-1_16,

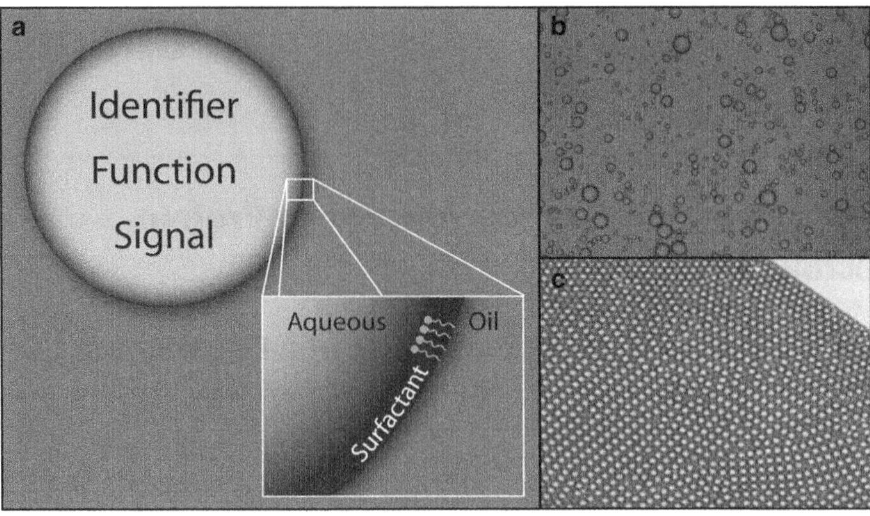

Fig. 1 Microdroplets as tools for protein research. (**a**) The aqueous compartment of a microdroplet serves to co-localize a functional protein with its identifying DNA sequence and an output signal. Photomicrographs of (**b**) polydisperse and (**c**) monodisperse droplets. Photograph (**c**) is reprinted from (30) with permission from John Wiley & Sons

Such emulsion droplets (with diameters between 10 and 50 μm) are emerging as an attractive tool for high-throughput research and are compatible with a wide range of biological and biochemical components and processes, from cell growth to DNA replication, in vitro or in vivo protein expression, and enzymatic turnover (1). They can be very rapidly formed and, due to their small size, their use requires minimal amounts of potentially valuable reagents.

Microdroplets have proven utility in the research laboratory in a variety of different applications including protein crystal growth (2), cell (3–11) or enzyme assays (12, 13), directed evolution (14–18), and DNA sequencing (19). As the uptake of experimental protocols involving droplets becomes more widespread, they will undoubtedly find even broader application than already demonstrated to date.

Generation of bulk emulsions is a batch process that is very rapid—10^{10} droplets can be formed in under 5 min—and requires little in the way of equipment, although the droplets formed using this method have a wide variation in size (Fig. 1b). Alternatively microfluidic droplets are produced in a continuous flow manner at rates lower than those of bulk emulsion, although still fast (typically $>10^6$/h), and are monodisperse (Fig. 1c, typical variation in volume of $<3\%$) (20, 21), but their use requires specialized equipment and a certain degree of technical expertise. One potentially significant advantage of producing droplets on a microfluidic platform is the possibility of combining separate steps on chip to carry out experiments requiring multiple manipulations (22).

In this chapter, we will describe both bulk emulsion and microfluidic methods for microdroplet production, including detail of a minimal equipment setup necessary for microfluidic droplet work. We hope that in this way, we will make basic droplet generation and manipulation steps readily available to a broad audience and envision that in the future, droplets or the compartmentalization principle will find use in many more areas of science.

2 Materials

2.1 Bulk Emulsion Droplet Generation

1. In vitro transcription/translation (IVTT) mix (e.g., RTS 100 *E. coli* HY kit from 5 PRIME or RiNA, see Note 1).
2. DNA template for in vitro expression [diluted sufficiently so that the majority of droplets contain no more than one copy of DNA (see Table 1)].
3. Oil and surfactant (e.g., mineral oil mix consists of 4.5% Span 80 and 0.5% Tween 80 in mineral oil, all of which are available from Sigma).
4. Omni International tissue homogenizer and Omni Tips Clear Plastic Homogenizing Probes (7 × 110 mm).

2.2 Microfluidic Droplet Generation

2.2.1 Basic Microfluidic Droplet Rig

1. Two syringe pumps (e.g., Chemyx Classic) (three are practical if two different aqueous phases are to be mixed at the point of droplet formation).
2. An optical lens with c-mount fitting (e.g., Navitar 12× Zoom).

Table 1
Concentrations required for single occupancy in droplets of varying sizes (based on the Poisson distribution given by the formula shown in Fig. 2)

Diameter (μm)	Volume	Concentration for 0.2 occupancy[a]	Concentration for 0.1 occupancy[b]	
500	65 nL	5 aM	2.5 aM	Microfluidic droplets (500–10 μm)
100	520 pL	630 aM	320 aM	
40	34 pL	10 fM	5 aM	
20	4.2 pL	79 fM	40 fM	Bulk emulsion droplets (20–1 μm)
10	520 fL	630 fM	320 fM	
1	0.52 fL	630 pM	320 pM	

[a]Giving 82% empty, 16% singly occupied, and 1.8% multiply occupied, i.e., 90% of occupied droplets are singlets
[b]Giving 90% empty, 9% singly occupied, and 0.5% multiply occupied, i.e., 95% of occupied droplets are singlets

3. A bright illumination source such as a white light LED array (Thorlabs LIU 004).
4. Firewire c-mount camera with low shutter time (such as Allied Vision Technologies Pike F-032B, shutter time <20 μs).

2.2.2 PDMS Device Preparation

1. Patterned silicon wafer (commercially available from, for example, Gesim, www.gesim.de) or access to a fully equipped photolithography suite.
2. PDMS monomer and curing agent (Sylgard 184, Dow Corning Corporation).
3. Vacuum chamber.
4. Oven for curing.
5. Plasma chamber.
6. Biopsy punch (1 mm, Kai Medical).
7. Aquapel (consumer product, widely available).

2.2.3 Droplet Production in Devices

1. Surfactant (fluorous tri-block copolymer can be prepared as previously described (23, 24), or purchased from Raindance™ Technologies or Sphere Fluidics).
2. HFE7500 oil (3 M).
3. Washed cells for encapsulation.
4. Density matching agent such as Percoll®, OptiPrep™, or sodium alginate (all available from Sigma-Aldrich).
5. Fine Bore Polythene Tubing, 0.38 mm ID 1.09 mm OD (Portex, Smiths Medical).
6. Syringes of suitable volumes for sample and oil, typically 100 μL and 2.5 mL, respectively. Glass syringes are preferable but if these are not available then disposable plastic syringes are adequate.

2.3 Downstream Processing of Samples

1. Microscopy slides for fluorescence analysis using micrographs—we find KOVA Glasstic Slides with hemocytometry counting grid to be ideal.
2. 1*H*,1*H*,2*H*,2*H*-perfluoro-1-octanol (PFO, Sigma-Aldrich) for breaking fluorous emulsions or diethyl ether (Sigma-Aldrich) for breaking hydrocarbon emulsions.
3. Centrifuge.

3 Methods

In principle any experiment that can be performed in micro-wells or -tubes can also be carried out in microdroplets. However, it is important to be aware of several caveats: small molecules may

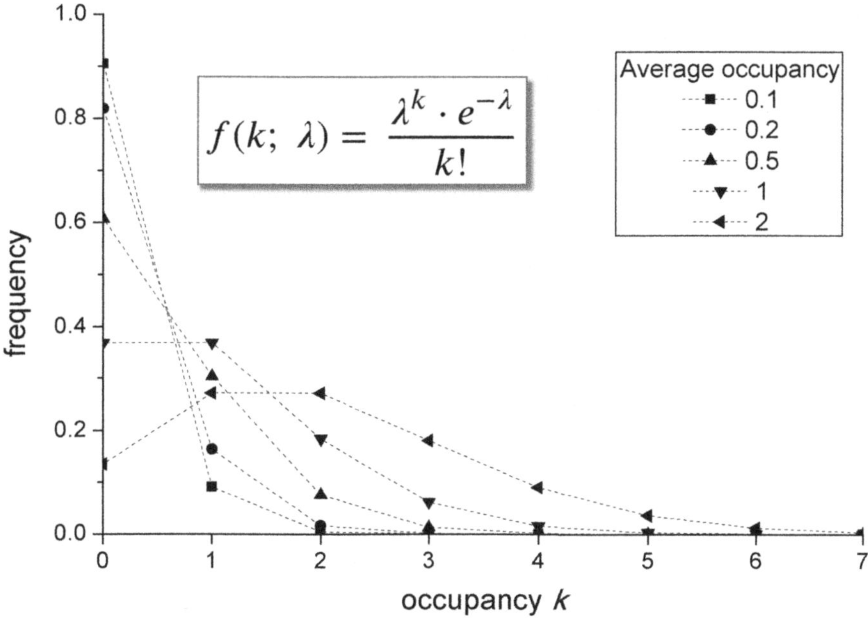

Fig. 2 Poisson distribution describing statistical frequency of different occupancies. In the Poisson equation, f is the frequency of any particular occupancy k when the average occupancy is given by λ

diffuse in and out of droplets and the interface between the aqueous and oil phases is not as inert as a plastic surface. These problems can often be addressed to some extent by the choice of surfactant and oil phase (e.g., fluorous vs. mineral oils (7, 23)) or by the addition of biomolecules such as BSA (25) or DNA (26, 27) that alter the interface character.

To express protein in microdroplets, two approaches have been employed: in the first approach, cells expressing protein are loaded into droplets (28). The protein can then be analyzed in cells (28), in the periplasm (12), displayed on yeast (14) or liberated from the cell by lysis at droplet formation (18). Alternatively, IVTT can be carried out from purified DNA template (16, 17, 26, 27, 29, 30). In either case it is important to be able to load single-protein sources into droplets, and this depends on Poisson distribution (see Fig. 2), with the target concentration defined depending on desired loading level and droplet sizes as outlined in Table 1. With bulk emulsion it is worth remembering that achieving solely singly occupied droplets is complicated by the polydispersity of droplet sizes—each twofold change in radius corresponds to an eightfold change in volume, so large droplets can easily dominate the total volume and end up multiply occupied.

In the following we will present detailed protocols for carrying out in vitro transcription/translation (IVTT) in bulk emulsion, and for producing monoclonally distributed cells in microfluidic droplets, but the droplet contents can be exchanged or altered according to need (see Note 2).

3.1 Bulk Emulsion Droplet Generation

There are a number of methods for producing bulk emulsions (17, 27, 29, 31–33), including vigorous stirring, vortexing, shaking, and extrusion. The method we describe here is very rapid and convenient, and uses only simple equipment (i.e., a tissue homogenizer).

1. Prepare the oil/surfactant mixture consisting of mineral oil (95% w/w), Span 80 (4.5% w/w), and Tween 80 (0.5% w/w).
2. Place 950 μL oil/surfactant in a 1.8 mL tube and cool on ice.
3. Add 50 μL of aqueous phase to the tube.
4. Homogenize at 5,000 rpm for 3 min ensuring the homogenizer tip is close to, but not touching, the bottom of the tube. It is advisable to keep the tube on ice during homogenization to prevent heating of the sample.
5. The emulsion is now formed and ready for further steps such as incubation to allow a reaction, such as cell-free protein expression, to take place.
6. When an experiment is carried out for the first time, it is important to estimate the droplet size. For large droplets (>3 μm) this can be conveniently done by measuring the diameter of a representative number of droplets using a light microscope (calibrated using a stage micrometer), but for smaller droplets it is necessary to use more advanced methods such as laser diffraction (34).

3.2 Microfluidic Droplet Generation

3.2.1 Basic Microfluidic Droplet Rig

The basic operations required for a microfluidic droplet platform are the ability to drive liquids into a microfluidic chip, and to be able to observe the point of droplet formation to monitor monodispersity and stability of droplet formation. There are many different manufacturers who produce suitable components, so the particular brands and models mentioned here are merely representative examples: these components work well and are familiar to the authors, but our selection should in no way deter researchers from substituting these with equivalents from other manufacturers. A minimal droplet platform is depicted in Fig. 3 and consists of two syringe pumps to form droplets from a single aqueous phase. A third pump is required if two aqueous phases are to be mixed in a ratio other than 1:1 at the point of droplet formation. The critical criterion for the selection of a syringe pump is the ability to pump smoothly even at low flow rates (i.e., <20 μL/h). Smooth flow is best provided by the smallest-capacity syringe as a small volume will result in a faster and hence smoother action of the screw drive on the pump. To observe droplet formation a computer-interfaced microscopy system is necessary. We use an optical lens coupled via a c-mount to a camera. As regards the optical setup, a camera capable of low shutter speeds (i.e., 20–50 μs) is necessary to produce images without excessive motion blur. An entry-level camera capable of recording video clips at 30 frames per second is not fast

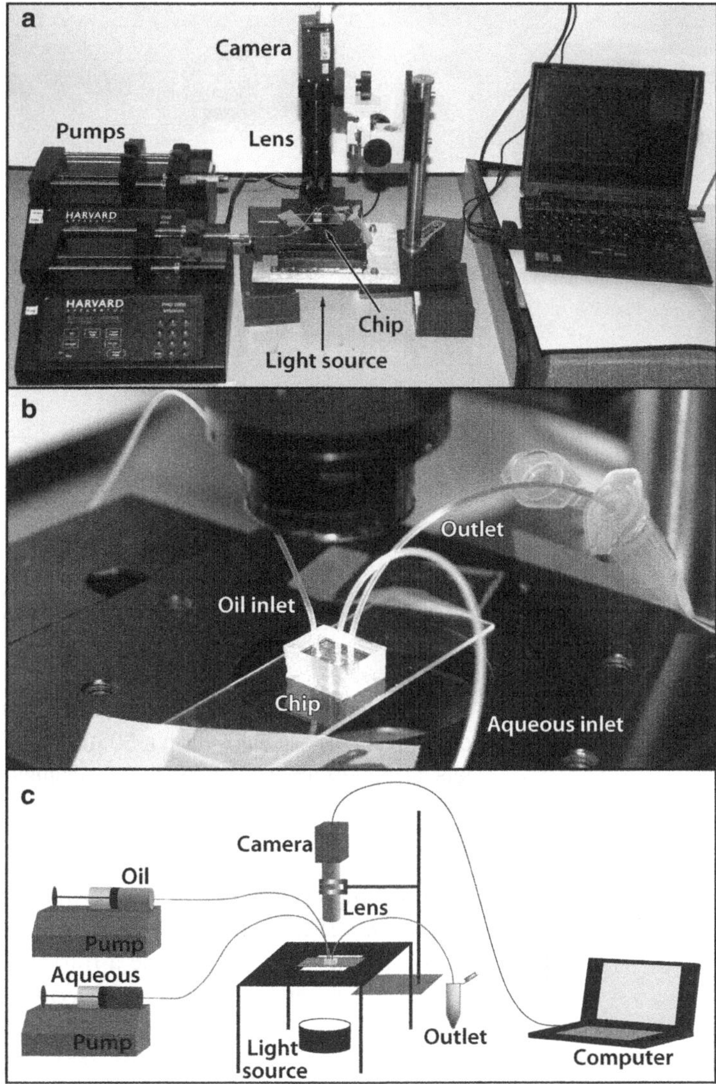

Fig. 3 A minimal microfluidic droplet setup. (**a**) Photograph of the setup, (**b**) a close-up view of the chip plumbed in and ready to use, and (**c**) diagrammatic outline of the same setup. The entire rig has a footprint of approximately 0.75 m^2 and a setup cost of around £5,500

enough to make good-quality video captures, but is sufficient for monitoring droplet formation with single pictures, if shutter times can be adjusted accordingly. A computer-connected camera enables recording of real-time images; any commercially available computer will be adequate for this task as long as it has the necessary inputs (USB or Firewire, depending on the camera). The final essential component is a bright light source to illuminate the chip—a bright LED light source is ideal. If a microscope is available, any setup will be suitable, as long as the camera can be connected via a c-mount. Inverted microscope setups are preferable because

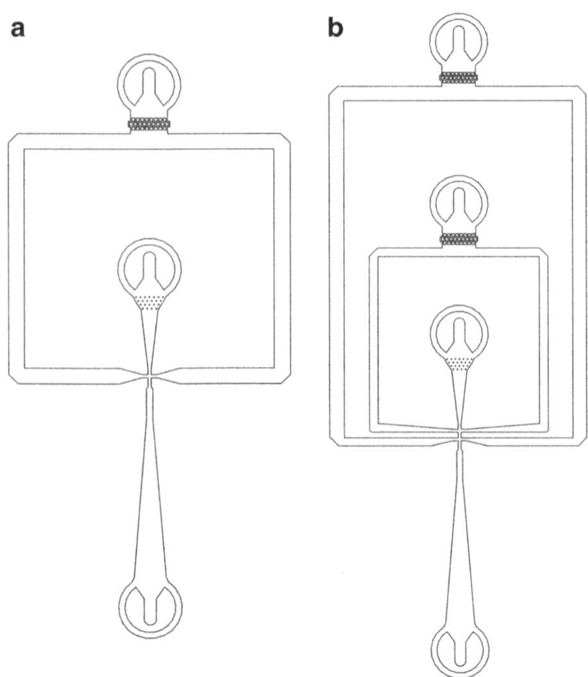

Fig. 4 Microfluidic chips for droplet production from either one (**a**) or two (**b**) aqueous streams. These chips have a 50 μm flow-focusing junction, and form droplets of approximately 50–90 μm in diameter (depending on the exact flow conditions)

tubing-chip connections do not interfere with higher magnification objectives. If dissemination quality videos of droplet formation are needed, one will have to use a high-speed camera (fps >1,000, see for example www.visionresearch.com).

3.2.2 PDMS Device Design and Manufacture

Prepare patterns using computer-aided design (CAD) software (e.g., the free program DraftSight). Simple sample designs for droplet formation at a flow-focusing junction with one or two aqueous solutions (Fig. 4) are available online from the author's institutional website at http://www.bio.cam.ac.uk/~fhlab/dropbase. Patterns with feature sizes larger than 5 μm can be printed on acetate film with a high-resolution inkjet printer (we use Micro Lithography Services Ltd, www.microlitho.co.uk).

With a printed pattern in hand, silicon wafers can be prepared using the following method. Alternatively they can be purchased pre-etched for around 350 Euros each, e.g. from Gesim (www.gesim.de).

To generate patterns with heights of approximately 25, 50, or 75 μm, access to a fully equipped photolithography suite, which will have all of the required equipment, is needed:

1. Spin coat a clean 3″ silicon wafer with SU-8 2025 photoresist to a thickness of 25, 50, or 75 μm. Different thicknesses are

achieved by adjusting the spin coating velocities and baking times (in the following section the first value describes the procedure for 25 μm, and in brackets follow the values for 50 and 75 μm channel heights, respectively):

(i) Initial spinning: 500 rpm for 5 s at an acceleration of 300 rpm/s

(ii) Second phase:

For 25 μm channel height at 3,000 rpm for 40 s at 300 rpm/s

For 50 μm channel height at 1,650 rpm for 40 s at 300 rpm/s

For 75 μm channel height at 1,000 rpm for 30 s at 300 rpm/s

2. "Bake" (i.e., place on a hotplate) the wafer for 1 (2, 3) minute(s) at 65°C, 2 (6, 9) minutes at 95°C, then a further 1 (2, 3) minute(s) at 65°C. This step ensures the complete evaporation of the solvent from the resin.
3. Expose your pattern onto the wafer using a mask aligner. Place the mask over the coated wafer and fix it to a blank glass mask in the tool. The exposure time required will depend on the brightness of the light source, but will typically be in the region of 3.5 s.
4. Bake the wafer after exposure for 1 min at 65°C and 1 (3, 6) minute(s) at 95°C.
5. Develop the wafer for 3–4 min in propylene glycol methyl ether acetate (PGMEA) to dissolve any non-cross-linked photoresist; then rinse the wafer well with isopropanol (IPA). An airbrush filled with PGMEA helps to develop small features.
6. Finally, bake the wafer for 1 min at 170°C to fully dry the SU-8 2025.

Using a silicon wafer carrying a positively etched pattern of the desired channel geometry, preparation of PDMS chips is straightforward.

1. Pour PDMS monomer into a container and add 1/10 curing agent by weight. Mix thoroughly and pour over wafer in a Petri dish. The thickness of the PDMS should not exceed the length of the biopsy punch-tool used for introducing the tubing connection.
2. To remove dissolved gas from the PDMS, place the Petri dish in a desiccator and apply a vacuum until PDMS stops bubbling.
3. Bake overnight (or at least 4 h) at 70°C. Check before you start that the Petri dish you use can withstand the temperature.
4. Cut a square of PDMS around the chip using a scalpel and lift the etched PDMS out of the dish. Be careful not to break the

brittle wafer by applying too much downward force when cutting.

5. Punch holes for entry and exit channels using a biopsy punch (1 mm diameter).
6. Treat the chip with oxygen plasma to facilitate bonding and place etched side down onto a glass slide directly or a PDMS backing (a thin layer of PDMS prepared separately, most easily by following steps 1–3 above using an empty Petri dish to produce a 2–3 mm thick layer of PDMS). The conditions for plasma oxidation are dependent on many factors and have to be established first depending on the size and power of the plasma chamber, the oxygen source (air or pure oxygen), and saturation of the chamber. In our hands, we treat with plasma for 1–2 s to achieve good bonding. Once plasma oxidation conditions have been optimized for a particular apparatus, good bonding, in which the glass/PDMS or PDMS/PDMS connection is stronger than the PDMS itself, is routinely achieved.

   If an oxygen plasma source is not available PDMS/PDMS bonding can be achieved using incompletely cured PDMS. In this modification of the procedure the PDMS-replica is cured for only 1.5 h on a hotplate at 65°C, cut, punched with holes, and then directly bonded to a thin layer (5 g PDMS and 500 mg curing agent in a 9 cm Petri dish, cured for 20 min at 65°C) of PDMS. Finally the assembly is left on a hotplate at 65°C overnight or for at least 3 h to cure fully.
7. Bonded and cured chips can be directly used for emulsion generation after being treated at 90°C to evaporate any moisture that may have condensed in them. However, any contact of the aqueous phase with the channels of the device downstream from the flow-focusing junction will result in wetting and render the chip unsuitable for further monodisperse emulsion production. This can be avoided by carrying out surface modification of chips to render them hydrophobic and thus prevent wetting.

Before surface modification, the freshly plasma-bonded chips should be incubated for 20 min at 90°C to reduce the risk of delamination. Companies (e.g., Sigma-Aldrich) offer a wide range of silanes that can be used to match the channel surface to the desired carrier (oil) phase. For fluorous oils, consumer products for glass treatment such as Aquapel provide very good surface modification. The use of a syringe with attached tubing to force Aquapel through the channels of the PDMS chip is sufficient to afford the desired modification of the channel surface. The channels can then be blow-dried with air. Surface-modified chips remain functional for several months and can be stored at room temperature and used directly for emulsion production.

3.2.3 Droplet Production in Devices

Oil and surfactant choice is crucial for successful experiments in microdroplets. We recommend the use of a fluorous oil phase rather than hydrocarbon-based oil (see Note 3). This preference is based on the idea that a fluorous oil is neither hydrophobic nor hydrophilic (i.e., a "third phase"), so small molecule escape from the droplets due to solubility in the oil is minimized. However, other mechanisms of small molecule escape (e.g., micromicellarization) may exist, so leaking will have to be assessed on a case-by-case basis with every new small molecule in question (see Note 4). Furthermore, the excellent compatibility of fluorinated oil with biomolecules (35) and its ability to form stable monodisperse droplets in conjunction with suitable surfactants (7, 23, 24) renders fluorous oil-based emulsions the system of choice. Block copolymer surfactants have been frequently used (9, 14, 36) and can be prepared following the protocols of Holtze and coworkers (23, 24) (and are commercially available from Raindance, albeit as part of a very expensive kit, or from Sphere Fluidics as PicoSurf).

The following protocol outlines a method for producing monodisperse microdroplets using a single aqueous phase and a fluorous oil phase.

1. Clean (see Note 5) and securely tape the PDMS chip to the microscope stage and focus the microscope on the flow-focusing junction.
2. Fill a 2.5 mL syringe with oil phase (e.g., HFE7500 + 0.5% w/w block copolymer fluorous surfactant). Plastic syringes are adequate but gastight glass syringes are superior due to their greater responsiveness.
3. Fill a 100–250 μL glass syringe with aqueous phase, ensuring that the concentration(s) of cells is appropriate to give singly occupied droplets for the expected droplet size (see Table 1). When loading cells, the density of the aqueous phase should be matched to cell density using Percoll®, OptiPrep™, alginate, or similar agents to prevent the cells from settling in the syringe (see Note 6).
4. To every syringe attach tubing of sufficient length to reach from the syringe pump to the PDMS chip as outlined in Fig. 3a, b (see Note 7). Connect a piece of tubing to the outlet on the chip and place the other end in a collection tube (see Note 8).
5. Place the syringes in the syringe pumps (see Note 9) and start flowing at high rates to fill the tubing. When the tubing is filled, reduce the flow. Connect the oil phase to the chip first and allow the oil to fill the channels, then connect the aqueous phase. Suggested initial flow rates for the chip design shown in Fig. 4 are 4,000 μL/h for the fluorous and 2,000 μL/h for the aqueous phase.
6. Observe the flow-focusing junction. Polydisperse droplets may form at the junction by jetting (Fig. 5a, b) when the

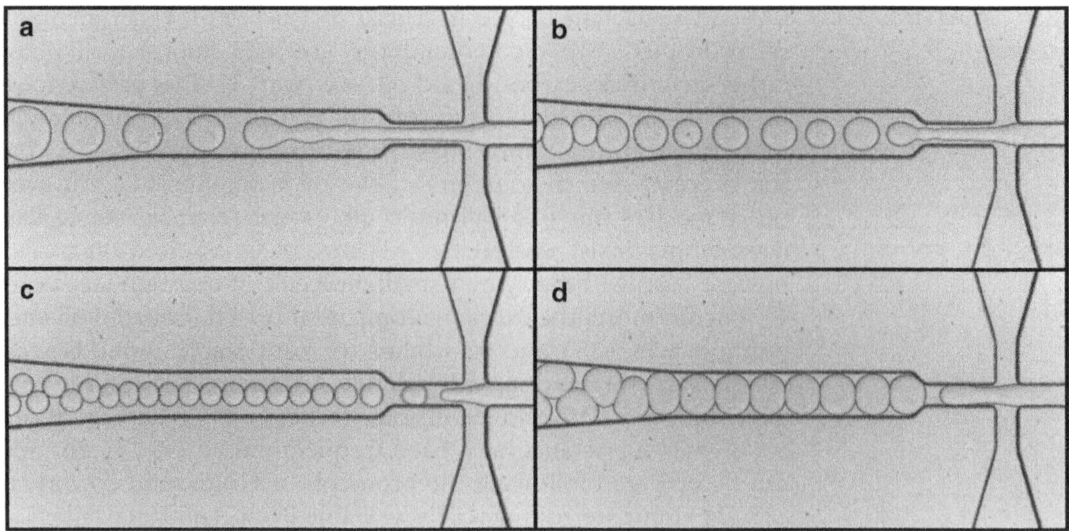

Fig. 5 Droplet production in the 50 μm microfluidic chip shown in Fig. 3 (with 50 μm channel height) using fluorous oil with 0.5% EA surfactant and distilled water at the indicated flow rates. (**a**) Jetting—flow rates are too fast so the aqueous phase is forming a constant stream with irregular droplet formation at the end of the jet (aqueous 3,000 μL/h, oil 4,000 μL/h). (**b**) Reduction of the aqueous flow rate reduces the jetting but not completely and the droplets formed are still polydisperse (aqueous 2,750 μL/h, oil 4,000 μL/h). (**c**) Formation of small monodisperse droplets (approximately 50 μm diameter; aqueous 2,000 μL/h, oil 4,000 μL/h). (**d**) Reducing the oil flow rate leads to formation of larger monodisperse droplets (approximately 90 μm diameter; aqueous 2,000 μL/h, oil 1,000 μL/h). Please note that the flow rates given here are for reference only—different solutions will require different flow rates

aqueous phase reaches it. To remedy this problem, gradually reduce the flow of the aqueous phase until the jetting ceases and the formation of monodisperse droplets can be seen in the video. The system will require time to stabilize after any change in flow rate, so allow about 1 min for equilibration before making any further changes. It may also be necessary to alter the flow rate of the fluorous phase to achieve monodisperse droplet formation. The final flow rates that give stable droplet formation will depend on the exact make up of both the aqueous and oil phases, so may well be different for different experiments. Droplet size is dependent not only on the channel geometries of the chip, but can be further adjusted to some extent by varying the ratios of the oil and aqueous flow rates (Fig. 5c, d).

7. Once droplet formation has stabilized, allow time for the initially produced, polydisperse, droplets to leave the outlet tubing before beginning collection (see Note 10).
8. Ensure that the syringes do not run out of liquid volume as the syringe pump could crush them.

3.3 Downstream Processing of Samples

There are a number of different possible ways in which microdroplets can be utilized, and the downstream processing steps will need to be tailored to the individual experiment, but some common tools are outlined here. As a general principle, we recommend that when carrying out an experiment for the first time, it is worthwhile producing a parallel sample without emulsification as a control to check that the encapsulation has not interfered with the biological experiment in any way.

(a) *Storage on chip*: an on-chip reservoir can accommodate large numbers of microfluidic droplets (as many as 10^6) and allow their continuous observation under a microscope (30). An alternative approach is to use a delay line to incubate moving droplets on chip (37).

(b) *Off-chip incubation and subsequent fluorescence analysis*: samples can be conveniently incubated off-chip and reinjected for analysis if prolonged incubation is required (>1 h). Alternatively, moderate numbers of droplets ($<10^4$) can be examined by fluorescence microscopy, conveniently carried out using disposable KOVA Glasstic hemocytometry slides. Subsequent image analysis can be carried out using the open-source image analysis software ImageJ (35).

(c) *Sample de-emulsification*: the emulsion can be broken to return the droplet contents to a single aqueous phase. This is useful for panning purifications during directed evolution of ligand-binding proteins. The key to any panning effort is to retain the genotype (via the linked phenotype) that is afforded by compartmentalization when the emulsion is broken. Several in vitro display systems make use of this principle (38–42), for example SNAP (26, 27), STABLE (40), or MHaeIII (39) display.

1. To break a fluorous oil emulsion, remove as much fluorous oil as possible and add 50 μL 1*H*,1*H*,2*H*,2*H*-perfluoro-1-octanol (PFO). Mix the sample thoroughly by vortexing, then spin down in a centrifuge to settle the phases. The aqueous phase will rest on top of the PFO/HFE7500 layer (and below the mineral oil if this was added). Remove the aqueous phase, and if more emulsion is visible, add another aliquot of PFO, remix and settle to recover the remaining aqueous material.
2. Similarly, mineral oil and related alkane-based emulsions can be broken by extraction with diethyl ether as described in detail in (43).
3. Alternatively, to break a mineral oil emulsion, freeze the emulsion in a microtube and centrifuge at maximum speed for 5 min. Examine the oil/water interface, and if any white emulsion can be seen, repeat the freeze-spin process until there is no emulsion remaining (44).

3.4 Perspectives

The number of applications for which droplets are of use exceeds the scope of this chapter. Therefore, we will focus on the field of directed evolution and give various examples for both in vitro and in vivo protein expression in bulk-generated and microfluidic droplets. In this way, we would like to highlight the current state of the art and give the reader a feel for the limitations and possibilities encountered to date.

In addition to the display systems mentioned above (Subheading 3.3, step (c)) developed for the selection of binders, cell-free methods in bulk droplets have been applied to enzyme evolution. Griffiths and Tawfik have evolved a DNA modifying methylase (17) and a phosphotriesterase (using microbeads as a genotype–phenotype linkage) (32). After expression of proteins and attachment to beads, the latter are re-emulsified with enzyme assay buffer and the formed product attaches to the beads. Beads are then sorted for product formation (32). In addition, a β-galactosidase was evolved by fluorescence-activated double-emulsion sorting (16).

Similarly, single cells have been encapsulated in bulk droplets and sorted for increased thiolactonase activity after double emulsification (15). Another successful system involving the bulk emulsification of single cells is compartmentalized self-replication (CSR) developed by Holliger et al., where a library of polymerases was expressed in *E. coli* and co-compartmentalized together with PCR reagents (45). Active variants replicate their encoding gene, whereas inactive ones cannot produce offspring. Therefore, the enrichment of a certain variant is directly coupled to its selective advantage. CSR has been used to evolve polymerases with relaxed substrate specificities capable of processing damaged DNA or unnatural oligonucleotides (46, 47).

Regarding IVTT in microfluidic droplets, first proof-of-principle experiments have been reported such as the successful expression of GFP from single genes (30) or the expression of β-galactosidase after DNA amplification from single genes and subsequent fusion with droplets containing IVTT mix (48).

Since 2010, the first directed evolution studies using microfluidic droplets have been reported. Agresti et al. reported the experimental evolution of a tenfold-improved horseradish peroxidase that was displayed on yeast after two rounds (14). By co-encapsulating fluorogenic substrate and single yeast cells displaying multiple copies of the enzyme on their surface, they were able to sort droplets for product formation directly on chip. Paegel and Joyce reported the evolution of RNA enzymes with RNA ligase activity in the presence of an inhibitor over five rounds (49). Kintses et al. evolved a hydrolase using cell-lysate assays in droplets and achieved a sixfold improvements in both activity and expression level (18).

Over the last two decades, droplet techniques have progressed from proof-of-principle experiments to tools for actual experiments to commercially exploited methods in some areas (e.g., 454 sequencing). We believe that as the field grows and methods mature, droplets will become available and be taken up by a wider range of scientists for various applications in protein nanotechnology.

4 Notes

1. We routinely use RTS 100 *E. coli* HY kit for cell-free protein expression (5 PRIME or RiNA). This kit is based on a bacterial lysate and therefore nucleic acid and protein degradation needs to be taken into account. We recommend keeping reactions on ice whenever possible to minimize this effect. Other kits are available from various providers and can in principle be used for protein production in emulsions, but the expression level and functionality of the protein should be checked. A system consisting of purified protein components, PURExpress, is available from NEB.
2. We advise the reader to carefully consider the requirements of his or her experiment and choose the simplest system available (e.g., bulk over microfluidic droplets when a quantitative read-out is not necessary) to obtain the maximum output with the minimum input of time and effort.
3. The method described in this chapter for microfluidic droplet production utilizes a fluorous oil/surfactant system; however, a number of alternatives involving hydrocarbon oil and non-fluorous surfactants are available, e.g., hexadecane containing 3% Span80 (50). The main differences between fluorous and hydrocarbon oil phases are as follows:
   (i) Small molecules may leak from droplets into the oil phase at differing rates, although there are ways of minimizing this effect (25).
   (ii) Droplet coalescence and Ostwald ripening can be controlled while the droplets are in devices (25), but taking droplets offline presents an additional challenge to droplet stability. In our hands, hexadecane/3% Span80 is most likely of the hydrocarbon oil mixtures to remain stable during off-line operations.
   (iii) The different physical properties of hydrocarbon vs. fluorous oils require readjustment of flow rates, e.g., to avoid jetting (as outlined in 3.3.3).
   (iv) There is a much greater choice of commercially available non-fluorous surfactants and oils and their purchase prices are much lower.
4. To check for compound leakage from droplets, the simplest method is to produce empty droplets and droplets containing the compound of interest, and then mix these two populations of droplets. After storage of the mixed droplets under

conditions of the experiment, check that there are still two distinct populations of full and empty droplets.

5. Scotch tape is useful for cleaning dust, oil, and fingerprints from PDMS.

6. To density-match cells, make a series of dilutions of your density-matching agent in water or buffer in microtubes. Add a small volume of concentrated cells to each tube and centrifuge at maximum speed for 10 min. Examine the tubes—in some the cells will have pelleted at the bottom and in others the cells will float. The correct density is midway between these concentrations.

7. Tubing is most easily cut using a clean-cut polymer tubing cutter. The easiest way to insert a needle into tubing is to hold the needle with the bevel facing up and hold the tubing in a pair of tweezers with the curvature down. Slowly guide the tubing onto the needle using the tweezers. To facilitate fitting the tubing onto the syringe needle, the tubing is prepared by first inserting a 25 G×5/8 needle and rotating it a few times to slightly stretch the tubing.

8. To prevent the outlet tubing falling out of the collection tube, it is useful to use a microtube with a small hole punched in the lid for collection. For convenience, a cap can be removed from a second tube and used repeatedly when the collection tube is changed.

9. If the biological sample is temperature sensitive, laboratory gloves filled with ice and tied off at the wrist provide convenient cooling packs to wrap around samples.

10. To maximize stability and facilitate recovery of microdroplet samples, the collected emulsion can be overlaid with a thin layer of mineral oil. This layer prevents sample evaporation and minimizes microdroplet adherence to the walls of the container.

Acknowledgments

This work was supported by the EU Marie-Curie network ProSA and the BBSRC. S.R.A.D. and M.F. are EU Marie-Curie fellows. F.H. holds an ERC Starting Investigator Grant.

References

1. Schaerli Y, Hollfelder F (2009) The potential of microfluidic water-in-oil droplets in experimental biology. Mol Biosyst 5: 1392–1404
2. Chen DL, Gerdts CJ, Ismagilov RF (2005) Using Microfluidics to Observe the Effect of Mixing on Nucleation of Protein Crystals. J Am Chem Soc 127:9672–9673
3. Huebner A, Bratton D, Whyte G, Yang M, deMello AJ, Abell C, Hollfelder F (2009) Static microdroplet arrays: a microfluidic device for droplet trapping, incubation and release for enzymatic and cell-based assays. Lab Chip 9:692–698
4. Huebner A, Srisa-Art M, Holt D, Abell C, Hollfelder F, Demello AJ, Edel JB (2007)

Quantitative detection of protein expression in single cells using droplet microfluidics. Chem Commun 2007:1218–1220

5. Hufnagel H, Huebner A, Gulch C, Guse K, Abell C, Hollfelder F (2009) An integrated cell culture lab on a chip: modular microdevices for cultivation of mammalian cells and delivery into microfluidic microdroplets. Lab Chip 9:1576–1582
6. Köster S, Angile FE, Duan H, Agresti JJ, Wintner A, Schmitz C, Rowat AC, Merten CA, Pisignano D, Griffiths AD, Weitz DA (2008) Drop-based microfluidic devices for encapsulation of single cells. Lab Chip 8:1110–1115
7. Clausell-Tormos J, Lieber D, Baret J-C, El-Harrak A, Miller OJ, Frenz L, Blouwolff J, Humphry KJ, Köster S, Duan H, Holtze C, Weitz DA, Griffiths AD, Merten CA (2008) Droplet-Based Microfluidic Platforms for the Encapsulation and Screening of Mammalian Cells and Multicellular Organisms. Chem Biol 15:427–437
8. Brouzes E, Medkova M, Savenelli N, Marran D, Twardowski M, Hutchison JB, Rothberg JM, Link DR, Perrimon N, Samuels ML (2009) Droplet microfluidic technology for single-cell high-throughput screening. Proc Natl Acad Sci USA 106:14195–14200
9. Shim JU, Olguin LF, Whyte G, Scott D, Babtie A, Abell C, Huck WTS, Hollfelder F (2009) Simultaneous determination of gene expression and enzymatic activity in individual bacterial cells in microdroplet compartments. J Am Chem Soc 131:15251–15256
10. El Debs BW, Tschulena U, Griffiths AD, Merten CA (2011) A competitive co-cultivation assay for cancer drug specificity evaluation. J Biomol Screen 16:818–824
11. Shim J-u, Patil SN, Hodgkinson JT, Bowden SD, Spring DR, Welch M, Huck WTS, Hollfelder F, Abell C (2011) Controlling the contents of microdroplets by exploiting the permeability of PDMS. Lab Chip 11:1132–1137
12. Huebner A, Olguin LF, Bratton D, Whyte G, Huck WTS, deMello AJ, Edel JB, Abell C, Hollfelder F (2008) Development of quantitative cell-based enzyme assays in microdroplets. Anal Chem 80:3890–3896
13. Damean N, Olguin LF, Hollfelder F, Abell C, Huck WTS (2009) Simultaneous measurement of reactions in microdroplets filled by concentration gradients. Lab Chip 9:1707–1713
14. Agresti JJ, Antipov E, Abate AR, Ahn K, Rowat AC, Baret JC, Marquez M, Klibanov AM, Griffiths AD, Weitz DA (2010) Ultrahigh-throughput screening in drop-based microfluidics for directed evolution. Proc Natl Acad Sci USA 107:4004–4009
15. Aharoni A, Amitai G, Bernath K, Magdassi S, Tawfik DS (2005) High-throughput screening of enzyme libraries: thiolactonases evolved by fluorescence-activated sorting of single cells in emulsion compartments. Chem Biol 12: 1281–1289
16. Mastrobattista E, Taly V, Chanudet E, Treacy P, Kelly BT, Griffiths AD (2005) High-throughput screening of enzyme libraries: in vitro evolution of a beta-galactosidase by fluorescence-activated sorting of double emulsions. Chem Biol 12:1291–1300
17. Cohen HM, Tawfik DS, Griffiths AD (2004) Altering the sequence specificity of HaeIII methyltransferase by directed evolution using in vitro compartmentalization. Prot Eng Des Sel 17:3–11
18. Kintses B, Hein C, Mohamed MF, Fischlechner M, Courtois F, Lainé C, Hollfelder F (2012) Picoliter cell lysate assays in microfluidic droplet compartments for directed enzyme evolution. Chem Biol 19:1001–1009
19. Shendure J, Ji H (2008) Next-generation DNA sequencing. Nat Biotechnol 26:1135–1145
20. Umbanhowar PB, Prasad V, Weitz DA (2000) Monodisperse Emulsion Generation via Drop Break Off in a Coflowing Stream. Langmuir 16:347–351
21. Theberge AB, Courtois F, Schaerli Y, Fischlechner M, Abell C, Hollfelder F, Huck WTS (2010) Microdroplets in Microfluidics: An Evolving Platform for Discoveries in Chemistry and Biology. Angew Chem Int Ed Engl 49:5846–5868
22. Kintses B, van Vliet LD, Devenish SR, Hollfelder F (2010) Microfluidic droplets: new integrated workflows for biological experiments. Curr Opin Chem Biol 14: 548–555
23. Holtze C, Rowat AC, Agresti JJ, Hutchison JB, Angilè FE, Schmitz CHJ, Köster S, Duan H, Humphry KJ, Scanga RA, Johnson JS, Pisignano D, Weitz DA (2008) Biocompatible surfactants for water-in-fluorocarbon emulsions. Lab Chip 8:1632
24. Chen C-H, Sarkar A, Song Y-A, Miller MA, Kim SJ, Griffith LG, Lauffenburger DA, Han J (2011) Enhancing Protease Activity Assay in Droplet-Based Microfluidics Using a Biomolecule Concentrator. J Am Chem Soc 133:10368–10371
25. Courtois F, Olguin LF, Whyte G, Theberge AB, Huck WTS, Hollfelder F, Abell C (2009) Controlling the retention of small molecules in emulsion microdroplets for use in cell-based assays. Anal Chem 81:3008–3016
26. Stein V, Johnsson K, Hollfelder F (2007) A Covalent Chemical Genotype–Phenotype Linkage for in vitro Protein Evolution. Chembiochem 8:2191–2194
27. Kaltenbach M, Stein V, Hollfelder F (2011) SNAP Dendrimers: Multivalent Protein Display on Dendrimer-Like DNA for Directed Evolution. Chembiochem 12:2208–2216
28. Huebner A, Srisa-Art M, Holt D, Abell C, Hollfelder F, deMello AJ, Edel JB (2007)

Quantitative detection of protein expression in single cells using droplet microfluidics. Chem Commun (12):1218–1220

29. Tawfik DS, Griffiths AD (1998) Man-made cell-like compartments for molecular evolution. Nat Biotechnol 16:652–656
30. Courtois F, Olguin LF, Whyte G, Bratton D, Huck WTS, Abell C, Hollfelder F (2008) An integrated device for monitoring time-dependent in vitro expression from single genes in picolitre droplets. Chembiochem 9:439–446
31. Agresti JJ, Kelly BT, Jäschke A, Griffiths AD (2005) Selection of ribozymes that catalyse multiple-turnover Diels–Alder cycloadditions by using in vitro compartmentalization. Proc Natl Acad Sci USA 102:16170–16175
32. Griffiths AD, Tawfik DS (2003) Directed evolution of an extremely fast phosphotriesterase by in vitro compartmentalization. EMBO J 22:24–35
33. Stein V, Sielaff I, Johnsson K, Hollfelder F (2007) A covalent chemical genotype-phenotype linkage for in vitro protein evolution. Chembiochem 8:2191–2194
34. Miller OJ, Bernath K, Agresti JJ, Amitai G, Kelly BT, Mastrobattista E, Taly V, Magdassi S, Tawfik DS, Griffiths AD (2006) Directed evolution by in vitro compartmentalization. Nat Meth 3:561–570
35. Kaltenbach M, Devenish SRA, Hollfelder F (2012) A simple method to evaluate the biochemical compatibility of oil/surfactant mixtures for experiments in microdroplets. Lab Chip 12:4185–92
36. Schaerli Y, Wootton RC, Robinson T, Stein V, Dunsby C, Neil MA, French PM, deMello AJ, Abell C, Hollfelder F (2009) Continuous-flow polymerase chain reaction of single-copy DNA in microfluidic microdroplets. Anal Chem 81:302–306
37. Frenz L, Blank K, Brouzes E, Griffiths AD (2009) Reliable microfluidic on-chip incubation of droplets in delay-lines. Lab Chip 9:1344–1348
38. Bertschinger J, Grabulovski D, Neri D (2007) Selection of single domain binding proteins by covalent DNA display. Prot Eng Des Sel 20:57
39. Bertschinger J, Neri D (2004) Covalent DNA display as a novel tool for directed evolution of proteins in vitro. Prot Eng Des Sel 17:699
40. Doi N, Yanagawa H (1999) STABLE: protein-DNA fusion system for screening of combinatorial protein libraries in vitro. FEBS Lett 457:227–230
41. Yonezawa M, Doi N, Higashinakagawa T, Yanagawa H (2004) DNA display of biologically active proteins for in vitro protein selection. J Biochem 135:285
42. Yonezawa M, Doi N, Kawahashi Y, Higashinakagawa T, Yanagawa H (2003) DNA display for in vitro selection of diverse peptide libraries. Nuc Acids Res 31:e118
43. Kaltenbach M, Hollfelder F (2011) SNAP display: In vitro protein evolution in microdroplets. In: Jackson RH, Douthwaite JA (eds) Ribosome Display and Related Technologies: Methods and Protocols. Humana Press (Springer Protocols), London
44. Kiss MM, Ortoleva-Donnelly L, Beer NR, Warner J, Bailey CG, Colston BW, Rothberg JM, Link DR, Leamon JH (2011) High-Throughput Quantitative Polymerase Chain Reaction in Picoliter Droplets. Anal Chem 80:8975–8981
45. Ghadessy FJ, Ong JL, Holliger P (2001) Directed evolution of polymerase function by compartmentalized self-replication. Proc Natl Acad Sci USA 98:4552–4557
46. d' Abbadie M, Hofreiter M, Vaisman A, Loakes D, Gasparutto D, Cadet J, Woodgate R, Pääbo S, Holliger P (2007) Molecular breeding of polymerases for amplification of ancient DNA. Nat Biotechnol 25:939–943
47. Ghadessy FJ, Ramsay N, Boudsocq F, Loakes D, Brown A, Iwai S, Vaisman A, Woodgate R, Holliger P (2004) Generic expansion of the substrate spectrum of a DNA polymerase by directed evolution. Nat Biotechnol 22:755–759
48. Mazutis L, Araghi AF, Miller OJ, Baret JC, Frenz L, Janoshazi A, Taly V, Miller BJ, Hutchison JB, Link D, Griffiths AD, Ryckelynck M (2009) Droplet-based microfluidic systems for high-throughput single DNA molecule isothermal amplification and analysis. Anal Chem 81:4813–4821
49. Paegel BM, Joyce GF (2010) Microfluidic Compartmentalized Directed Evolution. Chem Biol 17:717–724
50. Jousse F, Farr R, Link DR, Fuerstman MJ, Garstecki P (2006) Bifurcation of droplet flows within capillaries. Phys Rev E 74:036311

Chapter 17

Label-Free, Real-Time Interaction and Adsorption Analysis 1: Surface Plasmon Resonance

Conan J. Fee

Abstract

A key requirement for the development of proteins for use in nanotechnology is an understanding of how individual proteins bind to other molecules as they assemble into larger structures. The introduction of labels to enable the detection of biomolecules brings the inherent risk that the labels themselves will influence the nature of biomolecular interactions. Thus, there is a need for label-free interaction and adsorption analysis. In this and the following chapter, two biosensor techniques are reviewed: surface plasmon resonance (SPR) and the quartz crystal microbalance (QCM). Both allow real-time analysis of biomolecular interactions and both are label-free. The first of these, SPR, is an optical technique that is highly sensitive to the changes in refractive index that occur with protein (or other molecule) accumulation near an illuminated gold surface. Unlike QCM (Chapter 18) SPR is not affected by the water that may be associated with the adsorbed layer nor by conformational changes in the adsorbed species. SPR thus provides unique information about the interaction of a protein with its binding partners.

Key words Surface plasmon resonance, Label-free detection, Biosensors

1 Introduction to Surface Plasmon Resonance

Surface plasmon resonance (SPR) is an optical technique by which one can follow the course of molecular interactions in real time. This and other related optical biosensor techniques have been regularly reviewed by David Myszka and Rebecca Rich, of the University of Utah, e.g., (1–4). SPR instruments detect changes in refractive index that occur through changes in mass within a very fine layer close to the sensor surface, typically using disposable sensors or "chips", which have functionalized surface layers that are prepared for the study of specific interactions. One or more molecular species are immobilized or "tethered" to the surface layer and binding partners in free solution are then presented to the tethered molecule. Flow or stirring is used to minimize mass transfer limitations so that the measured rate of interaction reflects actual binding kinetics rather than mass transfer of solutes between the bulk solution

Juliet A. Gerrard (ed.), *Protein Nanotechnology: Protocols, Instrumentation, and Applications*, Methods in Molecular Biology, vol. 996, DOI 10.1007/978-1-62703-354-1_17,

and the surface layer. Important information can be determined by SPR, including the association and dissociation kinetic constants, the analyte concentrations, the detection of specific species through biospecific responses, and the gathering of circumstantial evidence in support of proposed binding mechanisms, including epitope mapping or supramolecular complex formation. Measurement of binding kinetics over a range of constant temperatures also allows thermodynamic quantities to be determined. With present technology, SPR can measure association constants, k_a, over the range 100–10^8 M^{-1} s^{-1}; dissociation constants, k_d, over the range 1–10^{-6} s^{-1}; and equilibrium constants, K_D, over the range 100 mM to 1 pM.

The lowest detectable molecular weight is currently claimed to be around 100 Da in the most sensitive instruments but this should not be considered routinely feasible and it depends strongly upon the molecular weight of the tethered species. The maximum SPR response during analyte interaction is directly related to the ratio between the analyte (free solution) and the ligand (tethered) molecular weights. Therefore, one should not expect to be able to routinely detect a single sugar analyte in free solution binding to an antibody ligand tethered to the chip surface, although reversing the analyte/ligand pairing may well yield useful results.

The principles of SPR are described in more detail below but essentially the phenomenon depends on detecting the conversion of energy from light incident on a thin layer of metal that is located between two media of differing refractive indices (5). Normally, for biointeractions, the two media are comprised of a glass prism, onto which a thin layer of gold is coated, and an aqueous flow cell. Light is directed up through the prism from below and the intensity of reflected light from the gold layer is detected by optical detectors on the other side of the prism (Fig. 1). Several different modulations to the light characteristics in response to binding events near the gold surface may be detected, with two important ones being *angular modulation* and *wavelength modulation*. Light intensity and phase changes can also be detected. In *angular modulation*, part of the energy from monochromatic light incident on the metal at a specific angle is converted into an electron wave that travels along the gold surface (the surface plasmon) on the aqueous side and this angle is detected by measuring a localized drop in the reflected light intensity.

Because the specific angle of the incident light that creates the surface plasmon wave depends critically upon the difference in refractive index between the glass and aqueous sides of the gold layer, when a protein binds on the aqueous side of the gold, the change in local refractive index causes a change in the angle of light energy that is coupled to the surface plasmon wave. Thus, the angle of lowest reflected light intensity detected by the optical detectors changes in response to the amount of mass coupled near the gold surface (Fig. 2).

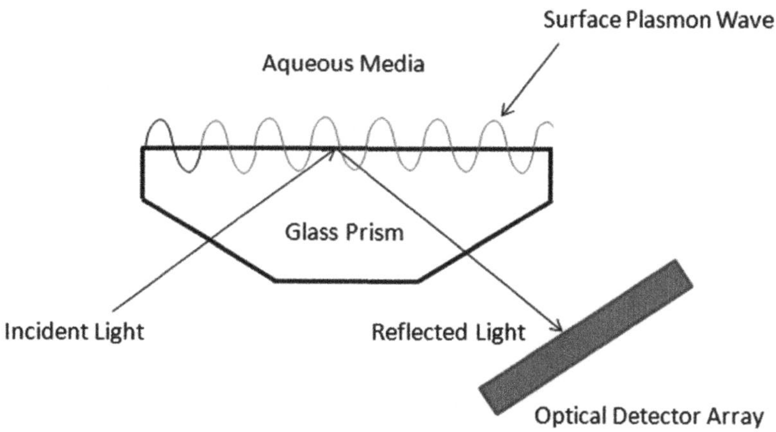

Fig. 1 Surface plasmon resonance configuration for biomolecular interactions

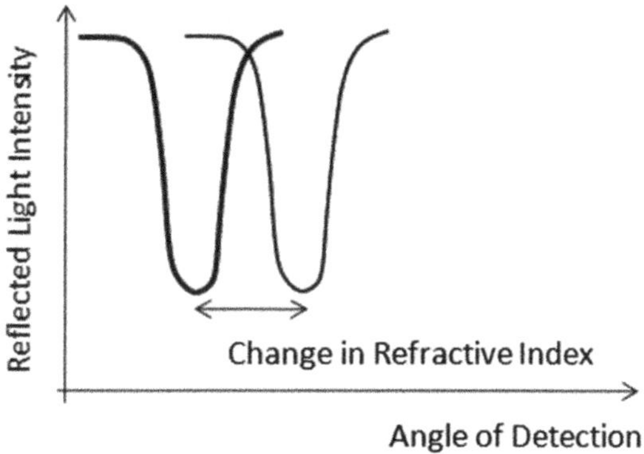

Fig. 2 Change in angle of minimum detected light intensity with change in refractive index

In *wavelength modulation*, different wavelengths of light are used to excite the surface plasmon, while the angle of incidence remains constant. The wavelength of light that has the strongest coupling to the plasmon is detected and this changes in response to changes in refractive index. For simplicity, angular modulation is assumed to be the measured quantity in the subsequent description but analogous phenomena occur for other modes of light modulation.

The actual angles involved and the extent to which they might change in response to binding near the surface will depend upon the particular instrument design (glass refractive index, metal type and thickness, surface coating, frequency and wavelength of incident light, etc.) so normally the change in angle is plotted in terms of "response units" (RU) against time, to give a "sensorgram" (Fig. 3 and Eq. 1). Typically, instruments are calibrated internally

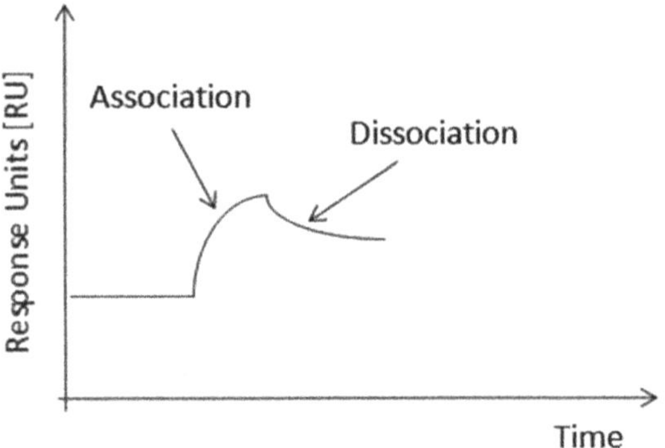

Fig. 3 The SPR "sensorgram"

so that a change of 1,000 RU nominally indicates the addition of 1 ng/mm^2 of bound protein (6). (It is worth noting here that if the surface coating layer on the chip in which the interaction occurs is nominally 100 nm thick, this corresponds to a bound protein concentration of 10 mg/mL per 1,000 RU, so a response of 15,000 RU corresponds to an effective concentration of 150 mg/mL at the chip surface!):

$$\Delta \text{RU} = m\Delta\eta, \tag{1}$$

where m is an instrument-specific constant that relates the "response" of that particular instrument to the change in refractive index, η.

SPR does not require labeling of the molecules involved but, as noted above, at least one partner in the interaction must be immobilized on or near the sensor chip surface. The absence of labeling means that the interactions of the native form of a molecule can be studied, without the potential interference that might arise from a covalently attached moiety such as a fluorophore. However, this is strictly true only for the molecule(s) in free solution, as the tethered molecule must, necessarily, be constrained to the chip surface by a stable linker. Commonly, for proteins, this is achieved through amine coupling, which may result in multiple tethering positions through multiple primary amine groups on lysine residues. Thus, the tethered molecule may not be in its fully native free-solution form and its binding site may be sterically hindered and/or conformationally constrained. For affinity detection or concentration assays, this is not important because either the rate of binding is irrelevant or, in the latter case, is implicitly accounted for by using a calibration curve. However, the possible effects of tethering must be considered when quantifying binding kinetics and, ideally, a single tethering site used that is as far removed as possible from the binding site and that allows full conformational freedom in the tethered molecule.

2 SPR Instruments

Before going through each of the steps involved in SPR assays, it is useful to briefly describe the configurations of several contrasting commercial instruments from three manufacturers: Biacore (GE Healthcare), Bio-Rad, and ForteBio. (Although it should be noted here that while the ForteBio instrument is an optical sensor, it is not actually an SPR device. However, its operation and the data it produces are very similar to an SPR.) The manufacturers chosen are by no means exclusive descriptions and the discussion is deliberately broad, as the technology is evolving and there is little point in providing extensive detail regarding systems that might be superseded. A table summarizing the 2008 peer-reviewed literature lists no fewer than 57 individual SPR instruments from some 25 suppliers and 14 non-SPR optical biosensor instruments from six suppliers (4).

2.1 Biacore

From 1990 until around 2007, it is fair to say that Biacore™ SPR instruments were synonymous with surface plasmon resonance sensing for biomolecular interactions. Indeed, many peer-reviewed papers prior to 2007 used the phrase "Biacore analysis" as a synonym for "SPR analysis". The market dominance of Biacore during that time meant there was little ambiguity in the use of such a term but the availability of alternative commercial SPR systems from other manufacturers means that there is no justification for its use today.

That said, Biacore, acquired by GE Healthcare (Uppsala, Sweden) in 2006, certainly set the standard for SPR analysis during their first 17 years or so of operation and remains the market leader at the time of writing. The company at this time offers no less than eight current commercial models for a range of applications, from a single sample system (the *Biacore J*) through to array systems (e.g., the *Biacore 4000*) that can measure over 2,000 interactions per hour with parallel sample injection.

Although it must be stressed that several configurations are available, a typical automated SPR instrument such as the *Biacore 3000* comprises an autosampler, a microfluidics device, a series of consumable sensor chips, and the optics system.

The autosampler system is essentially a temperature-controlled liquid handler that allows the input of all reagents for sensor chip surface activation, surface immobilization, analyte injection and, in some cases, recovery of eluted analyte, and automated preparation of matrix-assisted laser desorption-ionization time-of-flight (MALDI-TOF) mass spectrometry plates. The sensor surface has independent temperature control for interaction analysis. As described below, activation of the chip surface often involves the mixing of reagents that are prone to rapid hydrolysis, so a feature of the autosampler is the ability to keep these components separate until needed and then mix them in the autosampler system immediately prior to injection. The system control software is versatile in that it allows step-by-step command level control of the autosampler and injection conditions, etc.

The microfluidic device normally continuously injects running buffer from one of two independent buffer reservoirs, at flow rates between 5 and 100 μL/min. The device in the Biacore 3000 instrument incorporates the optics (prism) and forms four flow channels on the sensor chip by clamping the microfluidics head onto the sensor chip surface. Flow can be independently directed through each channel. Generally, one channel must be used as a noninteracting reference (control) for subtraction of responses due to bulk shift (change in refractive index between the sample solution and the running buffer) and nonspecific adsorption.

A wide range of sensor chips is available, including plain gold, carboxymethyl dextran-coated chips with varying carboxyl group densities and/or dextran layer thicknesses, streptavidin, lipid bilayers, and membranes, and nitrilotriacetic acid (NTA) chips suitable for capture of His-tagged molecules. Although the stability of an immobilized biomolecule will be unique to the particular molecular species involved, the immobilization method used, and the conditions to which it has been subjected, the sensor chips can often be reused many times. The sensor chips themselves are relatively robust and can retain activity after ejection from the system and refrigerated storage in either wet or dry state. An advantage of incorporating the optics, including the prism, within the instrument is that these are unlikely to be contaminated if suitable cleaning and system operation protocols are followed.

2.2 Bio-Rad

Bio-Rad Laboratories (Hercules, CA) currently produce a single multiplex SPR instrument, the ProteOn™ XPR36. Again, the instrument consists of a temperature-controlled autosampler, a microfluidics system, and a series of sensor chips that are docked into the optics/fluidics systems. The sensor surface has independent temperature control. The microfluidic system forms six flow channels when the chip is docked and it rotates 90° so that a 6×6 array is formed at the intersections (spots) formed between the channels in each orientation. A charge-coupled device (CCD) camera views the reflected light from the entire array area so that the instrument response can be followed independently not only at the 36 interaction spots but also between them. These latter "interspots" can be used as controls, eliminating the need to use a channel for subtracting bulk shift and nonspecific adsorptions from the raw signal.

The autosampler consists of six sample needles that move in unison and draw up samples or surface activation reagents from vials or a microtiter plate as a set. Channels can be left unused simply by placing running buffer in the corresponding sample vials/wells. By careful planning of the flow direction and the samples applied, the system allows a single channel to be immobilized with up to six ligands or replicates, or all six channels to be immobilized with 36 different ligands. Six analyte samples (or any combination of analytes and running buffer blanks) can then be flowed over the array and the interactions measured on any of the 36 spots.

A point of difference for the Bio-Rad system is that the sensor chips themselves incorporate the prism. When docked, the system initializes the optics by normalizing the background signal across the entire CCD array. The initial normalization is carried out in air because the chip is assumed to be completely dry but when re-docked after having been ejected, the user chooses whether to use the original normalization data or to remeasure the background using a glycerol solution to match the refractive index of the glass (assuming that the chip may be wet). The sensor chips may, again depending on the stability of the immobilized biomolecule(s), be reused many times if left docked in the instrument but care must be taken when storing the chips external to the instrument to avoid contaminating the prism surface. Thus, the Bio-Rad sensor chips should be stored dry, in contrast to the Biacore sensor chips, which can be stored either immersed in buffer or dry. Bio-Rad currently supplies chips that are coated with a carboxylated alginate layer with three levels of carboxyl group density, a neutravidin chip for capturing biotinylated molecules and a tris-nitrilotriacetic acid (NTA) chip for his-tagged proteins.

2.3 ForteBio

The ForteBio Octet system is significantly different from the two instruments described above in two major ways. First, it does not use microfluidics but instead uses eight or sixteen parallel, disposable "dipstick" optical tips, immersed in a shaking 96- or 384-well microtiter plate, respectively. One advantage of this approach is that there is no restriction on the use of solids-laden samples because the analyte samples do not pass through microchannels on the sensor tips but the microtiter plates are simply shaken to induce sufficient bulk mass transfer next to the stationary tip surfaces.

Second, although the system exploits a change in the modulated properties of reflected light and is thus an optical sensor, it is not actually a surface plasmon resonance system. Nevertheless, the response of the device to surface binding events is so similar to SPR, that in all practical senses the user can regard it as a comparable system i.e., for real-time, label-free quantification of biomolecular interactions. However, the difference in the underlying physical phenomenon exploited by this system is significant.

The Octet system analyzes the interference pattern of white light reflected from two surfaces: a layer of immobilized protein on the biosensor tip, and an internal reference layer (Fig. 4). Binding of an analyte to an immobilized ligand will increase the thickness of the surface layer and result in a change in the interference pattern and therefore a change in the wavelength of the lowest light intensity. This is then converted to a sensorgram in a similar way to

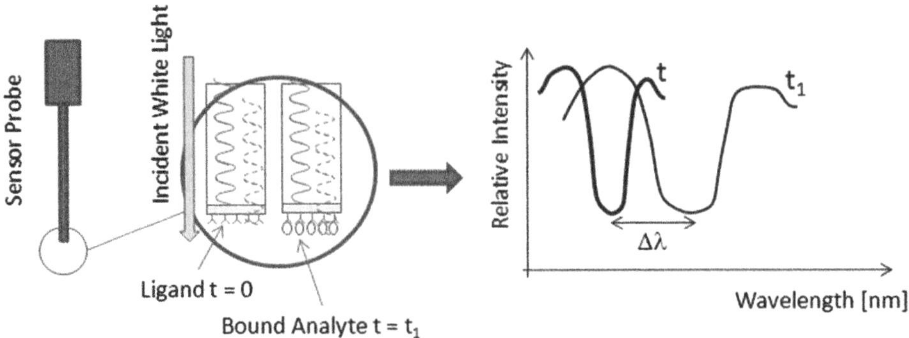

Fig. 4 The ForteBio system

the SPR systems described above. The response is affected only by bound molecules and not by changes in refractive index or bulk solution properties. The advantage is that bulk shift in the signal does not occur. However, a control is still required to account for nonspecific interactions.

The autosampler system moves the sensor probes (8 or 16) in unison between programmed microtiter well locations, so can allow for surface activation, immobilization, washing, analyte binding, regeneration, etc. The microtiter plates are housed in a temperature-controlled compartment.

Because the instrument response depends on a change in *thickness* of the adlayer on the end of the sensor tip, unlike SPR sensor chips, the tips are limited to a monolayer of immobilized ligand and detect a monolayer of bound analyte. The response signal from an SPR sensor chip, on the other hand, can be boosted, at least for concentration analysis if not kinetics analysis, by using a higher density of immobilized ligand within a deep surface layer to increase the level of binding achieved and thereby increasing the total change in refractive index that occurs with binding. A series of preprepared immunoglobulin probes, immunoglobulin-binding probes (e.g., protein A), and probes suitable for amine coupling or His-tagged protein capture and streptavidin and hydrophobic surfaces are available.

3 The SPR Method

The SPR method generally comprises the following steps, designed first to immobilize a ligand to the chip surface, then to measure real-time interactions with analyte(s) in free solution, and, finally, to regenerate the ligand for subsequent experiments and to process the interaction data:

1. Pre-concentration (not required for affinity immobilization techniques such as streptavidin-biotin or His-tag-NTA).

2. Chemical activation of the sensor chip surface.
3. Surface immobilization (tethering) of one or more binding partners.
4. Deactivation of residual activated sites on the sensor chip surface.
5. Surface stabilization to remove weakly bound material.
6. Interaction with the analyte(s). Sequential injection of one or more binding partners and measurement of the response during binding (association), followed by a return to the running buffer to measure the response during dissociation (Fig. 3).
7. Regeneration—removal of residual binding partner(s) from the ligand at the end of the dissociation phase.
8. Repetition of steps 5–7 for subsequent experiments (or steps 3–7 if ligand is to be completely removed during regeneration, as in affinity capture immobilizations).
9. Processing of raw data and appropriate analysis and presentation.

With a new sensor chip, a preconditioning step prior to amine coupling can be carried out with a series of dual injections of 100 mM HCl, 50 mM NaOH, and 0.5% sodium dodecyl sulfate (SDS), for 5 or 6 s each, at a flow rate of 100 μL/min (about 10 μL each injection). This rewets and cleans the sensor chip surface, giving a good, stable response prior to further work.

3.1 Pre-concentration

The objective of pre-concentration experiments is simply to determine the most effective pH and concentration at which to immobilize the target molecule by following the SPR response to electrostatic interactions between the molecule and the sensor chip surface. The usual starting point for interaction analysis is to tether one of the binding partners via carboxyl groups located within a stable polymer coating (e.g., carboxylated dextran or alginate) of about 100 nm thickness on the sensor chip. The carboxyl groups are active in the de-protonated form, i.e., when the pH in the surface layer (which can differ from the bulk phase) is above the pK_a of the carboxyl groups (usually around pH 3.5). Above this pK_a, the carboxyl groups have a single negative charge that is reactive but will repel other negatively charged species, so for instance, if the pH is above a protein's isoelectric point (pI), the protein will not easily penetrate into the surface layer on the sensor chip and binding will be low. To identify the most efficient binding conditions, the molecule of interest (for proteins, 20–200 μg/mL) should first be injected over the chip at a range of pH values to determine the maximum SPR response (Fig. 5). For a protein, the further the pH is below its pI, the more positively charged it will be but the sensor chip surface will also be more protonated, reducing

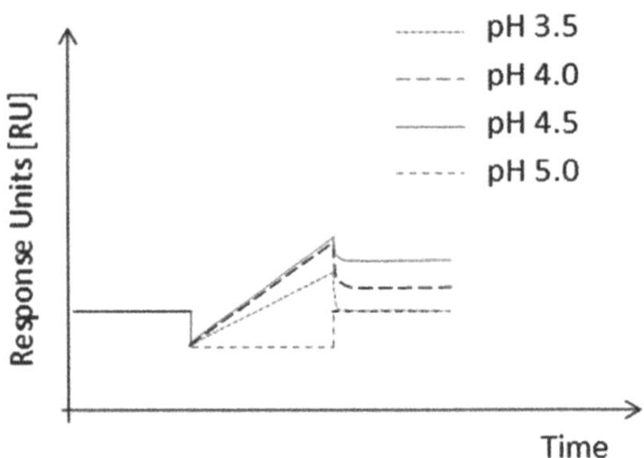

Fig. 5 Pre-concentration runs showing that the optimum immobilization in this hypothetical case would be one carried out at pH 4.5

its reactivity. Also, some proteins are not stable at low pH. Therefore, there is likely to be an optimal pH for binding that is best determined experimentally through pre-concentration experiments. However, immobilization can often be carried out successfully, if not optimally, at a midpoint between pH 3.5 and the protein's pI. The salt concentration of the buffer should be kept below 100 mM (NaCl) during pre-concentration to avoid masking the attractive electrostatic charges. Experiments with a range of protein concentrations should also be carried out at the optimal pH prior to immobilization to obtain a quantitative estimate of the likely response during immobilization. This is important, as the response range should be appropriate for the intended use (a high response (>1,000 RU) to maximize the lower limit of detection for concentration analyses and a low response (<250 RU) to minimize mass transfer effects during kinetic analyses.)

3.2 Chemical Activation of the Sensor Chip Surface

Generally, before the first binding partner is immobilized, the SPR chip surface must be chemically activated to allow coupling. A number of specially prepared surfaces such as streptavidin or metal chelating (NTA) chips may not require this step and are available from some suppliers for capture of biotinylated or His-tagged proteins, respectively. However, the most common approach is to tether one of the binding partners to carboxyl groups within the polymer coating on the sensor chip. One can regard the polymer coating on SPR chips more as a "tangled fishing line" rather than a uniform monolayer so that the carboxyl groups are located in space (Fig. 6a), much as they would be in solution, as opposed to tight orientation on a crowded surface (Fig. 6b).

Activation of the carboxylated surface with *N*-hydroxysuccinimide (NHS) provides a good leaving group on the carboxyl groups suitable for spontaneous formation of various bonds. Such activation can be carried out using a mixture

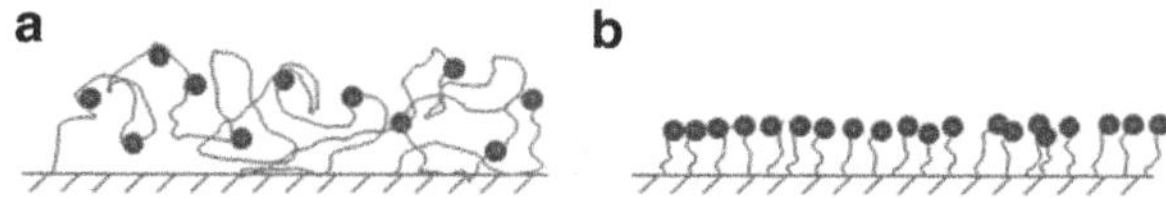

Fig. 6 The polymer coating on an SPR sensor chip is more like the "tangled fishing line" in (**a**) than the uniform monolayer in (**b**)

of *N*-ethyl-*N'*-(dimethylaminopropyl)carbodiimide (EDC or EDAC) and NHS or *N*-hydroxysulfosuccinimide (sulfo-NHS). The simplest approach is to form an amide bond directly with the primary amines on the lysine residues of proteins. However, the chemistry is quite versatile, so after NHS activation, subsequent exposure to a cystamine in the presence of dithiothreitol will produce a thiol group for coupling to amine or carboxyl groups, 2-(2-pyridinyldithio)ethane-amine (PDEA) will produce a reactive disulfide for coupling to free cysteines, and hydrazine will allow aldehyde coupling.

Direct coupling of proteins to EDC/NHS-activated carboxyl groups is straightforward but is nonspecific and may lead to multiple coupling, thereby constraining molecular flexibility. The orientation of coupling is also likely to be somewhat random, so the resulting adlayer will probably comprise a mixture of fully active binding molecules and others wherein the binding sites are partially or fully blocked. Maleimide linkages may be useful in the fortunate case where a single free cysteine is available on the protein of interest and this will almost certainly result in a uniform orientation and single-point tethering. Alternatively, the introduction of a thiol group on the sensor chip surface may be useful if a reactive disulfide group can be introduced onto the target protein first via an amine or carboxyl group. If a free cysteine is not present on the target molecule, disulfide bonds can be broken by a reducing agent under non-denaturing conditions so that these may form a new disulfide bond with reactive disulfide or maleimide but care must be taken that the activity (structure) of the target molecule is not compromised with this approach. Aldehyde coupling may be useful to link to carbohydrate groups (on glycoproteins, for example) if they are prepared first by oxidizing the *cis*-diols of sugar groups with periodate.

The degree of surface activation will affect the amount of ligand immobilized on the surface and can be controlled by the concentration of EDC/NHS and the time of exposure to the reagents. Typically, for concentration analysis, one aims to maximize immobilization so a high activation reagent concentration and a long activation period will be used. For kinetic analysis, the minimum immobilized species that will give a sufficient signal should be immobilized to avoid mass transfer limitations during analyte binding. Thus, in the latter case, a dilute reagent and/or a short activation period is recommended. Surface activation conditions for these

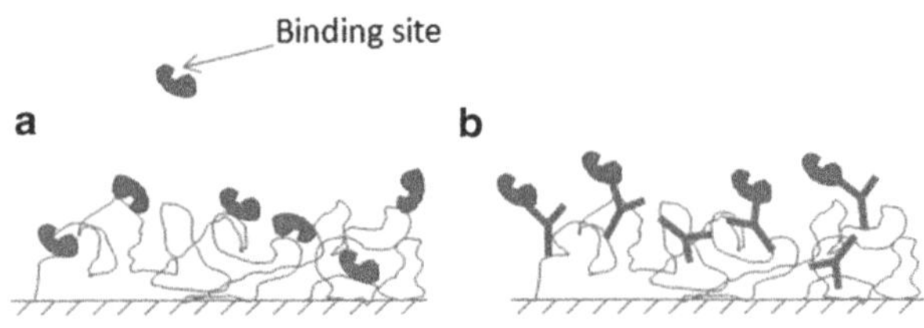

Fig. 7 Orientation of tethered molecules (**a**) random via direct coupling and (**b**) more uniform via indirect coupling

applications should be determined experimentally but good starting points are as follows:

For concentration analysis
A mixture of 200 mM EDAC and 50 mM NHS, injected for about 30 s, using a medium- or high-density sensor chip.

For kinetics analysis
A mixture of 25 mM EDAC and 5 mM NHS, injected for 30 s, using a low-density chip.

A reference channel should be prepared identically, in parallel.

3.3 Surface Immobilization (Tethering) of One or More Binding Partners

For amine coupling, the molecule of interest should be passed over the chip at a concentration in the range 20–200 μg/mL as soon as possible after surface activation to avoid loss of surface reactivity through hydrolysis of the NHS-activated carboxyl groups. The optimal pH should have been determined through preconcentration experiments and the salt concentration should normally be kept below 100 mM (NaCl equivalent) to avoid masking the electrostatic interactions that attract the target species into the surface layer. Typically, a 3–5 min injection at a flow rate of 25–50 μL/min would be used, requiring a total of about 2–50 μg of the target molecule. (A little excess volume (50 μL) in the sample vials/wells might be required to avoid taking up air into the sample needles in most microfluidics-driven systems.) For kinetic analysis, the aim should be to keep the total response of the analyte below about 250 RU to avoid kinetics being complicated by surface crowding. Coupling temperature can be increased to maximize immobilization or lowered to decrease the binding.

Spontaneous coupling via primary amines should then occur. The result of this *direct coupling* will be the immobilized target tethered in essentially random orientations, wherein the direction of the binding site relative to the surface and perhaps even the access to the binding site are variously distributed (Fig. 7a). This heterogeneity can sometimes lead to complex kinetic behaviors, for example if binding sites on some tethered molecules become accessed only at the highest analyte concentrations. A more uniform surface orientation may be possible through *indirect coupling* by first tethering an antibody to the target molecule to the surface via amine coupling and then capturing the target molecule in a subsequent injection (Fig. 7b). This approach is also likely to result in a highly specific tethering of the target molecule even from a complex mixture so can be useful where prior purification is difficult. Similarly, for affinity surfaces such as streptavidin or NTA, the tethering orientation will be uniform, according to the position of the biotin or His-tag on the ligand, respectively, and direct, selective tethering even from complex mixtures such as lysed cells (after solids removal for systems that use microfluidics) is possible.

The choice of which molecule of an interacting pair should be immobilized and which should be the analyte should be considered carefully. The SPR response during analyte binding will, at least to a first approximation, equal the response during ligand immobilization multiplied by the ratio of analyte to ligand molecular weights. Therefore, one would expect a low molecular weight analyte to produce a very small response if the immobilized ligand has a high molecular weight and is bound sparingly. In this case, one might improve the response by immobilizing the low molecular weight molecule and using the high molecular weight partner as the analyte. If the analyte is multivalent and cross-links the ligand molecules, the surface conformation of the latter may change during interaction and affect the response. In this case the immobilization/analyte choice should be reversed.

The reference channel or sensor should be subjected to running buffer during this step.

3.4 Deactivation of the Sensor Chip Surface

Any unreacted, NHS-activated carboxyl functional groups must be deactivated after immobilizing the target ligand, so as to avoid covalent attachment of the analyte during subsequent interaction analysis. This step is straightforward and can be achieved with a 30-s injection of 1 M ethanolamine to cap the residual NHS-activated carboxyl groups.

The control channel should also be subjected to this step.

3.5 Surface Stabilization

During the above immobilization steps, the surface may have attracted some non-specifically bound species or there may be weakly tethered components, both of which may result in a downward drift in response during subsequent interaction steps. It is best, therefore, to run running buffer over the surface until the response reaches a steady state. In Subheading 3.7, below, regeneration of the sensor chip surface after an interaction is described and it is recommended that at least one regeneration step be carried out to remove any weakly bound ligand prior to following interactions. As described below, particularly when using a new immobilized species, one should always check the stability of the immobilized ligand to the regeneration conditions and adjust these if necessary.

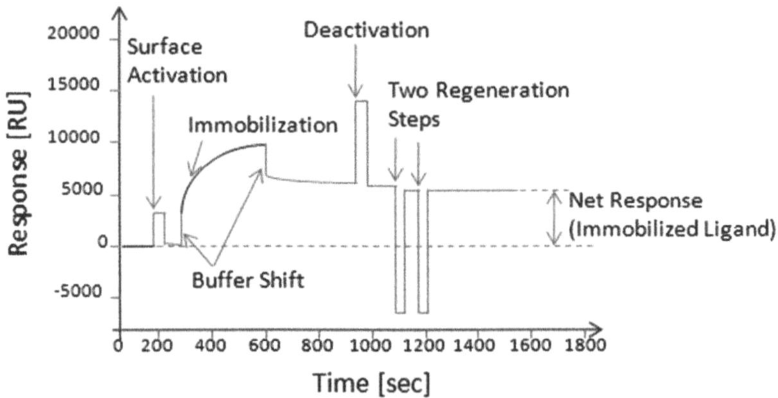

Fig. 8 A typical sensorgram during surface activation, immobilization, deactivation and two regeneration steps, showing a stable net response

At the end of all steps from Subheadings 3.2 to 3.5, the overall sensorgram should have a form similar to that in Fig. 8.

3.6 Interaction with the Analyte(s)

A running buffer should be chosen to stabilize the proteins but not interfere with protein binding. For example, Tris (which contains primary amines) should be avoided when using amine coupling. Generally, any physiologically relevant buffer with a little surfactant added to reduce non-specific interactions will suffice. Two recommended running buffers are HEPES (4-(2-hydroxyethyl)piperazine-1-ethanesulfonic acid) buffered saline (HBS, 10 mM HEPES containing 0.15 M NaCl, 3.4 mM EDTA, and 0.005% surfactant P20 at pH 7.4) or phosphate buffered saline (PBS) containing 0.005% Tween 20, pH 7.4. In some cases, particularly where the analyte comes from a very complex or challenging sample matrix such as whey (7), additional salt can be added to further suppress nonspecific interactions (unlike during covalent immobilization, when high salt will decrease the penetration of the molecule of interest into the surface layer).

Analyte samples and blanks (controls) should be prepared in the running buffer so that nonspecific interactions and bulk shifts can be subtracted and net changes in responses determined.

Prior to studying an interaction, it is recommended that several blank injections of running buffer be made immediately before analyte injection. In theory, as running buffer is constantly being run over the chip surface while docked, there should be no difference between the baseline SPR signal in between steps and the signal during an injection (as long as the same buffer source is used for both). However, SPR instruments are extremely sensitive, and changing the source of running buffer from background flow to injected samples can result in a slightly different response between the immobilized surface and the blank control. Several blank injections can "settle" the surface down, in practice.

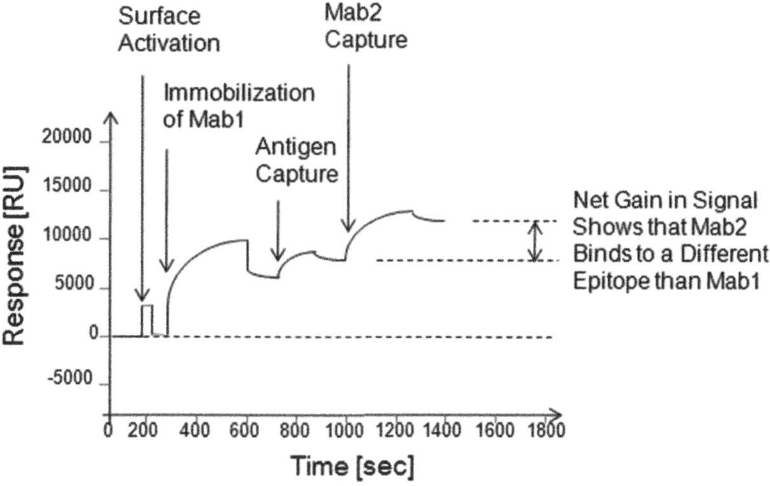

Fig. 9 Epitope mapping by SPR

(As a practical aside, and one which is generally true when working with highly sensitive analytical instruments, one should *never* lean on the instrument or even the bench upon which it is sitting while carrying out an experiment. The optics within these instruments are very sensitive to alignment and a detectable shift in SPR response may result if one leans on the top of an instrument or against the bench upon which it is sitting!)

There are several levels of data one might wish to collect by SPR. The first is a simple "yes/no" answer to either the question of whether or not two proposed binding partners actually interact or the question of whether or not an analyte known to interact with the ligand is present in a given sample solution. Neither of these analyses requires extensive quantification and generally it is only a matter of subtracting the appropriate reference channel data to eliminate nonspecific binding and buffer shift to determine whether there appears to be binding.

Likewise, epitope mapping relies on detecting additive responses to the stepwise creation of complex structures but not necessarily quantitative determination. In epitope mapping by SPR, one can determine, without the use of labels, whether different antibodies bind to different parts of the antigen. The technique involves immobilizing the first monoclonal antibody (Mab1) on the chip surface via amine coupling, binding the antigen to the bound Mab1, and then injecting a second monoclonal antibody (Mab2), also known to bind to the antigen. If the two antibodies bind at the same or a closely located epitope, Mab2 will be blocked and there will be no increase in signal after its injection. If the two Mabs bind to different epitopes, this will be seen by an increase in the signal (Fig. 9).

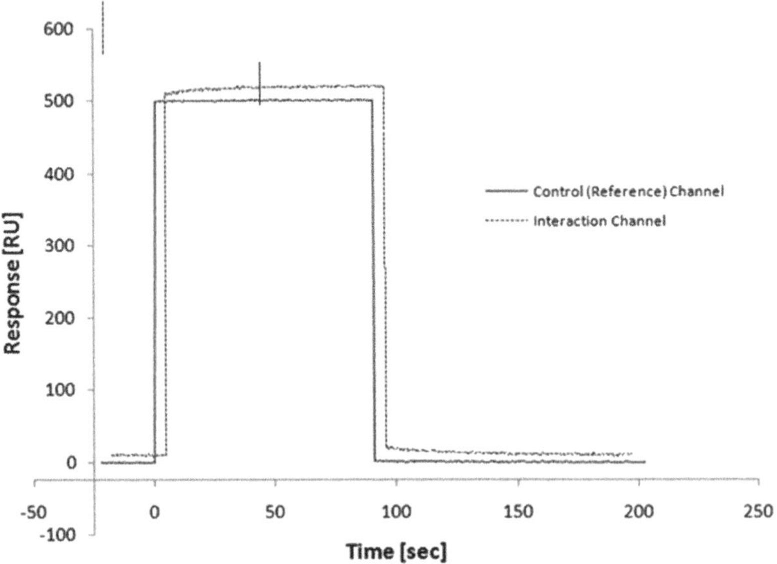

Fig. 10 Typical "raw" interaction and reference channel sensorgrams, requiring data processing before the SPR responses due to specific binding, can be determined (Fig. 10)

Quantitative data are required for concentration analysis. In this case, a calibration series must be prepared using an appropriate serial dilution of standard and the responses of analyte samples compared with this for quantification. This type of analysis requires, above all, consistency of operation to ensure the unknown analyte(s) can be quantitatively compared to the standard curve. The main constraint is to ensure that the data are reproducible and that the surface does not become saturated during binding; otherwise differences between analyte concentrations cannot be distinguished. This, however, should be an obvious constraint and as long as the samples are diluted to bring the SPR responses to within those of the standard curve data, all should be well.

Quantitative analysis of binding kinetics requires the most stringent planning and control of the experiments, careful consideration of the data quality, and elimination of artifacts. Obtaining the correct shapes of the association and dissociation curves is crucial and repeatable sensorgrams should be collected, preferably spanning the range between zero and saturation of the immobilized ligand.

An example of a typical raw interaction sensorgram and the corresponding control or reference channel sensorgram is given in Fig. 10. Note that there is a large step change in the response of both channels, due to the difference in the bulk refractive index of the analyte sample compared to the running buffer; the two curves are slightly offset from one another, which may arise from minor

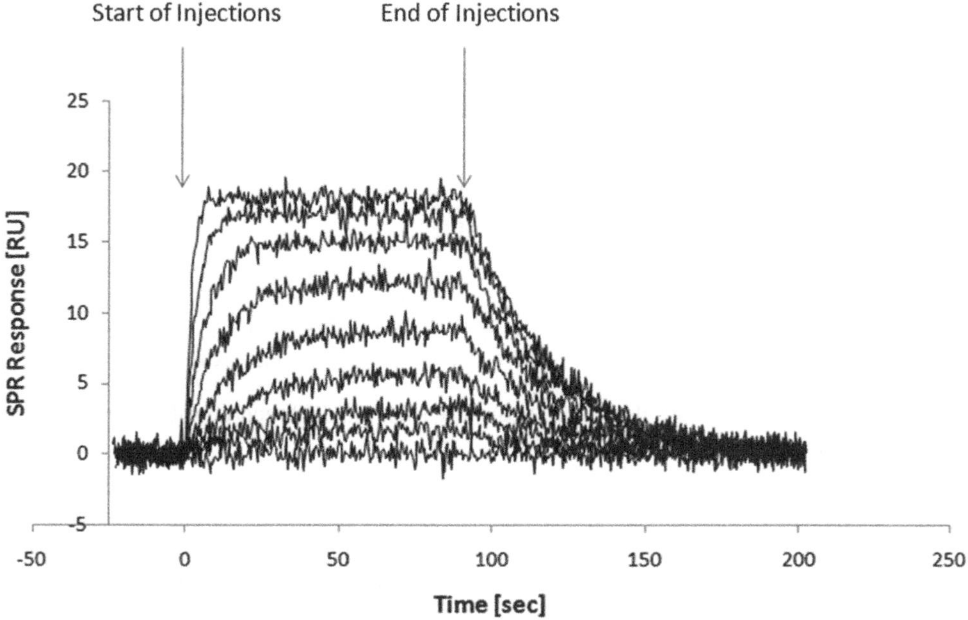

Fig. 11 A series of sensorgrams for an interaction analysis after processing of raw data (simulated using Clamp software)

differences in flow paths and baseline signals between channels; and there are a couple of spikes (artifacts) in the curves. Initial data processing is required to vertically and horizontally align the interaction sensorgrams with the reference channel sensorgram, to delete the spike artifacts, and to then subtract the reference channel from each interaction channel sensorgram. Although the scale of the buffer shift will usually obscure this, the reference channel does not only undergo buffer shift but also undergo some association and dissociation due to nonspecific interactions. The scale also obscures the contribution of noise to the signal. The result, including a number of interaction sensorgrams, rather than just the single one shown in Fig. 10 for clarity, is shown in Fig. 11. Note that the noise in the sensorgrams is much more evident. The data for these two figures were generated using Clamp, a data fitting, and modeling software package obtained from the Center for Interaction Analysis at the University of Utah (http://www.cores.utah.edu/interaction/software.html).

The curves in Fig. 11 represent a good set of data for kinetic analysis. The shapes of the association and dissociation curves indicate simple binding; the sensorgrams cover the range from zero response (a blank analyte sample) through to a well-saturated equilibrium condition, with an even distribution over that range; the responses are small but clearly distinguishable (ideal for kinetic analysis, whereas large responses may be subject to mass transfer effects); and the dissociation curves have clear curvature. Ideally,

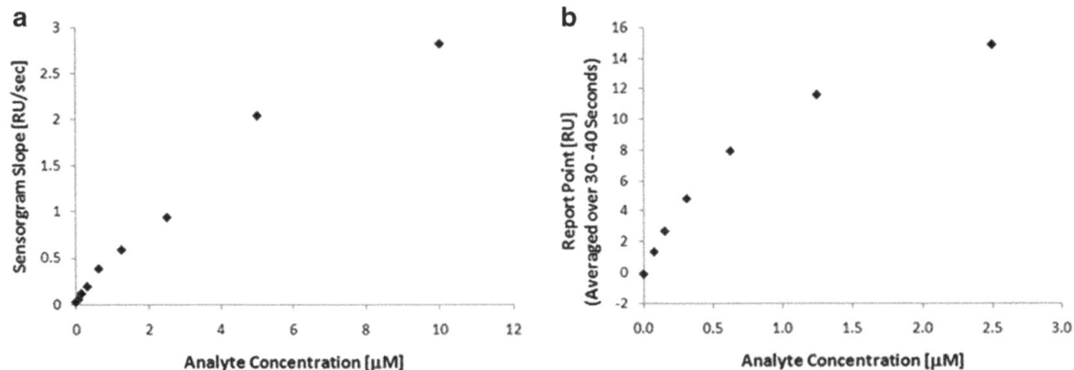

Fig. 12 Standard concentration curves derived from (**a**) initial slopes and (**b**) report points of average responses between 30 and 40 s from the data in Fig. 11

one should obtain at least triplicate data sets for the sensorgrams at each concentration. Somewhat unusually, in this case, the sensorgrams all appear to return to the baseline during the rather brief dissociation phase—this is not typical of all interactions, by any means, and usually a regeneration step is required, as described below, to remove the analyte completely between interactions.

For concentration analysis, there are two methods of quantifying the analyte concentration. In the first method, the initial slopes of the sensorgrams during association are determined and used to determine concentration by comparing against a standard curve of initial sensorgram slope versus concentration (8). In the second method, a "report point" is used, in which the SPR response is taken during the association at a set time after injection—the higher the analyte concentration, the higher the response should be at the report point, provided that the surface has not become saturated (7). Usually an average of the SPR response over a short time interval around the report point is used to smooth out noise. Fig. 12 shows the standard curves generated from the data in Fig. 11, using (a) initial slopes of the sensorgrams and (b) the average responses over the period 30–40 s as report points. Note that the initial slopes give a reasonable standard curve over a greater concentration range than the report points in this case, where the data are not shown for the latter above 3 µM because the responses flatten out at higher concentrations. This should not present a limitation if more concentrated analyte samples can be diluted to give report point values within the standard curve range.

In the simplest case, we can consider simple one to one binding to be of the form

$$A + B \underset{k_d}{\overset{k_a}{\rightleftarrows}} AB \tag{2}$$

and the kinetics to follow the form

$$\frac{\mathrm{d}[AB]}{\mathrm{d}t} = k_\mathrm{a}[A][B] - k_\mathrm{d}[AB] \tag{3}$$

where (A) is the concentration of free, immobilized ligand, A; (B) is the concentration of free analyte, B; (AB) is the concentration of the analyte-ligand complex, AB; k_a is the association constant; and k_d is the dissociation constant.

Initially, when the analyte encounters the clean ligand surface, (AB) is zero, while there are plenty of ligand sites available so (A) may be assumed to be essentially constant for a short time. Thus, according to Eq. 3, the initial rate of change in (AB) (the slope of the sensorgram) is approximately directly proportional to (B), as shown in Fig. 12a. On the other hand, at the report point between 30 and 40 s, the ligand approaches saturation at the higher concentrations (i.e., (AB) is appreciable, while the free ligand sites available, (A), has decreased), so the SPR response is not as sensitive to differences in analyte concentration above 3 μM. Note that the upper useful concentration is given here only by way of an example and is unique to the hypothetical interaction simulated in Figs. 10 and 11—the actual analyte concentrations over which sufficient resolution can be obtained in any particular case will depend upon the particular binding pair under consideration, as well as the density of the immobilized ligand on the surface. For concentration analysis by either method, one would wish to use data far below ligand saturation, so maximizing the ligand density during immobilization is desirable.

At the end of the association period, the analyte sample is replaced by running buffer and the interaction monitoring continued for a given time. Accurate determination of k_d requires some curvature in the dissociation curve, which may require the extension of dissociation for many hours if k_d is very low. The limiting factor for measuring very slow dissociation ($k_\mathrm{d} < 10^{-6}\ \mathrm{s}^{-1}$) is signal drift, whereby the drift in baseline signal becomes significant compared to the change in response due to dissociation.

Early analyses of binding kinetics were carried out using linearization but this is unnecessary when computers are easily able to fit data by iterative means. Instruments usually come with proprietary software packages that not only drive the instrument during data collection but analyze and report association/dissociation kinetics. Two analysis software packages, Clamp and Scrubber 2, are available (both via http://www.cores.utah.edu/interaction/software.html), the latter being a particularly powerful package for manipulation and analysis of raw SPR data.

A system of differential equations can be derived from Eq. 2. Because the amount of free analyte in the injection is large compared to the number of ligand sites available for binding on the chip surface and we can assume that the supply of A to the chip surface is not limited by mass transfer, (A) is constant:

$$\frac{\mathrm{d}[A]}{\mathrm{d}t} = 0 \tag{4}$$

The change in immobilized ligand concentration is then

$$\frac{\mathrm{d}[B]}{\mathrm{d}t} = -k_\mathrm{a}[A][B] + k_\mathrm{d}[AB], \tag{5}$$

and the change in the analyte-ligand complex is given by Eq. 3, repeated below for convenience:

$$\frac{\mathrm{d}[AB]}{\mathrm{d}t} = k_\mathrm{a}[A][B] - k_\mathrm{d}[AB] \tag{3}$$

For equilibrium analysis, we can say that $\frac{\mathrm{d}[AB]}{\mathrm{d}t} = \frac{\mathrm{d}[B]}{\mathrm{d}t} = 0$, and therefore

$$k_\mathrm{a}[A][B] = k_\mathrm{d}[AB] \tag{6}$$

Amongst the variables in Eq. 6, (A) is known, as it is the concentration of the analyte injections we control, and (AB) is the measured (dependent) variable, while (B), $k_\mathrm{a,}$ and k_d are all unknown. Note that (A), (B), and (AB) are all functions of time. We may take a number of approaches to the determination of k_a and k_d. First, we recognize that (AB) is proportional to R, the response of the instrument to bound analyte at any time, so

$$[AB] = \omega R \tag{7}$$

where ω is a constant of proportionality. Also, $(B)_0$ is the total initial concentration of ligand sites available and because $(AB)_{\mathrm{max}} = (B)_{0,}$ then $(B)_0 = \omega R_{\mathrm{max}}$. As (AB) goes from 0 to $(AB)_{\mathrm{max}}$, (B) goes from $(B)_0$ to 0. Therefore

$$[B] = [B]_0 - [AB] = \omega(R_{\mathrm{max}} - R) \tag{8}$$

We can then write the kinetics in terms of R as

$$\frac{\mathrm{d}(\omega R)}{\mathrm{d}t} = k_\mathrm{a}[A]\omega(R_{\mathrm{max}} - R) - k_\mathrm{d}\omega R \tag{9}$$

Finally, we can eliminate the constant ω to express the kinetics entirely in terms of the two known variables (A) and R, the constant R_{max} estimated from the sensorgram curves and the two unknowns k_a and k_d:

$$\frac{\mathrm{d}R}{\mathrm{d}t} = k_\mathrm{a}[A](R_{\mathrm{max}} - R) - k_\mathrm{d}R \tag{10}$$

We can now make initial guesses at R_{max}, k_a, and k_d and then have the software iterate to fit the model in Eq. 10 to the experimental data by minimizing the sum of squares of the residuals between fitted and experimental data.

At equilibrium, $\frac{dR}{dt} = 0$, so solving for R_{eq} gives

$$R_{eq} = \frac{k_a[A]R_{max}}{k_d + k_a[A]} = \frac{[A]R_{max}}{K_D + [A]} \quad (11)$$

where K_D is the dissociation equilibrium constant $K_D = \frac{k_d}{k_a}$.

Integrating Eq. 10 with respect to time gives, during association,

$$R = \frac{[A]R_{max}}{K_D + [A]}\left[1 - e^{-(k_a[A]+k_d)t}\right] \quad (12)$$

Equations 10 or 12 can be used to fit sensorgrams taken over a range of (A) and the fit should be shown in published results to visually demonstrate that the fitted lines pass properly through the curves and that there are no artifacts or deviations from expected behavior. Equilibrium analysis determines K_D from a series of injections over a range of (A), where each curve should give the same value of K_D as long as they have reached a true plateau at the point for which the analysis is undertaken. Attaining a stable response at equilibrium is critical—equilibrium analysis *cannot* be applied if the association response is still increasing at the end of injection! The K_D values should also be consistent with the individual k_a and k_d values calculated from fitting Eq. 10 or 12.

During the dissociation phase, $(A) = 0$, so

$$\frac{d(R)}{dt} = -k_d R \quad (13)$$

Equation 13 is a typical first-order decay equation, characterized by a constant half-life. This means that the dissociation response curve is independent of (A) and will look identical around any given response level, R_0,

$$R = R_0 e^{-k_d t} \quad (14)$$

Curvature must be evident in the decay curve or else the decay is indistinguishable from a zeroth-order decay, $\frac{dR}{dt} = -k'$, where k' is a constant. Fitting of the dissociation curves by Eq. 14 to determine k_d is straightforward.

Figure 13 shows processed sensorgrams and fitted curves for all three systems described in Subheading 2 (9). It is always important that such curves and fitted lines are shown, rather than simply

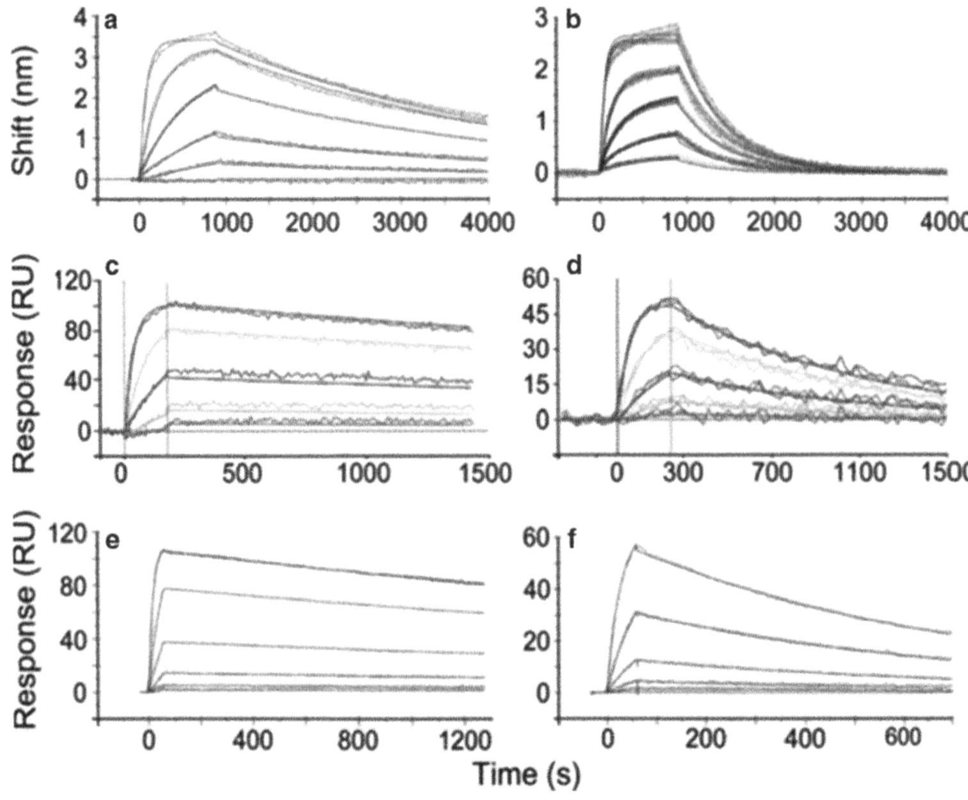

Fig. 13 Sensorgrams with fitted curves for interactions of an antibody fragment with rat (**a**, **c**, **e**) and human (**b**, **d**, **f**) calcitonin gene-related peptide alpha, measured on ForteBio Octet (**a**, **b**), Bio-Rad ProteOn (**c**, **d**), and Biacore 3000 (**e**, **f**) systems. Noisy and smooth lines distinguish measured data from simulated fits (reproduced from ref. 9)

a table of calculated association/dissociation of equilibrium constants. Ideally, a plot of residuals should also be shown but from Fig. 13, at least the reader can judge the goodness of fit for each sensorgram.

3.7 Regeneration

Regeneration is described here after the interaction step because it is routinely used between interaction cycles. Nevertheless, regeneration conditions must be determined prior to carrying out interaction analyses because they will critically affect surface stability and therefore repeatability. Usually, the analyte will not dissociate quickly enough to be completely dissociated by the end of an experimental run but it can be quickly removed by injecting an appropriate regeneration buffer to disrupt the intermolecular forces (Fig. 14). In the ideal case, regeneration removes all residual analyte but maintains analyte binding capacity of the surface from one interaction cycle to the next. In practice, a compromise is normally necessary between attaining sufficiently rigorous conditions for the effective removal of all residual analyte and nonspecifically bound materials from the ligand/surface and avoiding overly harsh

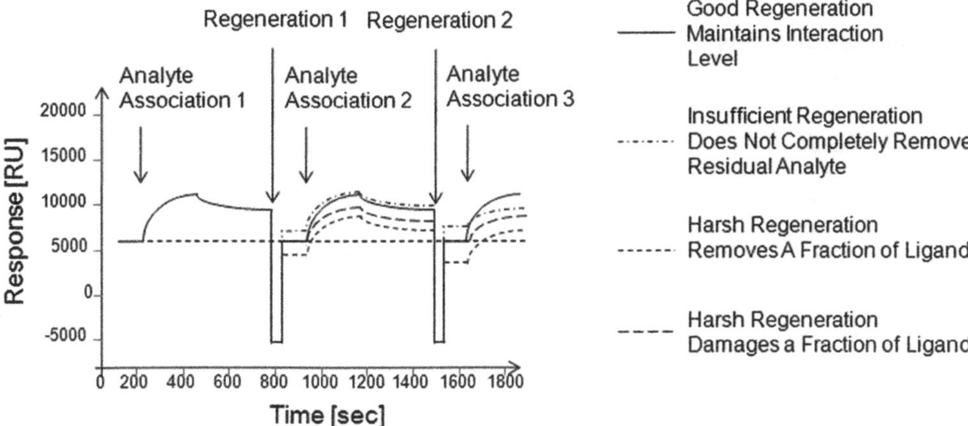

Fig. 14 Post-interaction regeneration of sensor chip surface

conditions that can remove ligand from the surface or damage the ligand's binding functionality. Figure 14 shows a number of scenarios where either too gentle a regeneration buffer results in insufficient removal of analyte, leading to an increase in baseline and gradual decline in analyte binding capacity, or too harsh a regeneration buffer results in either removal of ligand, causing a decrease in baseline and reduced analyte binding capacity or damage to the ligand binding functionality (without removal of mass), again reducing analyte binding capacity. It is important to establish the stability of the immobilized surface from one run to the next and the best conditions for regeneration must be determined experimentally by scouting a range of regeneration buffers. Start with gentle conditions and move to more stringent conditions if insufficient analyte/nonspecifically bound material is removed. Generally, the most significant variables are the regeneration buffer pH, salt concentration, presence of detergents or chaotropes, and the length of the regeneration buffer injection.

Table 1 lists a number of commonly used regeneration buffers for various strengths and types of interaction. A good starting point for regeneration for antibody-antigen interactions with immobilization by amine coupling of the ligand and providing that the ligand is stable at acid pH is to scout a range of 10 mM glycine-HCl buffers at pH values between 1.75 and 3, with 10 s injections. Regeneration of non-covalently immobilized ligands may require preparation of the immobilized surface between interaction cycles and care must be taken to attain as close as possible to identical surfaces for each run to ensure repeatability and comparison between interactions. As mentioned at the beginning of this section, regeneration conditions must be thoroughly scouted and a robust protocol determined before interaction studies begin. For concentration analyses, preparation of a new standard curve is recommended every ten cycles or so, depending on the stability of the surface under repeated interaction/regeneration runs. Running a

Table 1
Common regeneration buffers

Type of interaction	Acidic	Basic	Hydrophobic	Ionic
Weak	pH > 2.5 Formic acid HCl 10 mM Glycine–HCl	pH < 9 10 mM NaOH in HEPES or PBS	pH < 9 50% Ethylene glycol	1 M NaCl
Intermediate	pH 2–2.5 Formic acid HCl 10 mM Glycine–HCl H_3PO_4	pH 9–10 NaOH 10 mM Glycine/NaOH	pH 9–10 50% Ethylene glycol	2 M $MgCl_2$
Strong	pH < 2 Formic acid HCl 10 mM Glycine–HCl H_3PO_4	pH > 10 NaOH	pH > 10 25–50% Ethylene glycol	4 M $MgCl_2$ 6 M Guanidine chloride

few interaction/regeneration cycles before beginning to record critical data is worthwhile, to attain a steady state condition of the sensor chip surface whereby weakly bound ligand molecules are removed, labile ligands are destroyed, and a balance is reached between removal and binding of analyte and nonspecifically bound species with successive cycles.

3.8 Reproducibility of Experiments

Clearly, for a robust analysis, a statistically significant number of repeated measurements must be made to determine the certainty of results. There is no reason why a well-designed and executed sensor analysis should not produce excellent repeatability within an acceptable level of experimental error. One should *never* attempt to publish interaction behaviors that have not been verified by repeated observations because without this, it is not possible to tell whether there are systematic errors between injections of analyte, such as incomplete removal of analyte, loss of binding activity, or loss of ligand from the chip surface. One should also be very cautious about reporting "strange" kinetic behaviors in sensorgrams or over-interpreting unexpected deviations from the standard "sawtooth" shape of the association/dissociation curves shown here in Figs. 3, 9, 10, 11, and 13. Such deviations may well offer interesting engineering or scientific insights into mass transfer, nonspecific interactions, electrostatics, and so on but they are highly unlikely to be relevant to physiological binding events. For kinetic analysis of physiologically relevant biomolecular interactions, or detailed study of supramolecular assembly of proteins, the SPR should reliably allow a careful user to quantify k_a, k_d, and K_D.

3.9 Processing of Raw Data and Appropriate Analysis and Presentation of Data

Subheading 3.6 (Figs. 10 and 11) introduced the idea of processing of raw data. For kinetic analysis, each sensorgram in an analyte concentration interaction series must be aligned in both the "*y*" and "*x*" directions to ensure all interaction response curves start from the same point. Slight misalignments of raw data will inevitably occur, due to variations in injection times or tiny differences in flow path lengths between sample needles and the chip surface. Flow channels may undergo a slightly different degree of drift between injections, while small differences in the chip surfaces, extent of regeneration, and so on will mean each interaction starts from a slightly different response baseline. These must all be brought together to properly compare binding rates. Spikes, caused by transient air bubbles or electrical interference, must then be removed so that data can be fitted without distortions due to artifacts that do not represent the association/dissociation events. Most systems and independent analysis software packages have features that allow this to be done either automatically or under manual control.

3.10 Thermodynamics

Some information on binding, the water structure around proteins before and after association, and various relationships between enthalpic and entropic contributions to interactions and conformational changes in the proteins can be inferred from thermodynamic measurements. A discussion of these effects is beyond the scope of this chapter but the following relationships can be explored with a number of instruments in which sensor chip temperature can be systematically altered between runs.

Association and dissociation constants are functions of temperature, with both typically increasing with temperature according to an Arrhenius relation as follows:

$$\ln k = \ln A - \frac{E_A}{RT} \tag{15}$$

where k is the rate constant in question, A is the pre-exponential factor (as distinct from analyte A), E_A is the activation energy, and R is the universal gas constant. By varying the temperature between runs and plotting $\ln k$ versus $1/T$, one can determine both A and E_A. Also

$$\Delta G^\circ = RT \ln \frac{K_D}{C^\circ} \tag{16}$$

where ΔG° is the Gibbs free energy change associated with binding and C° is the standard state concentration (usually 1 M). The binding affinity includes both enthalpic (ΔH) and entropic (ΔS) contributions

$$\Delta G = \Delta H - T\Delta S \tag{17}$$

so combining Eqs. 16 and 17 and assuming that ΔH and ΔS are constant with temperature, one can use Eq. 18 below to plot $\ln\frac{K_D}{C^\circ}$ versus $1/T$ for runs made over a range of temperatures to determine both thermodynamic quantities:

$$\ln\frac{K_D}{C^\circ} = \frac{\Delta H}{RT} - \frac{\Delta S}{R} \tag{18}$$

However, unless the temperature range is quite narrow, neither ΔH nor ΔS will be constant so that the plot suggested by Eq. 18 will not be truly linear. A more accurate approach in that case is to fit the integrated form of the van't Hoff equation, which requires information regarding the change in heat capacity with binding (10).

4 Summary

Optical biosensors such as SPR offer the ability to follow association and dissociation kinetics in real time with label-free molecules. A wide variety of systems is available from a number of suppliers, with various immobilization chemistries. The most common starting point for immobilization is through amine coupling to carboxy groups, while other chemistries such as streptavidin-biotin or NTA-His-tag can offer biospecific tethering with the advantage of uniform orientation of the ligand. Interaction analysis requires exacting control of experimental conditions and careful attention to detail, together with the demonstration of reproducible results. The most important aspect of biointeraction kinetic analysis is that the fitted curves are shown to closely and reproducibly fit the experimental data across an appropriate range of analyte concentrations.

References

1. Rich RL, Myszka DG (2006) Survey of the year 2005 commercial optical biosensor literature. J Mol Recognit 19:478–534
2. Rich RL, Myszka DG (2007) Survey of the year 2006 commercial optical biosensor literature. J Mol Recognit 20:300–366
3. Rich RL, Myszka DG (2008) Survey of the year 2007 commercial optical biosensor literature. J Mol Recognit 21:355–400
4. Rich RL, Myszka DG (2010) Grading the commercial optical biosensor literature-Class of 2008: 'The Mighty Binders'. J Mol Recognit 23:1–64
5. Homola J, Piliarik M (2006) Surface plasmon sensors. In: Homola J (ed) Surface plasmon based senors. Springer, Berlin
6. Van Der Merwe PA (2001) Surface Plasmon Resonance. In: Harding SE, Chowdhry BZ (eds) Protein-ligand interactions: hydrodynamics and calorimetry. Oxford University Press, Oxford
7. Indyk HE, Filonzi EL (2005) Determination of lactoferrin in bovine milks, colostrums and infant formulas by optical biosensor analysis. Int Dairy J 15:429–438
8. Fee CJ, Billikanti JM (2009) Simultaneous, quantitative detection of five whey proteins in multiple samples by surface plasmon resonance. Int Dairy J 20:96–105
9. Abdiche Y, Malashock D, Pinkerton A, Pons J (2008) Determining kinetics and affinities of protein interactions using a parallel real-time label-free biosensor, the Octet. Anal Biochem 377:209–217
10. Yoo SH, Lewis MS (1995) Thermodynamic study of the pH-dependent interaction of chromogranin A with an intraluminal loop peptide of the inositol 1,4,5-trisphosphate receptor. Biochemistry 34:632–638

Chapter 18

Label-Free, Real-Time Interaction and Adsorption Analysis 2: Quartz Crystal Microbalance

Conan J. Fee

Abstract

In this chapter, a second biosensor technique is described: the quartz crystal microbalance (QCM). The quartz crystal microbalance is a physical technique that detects changes in the resonance frequency of an electrically driven quartz crystal with changes in mass. Unlike surface plasmon resonance (SPR), QCM is affected by both the water that may be associated with the adsorbed layer and by conformational changes in the adsorbed species, while SPR is insensitive to both effects. Thus QCM can both corroborate the findings of an SPR experiment and provide some complementary information. Also, the QCM surface is highly versatile and can range from plain quartz, through gold and other metal surfaces (e.g., titanium or stainless steel) to polymeric materials. Thus, the QCM technique has wide utility in tracking interactions with a variety of materials.

Key word Quartz crystal microbalance

1 Introduction: Quartz Crystal Microbalance

The quartz crystal microbalance (QCM) allows the real-time measurement of mass associated with adsorption of solution-phase analytes either directly on material surfaces or onto ligands immobilized on a surface, forming a so-called adlayer. The underlying principle of operation is that the resonant frequency of a quartz crystal excited by an oscillating electrical current (Fig. 1a) is related to the total mass of the crystal, which, to a first approximation, includes bound coatings and adsorbed materials as well as local solvent molecules, particularly water. More particularly, such a crystal can generally be coated with various materials, including gold, silver, or titanium, and a number of polymers so can be used, for example in biomedicine applications (1). A further refinement of the technique is made possible if the resonant frequency is followed immediately after ceasing the electrical excitement in order to measure the rate of decay in the signal or, indirectly, the dissipation of the energy in the oscillating system (Fig. 1b). This approach

Juliet A. Gerrard (ed.), *Protein Nanotechnology: Protocols, Instrumentation, and Applications*, Methods in Molecular Biology, vol. 996, DOI 10.1007/978-1-62703-354-1_18, © Springer Science+Business Media New York 2013

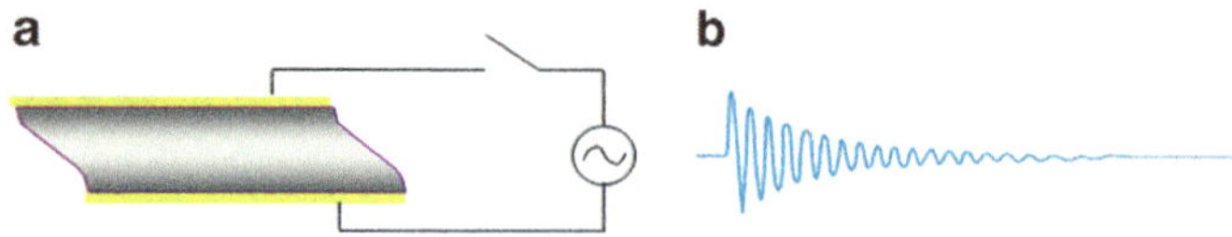

Fig. 1 (**a**) Cross-sectional representation of the strain of a quartz crystal driven by an alternating electric current when the circuit is closed. (**b**) Decaying resonant frequency when the circuit is opened

is known as QCM with dissipation (QCM-D), where the rate of decay in the signal is related in part to the viscoelasticity of the adlayer, such that a loosely bound, soft adlayer will tend to dampen the signal more than a tightly bound rigid adlayer. The 2001–2005 literature on protein interaction studies by QCM was reviewed by Cooper and Singleton, providing an introduction to the QCM technique, protein interactions, and describing a wide range of applications (2).

2 Principles of QCM-D

A quartz crystal, or more specifically the α-quartz form of SiO_2, is a piezoelectric material, i.e., one in which a mechanical strain (for example in response to compression or shear) will induce an electric field within the material. Conversely, the application of an electric field will induce a strain in the material. Thus, a thin slice (0.1–0.3 mm) of quartz crystal can be made to distort laterally through the application of an electric potential across electrodes coated onto two opposite faces of the slice. An oscillating current will cause the crystal to deform in alternate directions at a resonant frequency, f, given by

$$f = n\frac{v_q}{2t_q} = nf_0 \quad (1)$$

where n is an odd integer describing the fundamental frequency ($n=1$, $f_0=v_q/2t_q$) or harmonic overtones ($n=3, 5, 7\ldots$), v_q is the speed of sound in the quartz slice, and t_q is the thickness of the slice. The mass per area of crystal, m_q, is related to the thickness by $m_q=t_q\rho_q$ (kg/m^3), where ρ_q is the density of the quartz (3). Substituting for m_q in Eq. 1 and differentiating with respect to m_q gives

$$df = -\frac{f}{m_q}dm_q \quad (2)$$

If we can assume that an added mass is small compared to the total weight of the crystal; that an adlayer is adsorbed tightly with no slip; and that it is evenly distributed over the entire surface, then by letting $d \rightarrow \Delta$, we can obtain the Sauerbrey relation, which states that the frequency change is linearly related to mass change (4):

$$\Delta f = -\frac{n}{C}\Delta m \tag{3}$$

where $C = \frac{\nu_q \rho_q}{2f_0^2}$ is the mass sensitivity. For the quartz crystals used in QCM-D, ρ_q is typically 2,650 kg/m^3 and ν_q = 3,440 m/s, so the mass sensitivity for a crystal oscillating at a resonant frequency of 5 MHz (n=1) is about 17.7 ng/(cm^2 Hz) (3). In aqueous solutions, the frequency can be resolved to within 0.2 Hz, so that mass changes of the order 3–4 ng can be tracked by following changes in the fundamental frequency on a typical quartz crystal sensor of diameter 1 cm. In water, the QCM operating with a resonant frequency of 5 MHz will probe a region near the crystal surface about 250 nm deep (5). Clearly, the frequency change is amplified by the use of overtones so that sensitivity is increased three-fold by tracking the third overtone (n=3) and a crystal with a higher fundamental frequency will increase sensitivity by f_0^2, although, ultimately, sensitivity will be limited by the signal to noise ratio of the instrument and its ability to track higher frequencies.

Höök gives the useful examples that a monolayer of water on a 5 MHz crystal oscillating at its fundamental frequency gives a −1.4 Hz shift, while a monolayer of protein will (depending upon the molecular weight) result in a frequency shift of −20 to −80 Hz, easily followed by the QCM by a frequency counter (6).

One must also be careful, however, to control experimental conditions carefully, to incorporate suitable control experiments, and to analyze results thoroughly, as the assumption that the adlayer is tightly and rigidly adhered to the surface is highly questionable under many circumstances. Indeed, it is the recognition of the potential for violation of this underlying assumption that can lead to useful insights into adsorption and interaction behaviors through proper use of dissipation measurements.

Dissipation of energy can occur through viscous damping of the surrounding fluid, water bound to the adsorbed protein species or trapped between bound molecules on the crystal surface coating, and/or differential movement between the underlying oscillating quartz and a soft adsorbed layer (which may exhibit viscoelastic behaviors, especially adsorbed cells). In cases where viscoelastic behavior occurs, the Sauerbrey Eq. 3 will underestimate the true mass of the adlayer. The dissipation, D, is the inverse of the Q factor that describes the ability of a system to retain energy, and is directly related to the ratio of energy dissipated, $E_{\text{dissipated}}$, to that stored in the oscillating system, E_{stored}:

$$D = \frac{1}{Q} = \frac{E_{\text{dissipated}}}{2\pi E_{\text{stored}}} \tag{4}$$

Höök realized that when the driving power is switched off, the voltage or current decays as an exponentially damped sinusoidal

$$A(t) = A_{\downarrow}0e^{\uparrow}(t/\tau)\sin(2\pi ft + \varphi) \quad (5)$$

where ϕ is the phase angle of the signal, A_0 is a constant, and τ is the decay time constant of the signal. As dissipation is related to the decay time constant by $D = 1/\pi f\tau$, both f and D can be determined simultaneously by numerically fitting the measured voltage or current during the decay to Eq. 5.

An early example of the information that can be obtained by QCM-D is that of mussel adhesive protein, first adhering loosely to a gold surface (tracked by a negative Δf and a positive ΔD) and then being cross-linked to increase surface rigidity (tracked by a positive Δf and a negative ΔD), as shown in Fig. 2 (7).

Figure 3 shows the adsorption of β-casein on stainless steel before (a) and after (b) protection of the surface by grafting of poly(ethylene glycol) chains (8), showing that protein adsorption could be reduced 40% when using a combination of native poly(ethylene glycol) and polyethyleneimine.

3 QCM-D Experiments

Commercial suppliers offer a range of QCM sensors with various surface coatings. Gold is a common coating material, where the gold serves as the electrodes for the crystal. Plain quartz, silver, copper, chromium, titanium, stainless steel, various polymeric surfaces such as polystyrene or polycarbonate, and functionalized surfaces such as carboxylated, streptavidin, or nitrotriacetate are available. It is also possible to spin coat or vacuum sputter various materials onto a crystal, making QCM/QCM-D an extremely versatile technique for studying the adsorption of proteins, DNA, cells, viruses, etc., on many materials. Its use in biomaterial research is thus an obvious application, as is the development of biosensors or study of fouling of process equipment in the food, beverage, fermentation, and pharmaceutical industries.

Assuming that a suitable instrument is available for driving the crystal and monitoring the frequency/voltage/current changes, the experimental set-up for QCM-D is relatively straightforward. Tight temperature control is essential. A flow chamber is typically supplied through inert microtubing by a peristaltic pump at a flow rate of approximately 100 μL/min, although the actual rate will depend upon the dimensions of the crystal and flow chamber concerned. A low-pulse pump type is preferable to minimize pressure pulsations affecting the signals and while in principle the flow can

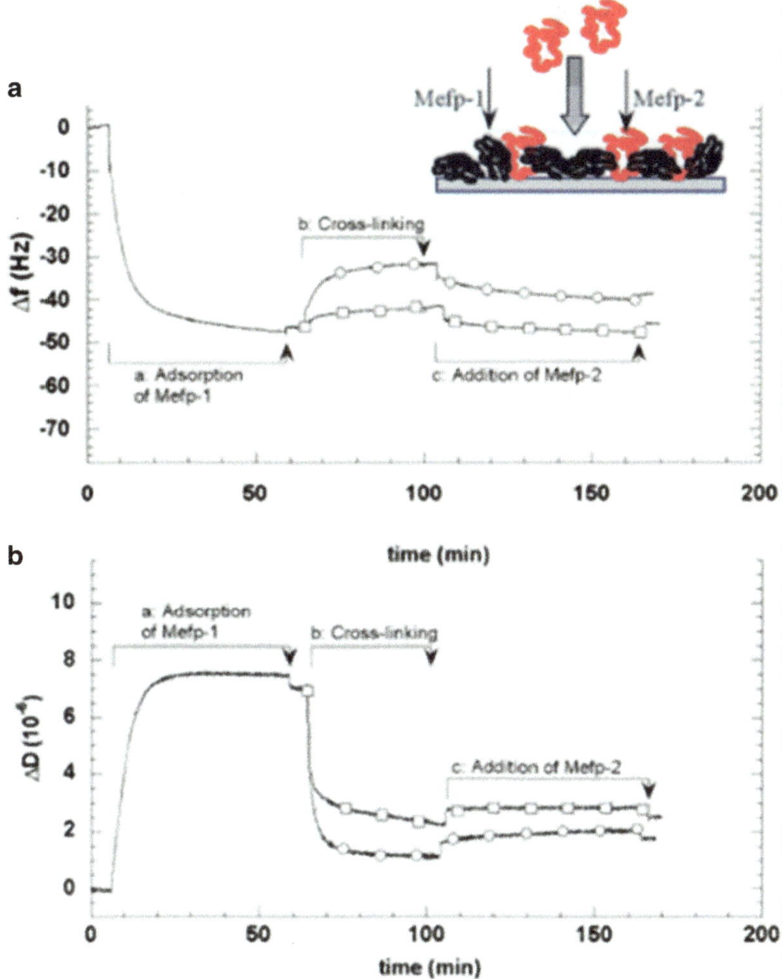

Fig. 2 (**a**) Changes in frequency, Δf, versus time upon adsorption of *Mytilus edulis* foot protein 1 (Mefp-1) (*solid line*) to the CH3-terminated (thiolated), nonpolar gold surface (*arrow* **a**). After rinsing, cross-linking was chemically induced (*arrow* **b**) using 1 mM $NaIO_4$ (*open circles*, *solid line*) or 10 mM Cu^{2+} (*open diamonds*, *solid line*). This was followed by rinsing and subsequent addition of 25 mg/mL Mefp-2 (*arrow* **c**) in the presence of 1 mM $NaIO_4$ (*open circles*, *solid line*) or 10 mM Cu^{2+} (*open diamonds*, *solid line*). (**b**) Changes in energy dissipation, ΔD, versus time for the experiments shown in (**a**). *Inset*: Schematic picture of the binding of Mefp-2 to free-surface patches in the adsorbed and cross-linked Mefp-1 layer (reproduced from ref. 7)

be either injected into the flow chamber from the upstream side or drawn through the chamber from the downstream side, the latter tends to result in smaller pressure pulsations.

A simple but effective approach is one in which the equilibration buffer is first run through the chamber until a steady state is obtained, then the pump stopped briefly and the upstream side of

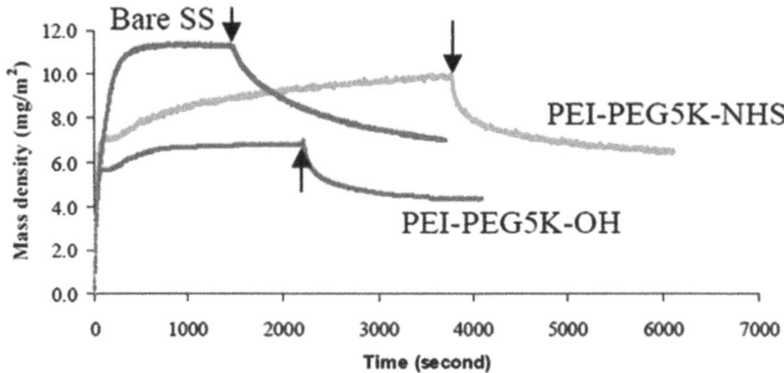

Fig. 3 Surface mass density of β-casein on bare stainless steel (SS) and SS treated with polyethyleneimine and either native 5 kDa polyethylene glycol (PEG) or PEG modified with *N*-hydroxysuccinimide (reproduced from ref. 8)

the flow tubing is swapped from equilibration buffer to the sample of interest, and the pump restarted. Outlet flow can be directed to waste, collected separately or recirculated to the inlet reservoir (beaker). Given that protein adsorption on the crystal surface is generally of the order of nanograms and will thus have little effect on the bulk inlet sample concentration for all but the most dilute of samples, recirculation is an effective means to minimize the amount of sample required. The process of briefly stopping the pump while swapping the inlet tubing to alternate sample reservoirs can be repeated as many times as desired to build up multiple layers or subject the crystal surface to various treatments. At the end of the experimental run or between steps, as desired, one should swap the inlet tubing back to the original buffer to check for reversibility of binding and to adjust for drift that may have occurred during the experiment. This also allows for the changes in signal that occur due to viscosity or conductivity rather than adsorption/desorption to be determined. This experimental approach, although it involves manual intervention to change samples, allows the user to decide when a change of inlet source might be useful by following the behaviors in real time, rather than preprogramming a series of steps through an automated liquid-handling system. For example, it may be difficult to decide in advance how long it might take to reach saturation of the surface, whereas it is a simple matter to determine this by observation.

Figure 4 shows the Q-Sense E4 system, with four parallel flow chambers (channels). Judicious use of a second channel through which the equilibration buffer and/or other buffers that are essentially the same as the samples but without the presence of adsorbing species are flowed is almost always important. A simple experiment in which one equilibrates the crystal, swaps to an adsorbing species

Fig. 4 The Q-Sense E4 system, showing four flow cells in a temperature-control chamber. The *right-hand* flow cell is *open* and a gold-coated quartz crystal ship is shown held in a pair of tweezers, ready for placement onto the "o"-ring-sealed flow chamber (reproduced from Q-Sense E4 manual)

and then back again to the original buffer may not strictly require use of a second control (or reference) channel, particularly for fast, completely reversible adsorption but without it, one cannot easily adjust for drift over the course of an experiment. It is advisable even in such simple cases but absolutely essential in many others for a reference channel to be used so that drift and buffer effects (changes in signal due to viscosity and conductivity changes between buffers) can be taken into account.

In practice, one must be careful to avoid microbubbles forming on the crystal sensor surface, which can occur due to changes in temperature, for example. The crystals themselves can develop microcracks and the fundamental frequency of the crystal in situ within the instrument is very sensitive to orientation and the distribution of forces when sealed into the flow chamber. Thus, it is important to degas buffers and check for imperfections on the crystal surface before use and it is generally not possible to obtain quantitatively comparable results after a crystal flow chamber has been opened or even loosened/tightened between runs.

Crystals should be thoroughly cleaned prior to use. The cleaning protocol should be developed and confirmed experimentally, particularly for unique surfaces, but for plain gold, the surface can usually be stripped of adsorbed species by soaking in Pirhana solution or a mixture of hydrogen peroxide (30%), ammonia (25%), and water at a ratio of 1:1:5 by volume for 5 min at 75°C and then rinsed consecutively with hexane, ethanol, and deionized water. Finally, prior to use, gold-coated crystals should be exposed to UV/ozone for 10–15 min. A minimum of 12 mW/cm^2 at 25 mm from a 185/254 nm lamp is recommended. The latter treatment may not be suitable for other materials, particularly polymeric coatings, on which free radicals may be generated during UV irradiation.

Signal processing (removal of artifacts, alignment of injections, subtraction of reference curves) and analysis of binding kinetics follow essentially the same principles as already described in Chapter 17 for SPR so will not be repeated here. It is essential though, just as was stressed for SPR experiments, that statistical reproducibility and goodness of modelling fits are shown for QCM/QCM-D experiments. Sauerbrey (non-viscoelastic) modelling may be possible if the dissipation value remains close to zero or $\Delta D/\Delta f < 0.05$ (where ΔD is expressed in terms of 1×10^{-6} and Δf is expressed in Hz) throughout the experiment. Viscoelastic behavior will induce an overtone-dependent Sauerbrey thickness (5), so deviations of D from zero or significant differences between the frequency shifts of different overtones indicate that viscoelastic modelling should be undertaken, using either the Voigt or the Maxwell model. The Maxwell model treats the behaviour of a material placed under strain as a spring and dashpot in series, which results in a model that responds immediately to stress through the spring element and then more slowly takes on a viscous component with permanent displacement through the dashpot. The Voigt model models behavior as a spring and dashpot in parallel, which results in a reversible strain with a good approximation of creep behavior. These and other models are reviewed in more detail by Ferreira et al. (9). The final model should be able to fit a wide range of overtones well, although the fundamental frequency signal ($n = 1$) is not generally very useful. Deviations of the model from some overtones may indicate nonoptimal parameters or a poor choice of model.

In addition to time series for f and D, it is possible to plot ΔD versus Δf and obtain potentially useful information on the changes in dissipation as a function of deposited mass. For example, it is possible to see differences in viscoelasticity with adsorbed mass either within a single run, for example as a result of multiple layer formation, water ejection, or cross-linking, or between adsorption runs made at, say, different pH values or temperatures. An example of this is shown in Fig. 5, where changes in slope are indicative of distinctly different processes occurring during adsorption (10). Figure 6 shows a transition from "mushroom" or "pancake" conformation to extended "brush" formation of polymer chains on a gold surface (11).

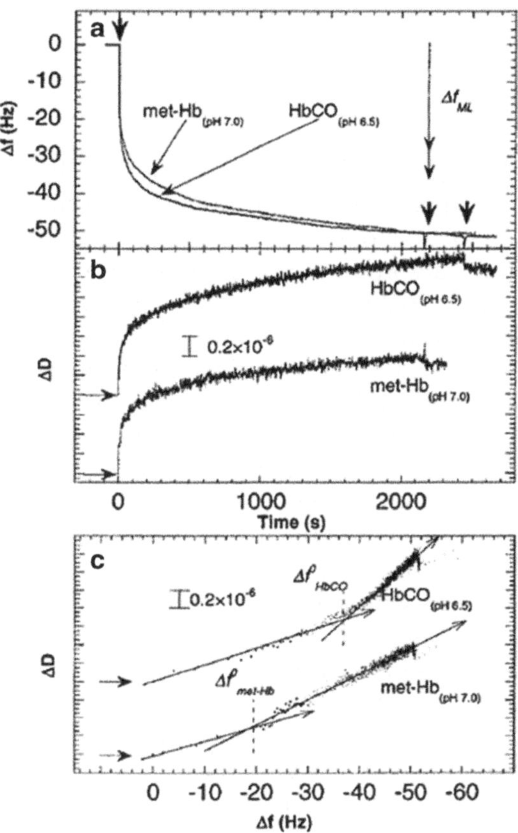

Fig. 5 Changes in frequency (**a**) and dissipation (**b**) as a function of time during adsorption of met-hemoglobin (met-Hb) and hemoglobin-CO (HbCO) at pH 7.0 and 6.5, respectively, to gold covered by a hydrophobic methyl-terminated thiol monolayer (10 mM Hepes). The proteins were introduced at $t=0$. The two *right arrows* in (**a**) indicate the times at which the two protein solutions were exchanged for pure buffer solutions. The *vertical double arrow* in (**a**) shows the predicted frequency change for a monolayer of Hb. Two limiting values are given (29 Hz $<\Delta f_{ML}<$ 35 Hz) depending on the orientation of the Hb molecules on the surface. (**c**) *D–f* plots using the data from (**a**) and (**b**). Note that the density of data points (equispaced in time) becomes smaller the faster the kinetics, in this type of plot. This explains the small number of data points near the origin, where the kinetics is fast. The *horizontal arrows* in (**b**) and (**c**) denote the starting points of the experiments, i.e., $\Delta D=\Delta f 5t=0$. The *arrows* drawn through the *D–f* graphs in (**c**) indicate the direction of time. Δf_0 denotes the break point of the *D–f* graph into two regimes (reproduced from ref. 10)

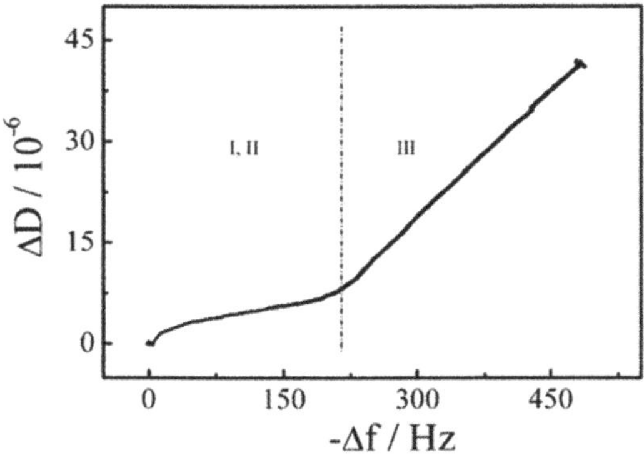

Fig. 6 ΔD versus Δf for the transition between various configurations of polymer chains on a gold surface, going from "mushroom" or "pancake" conformations in regions I and II to extended "brush" conformations in region III (11)

4 Summary

The QCM technique is able to track changes in mass on the surface of a quartz crystal, including on a thin surface coating, making it useful not only for studying molecular interactions but adsorption/desorption on a range of surface materials. Because the measured variable, resonant frequency, is related to mass, unlike optical techniques QCM can indicate the presence of bound and loosely associated water. For thin, rigid, evenly distributed adlayers, the Sauerbrey relation shows that a change in the resonant frequency of a quartz crystal is directly related to the change in mass. However, the assumption of a rigid layer is highly likely to be violated during protein adsorption and even more so for cell adsorption on surfaces because of the inherent viscoelasticity of the adsorbed species and the association of solvent (water) with the adsorbed species or trapped within the adlayer. In this case, viscous damping of the crystal oscillations will occur and it is necessary to allow for this by including a measurement of the energy dissipation of the crystal-adlayer system. QCM-D is a powerful technique that indicates when deviations from the Sauerbrey relation occur and provides additional information on the viscoelastic properties of the adlayer and thus insights into the nature of association during adsorption/desorption.

References

1. Hunter AC (2009) Application of the quartz crystal microbalance to nanomedicine. J Biomed Nanotechnol 5:669–675
2. Cooper MA, Singleton VT (2007) A survey of the 2001 to 2005 quartz crystal microbalance biosensor literature: applications of acoustic physics to the analysis of biomolecular interactions. J Mol Recognit 20:154–184
3. Höök F (2001) Development of a novel QCM technique for protein adsorption studies. In: Biochemistry and Biophysics, Chalmers University of Technology, Göteborg University, Göteborg, Sweden
4. Sauerbrey G (1959) Verwendung von schwingquarzen zur wagung dunner schichten und zur mikrowagung. Zeitschrift Fur Physik 155:206–222
5. Johannsmann D (2008) Viscoelastic, mechanical, and dielectric measurements on complex samples with the quartz crystal microbalance. Phys Chem Chem Phys 10:4516–4534
6. Kanazawa KK, Melroy OR (1993) The quartz resonator—Electrochemical applications. IBM J Res Dev 37:157–171
7. Fant C, Elwing H, Hook F (2002) The influence of cross-linking on protein-protein interactions in a marine adhesive: the case of two byssus plaque proteins from the blue mussel. Biomacromolecules 3:732–741
8. Ngadi N, Abrahamson J, Fee CJ, Morison KR (2008) QCM-D study on relationship of PEG coated stainless steel surfaces to protein resistance. Int J Chem Biomol Eng 1:126–130
9. Ferreira GNM, Da-Silva AC, Tome B (2009) Acoustic wave biosensors: physical models and biological applications of quartz crystal microbalance. Trends Biotechnol 27:689–697
10. Hook F, Rodahl M, Kasemo B, Brzezinski P (1998) Structural changes in hemoglobin during adsorption to solid surfaces: effects of pH, ionic strength, and ligand binding. Proc Natl Acad Sci U S A 95:12271–12276
11. Zhang GZ, Wu C (2009) Quartz crystal microbalance studies on conformational change of polymer chains at interface. Macromol Rapid Commun 30:328–335

Chapter 19

Atomic Force Microscopy for Protein Nanotechnology

Dmitry V. Sokolov

Abstract

This chapter introduces atomic force microscopy (AFM) as an important tool for protein nanotechnology. A short review of AFM-based imaging, mapping, and spectroscopy of protein samples is given. AFM imaging of β-lactoglobulin nanofibrils in air is demonstrated. Basic concepts of AFM are described. Protocols for β-lactoglobulin nanofibrils and multiwall carbon nanotubes (MWCNT) samples preparation are defined. The operation of the microscope is described using MWCNT and the NanoScope E instrument in contact mode as an example. Nanostructure manipulation based on AFM nano-sweeping is demonstrated.

Key words Atomic force microscopy, AFM, Contact mode, Quality imaging, Quality data acquisition, Carbon nanotubes

1 Introduction

Microscopy is a field of science focused on the research and development of methods and techniques for the characterization of features smaller in size than those normally visible to the human eye. This characterization is possible because the interaction of elementary particles with the surface and volume of the sample can be detected. The three most commonly used types of elementary particles, photons, electrons, and atoms, define the three largest branches of the microscopy: light, electron, and scanning probe microscopy (SPM). Depending on the purposes of the experiments, not only elementary but also nano- and microparticles can be used to form a probe. However, only SPM includes the word "probe" in its name.

The size of the features visualized by the probe depends on how small the probe particles are and how well they can be confined in space. The particles can either be focused into an approximately spherical volume or projected linearly to form a beam. The sample can be brought into the contact with a probing volume or beam, and the local physical or chemical properties can be measured in bulk or on a surface of the sample. The penetration depth of the

Juliet A. Gerrard (ed.), *Protein Nanotechnology: Protocols, Instrumentation, and Applications*, Methods in Molecular Biology, vol. 996, DOI 10.1007/978-1-62703-354-1_19, © Springer Science+Business Media New York 2013

probe depends on the permeability of the sample to the type of probe selected, and also on sample homogeneity.

Microscopy and spectroscopy investigate probe–sample interactions. Characterization of sample properties with spectroscopy is frequently performed at just one selected position, with no scanning of the probe across the sample, with one or two parameters varied in time. In contrast to spectroscopic measurements, scanning of the probe is required for the imaging and mapping of sample properties in space or along the surface. Spatial mapping of the sample properties is the primary goal of the microscopy. Microscopic data are recorded based on the assumption that during the acquisition time all the critical parameters are either fixed or fluctuating within the tolerance range, if not varied intentionally. The series of spectroscopic and/or microscopic data can be arranged in a matrix of one or a number of parameters or in a matrix of spectra. An image is frequently used to visualize the spatial distribution of just one parameter. That can be the intensity of transmitted light in conventional light microscopy, secondary or backscattered electrons in scanning electron microscopy or the height in SPM. However, visualization of the spatial distribution of spectra is rather difficult. Individual spectra from every pixel are often analyzed in comparison to reference spectra from known components. The distribution of the recognized components is successfully presented as a map. A mixture of different colors in every pixel of a map can provide a quantitative representation of the different components. Imaging techniques provide data visualization in the form of plane and stereo images, free hand rotated 3D models, which can be still or changing with time (1).

Scanning microscopy requires the systematic movement of a probe through the sample (see Note 1), and the measured properties are recorded in space and time. Imaging frequently requires the capture of data points at regular intervals of time, and scanning is the easiest way to collect regular spatial data from a probe scanning along a line at constant velocity (see Note 2). A series of lines can form a plane, and a series of planes can build a volume. Scanning requires a probe to scan, and probe microscopy assumes that the images are achieved by the mean of scanning through the sample volume. However, only scanning probe microscopy has historically combined the terms "scanning" and "probe" together. SPM is probably the only technique where the probe is made of a physically stable crystal, fiber, nanotube, or metal wire, and is therefore readily visible through a light microscope. The probing volume is defined by the sharpness of the tip of the probe, where a single or a small group of atoms is brought into contact with the sample. Depending on the application, the tip can be made as sharp as about 2 nm in diameter or as broad as a microparticle or a single cell attached to the tipless cantilever.

SPM can be used to directly measure the nanoscale physical and chemical properties of macromolecules, with up to molecular and atomic resolution. SPM can be divided into three groups according to the nature of the interaction between the probe and the sample: scanning tunnelling microscopy (STM), atomic force microscopy, and scanning ion-conductance microscopy (SICM). STM is based on the measurement of the current tunnelling through a few nanometers-wide gap between the probe tip and a conductive sample surface. AFM detects the attractive and repulsive intermolecular forces acting at a similar distance and is therefore capable of imaging a wider variety of samples. SICM utilizes an ion flux through a hollow probe to maintain a constant gap between the probe and a soft nonconductive sample surface. SICM is applicable but not limited to real-time measurement of ionic current through membranes in liquid environments with sub-micrometer lateral resolution. Since its invention by Binnig et al. in 1986 (2) AFM has grown into a distinct discipline that currently encompasses more than 30 different methods and techniques. STM and SICM groups are considerably smaller.

The quality of the SPM image depends on how well the gap between a probe tip and a sample is controlled at every position across the sample surface. Operation of the microscope frequently requires the processing and display of a number of data channels simultaneously. SPM is an example of multimodal or multichannel microscopy techniques. One of the modes of SPM is traditionally used for maintaining the probe at the selected conditions, e.g., at a specified distance from the surface, while the signals from other modes are collected for further analysis of surface properties. Obtaining quality data on properties of the surface depends on the accuracy of control of the gap between the probe and sample. Knowledge of the fundamental concepts, physics of probe–sample interaction, and possible artifacts are required to enhance the outcome of research based on SPM. The theory and application of SPM to the characterization of proteins and other biological samples are described in more details in several books (3–5).

This chapter focuses on the AFM technique (see Note 3) and is written using the example of the NanoScope E instrument because it is generic and more suitable for describing the general concepts of atomic force microscopy compared to the Nanosurf and other instruments with pre-aligned optics and advanced automation. Protocols developed for the NanoScope E can be applied to other types of the microscopes.

Nanotubes are selected as a model for imaging macromolecules due to higher stability of the samples and simplicity of the sample preparation. Principles of image acquisition for nanotubes are similar to those required for imaging other kinds of macromolecules.

Procedures and protocols for the acquisition of images of macromolecular structures, such as protein nanostructures, are described below in detail. Nearly all of the advanced AFM-based analytical techniques require precision locating of the features of interest and precise positioning of the probe in respect to the selected features. AFM imaging and mapping is the most convenient spatial referencing tool and is the most common basis for further analysis (see Note 4). Quantitative nanomechanics, mapping of mechanical and chemical properties of protein nanostructures, single-protein atomic force microscopy (6), and some of other advanced applications of AFM listed in Table 1 are based on the imaging capabilities of AFM. Imaging and characterization of proteins can be performed in controlled air and liquid environments.

Depending on the purpose of the research samples may require very little preparation. In the experiment that generated the data for Fig. 1 the β-lactoglobulin samples were deposited on the surface of mica and imaged in contact mode under ambient conditions. Contact mode is the basic but probably the most reliable mode of operation of AFM. The experiment described in detail below took 2–3 h including deposition of the samples on the substrate, sample rinsing, optimization of the imaging conditions, and acquisition of images.

2 Materials

2.1 Multiwall Carbon Nanotubes

1. Multiwall Carbon Nanotubes (MWCNT) (06-0470, Strem Chemicals, Newburyport, USA).
2. *N*-Methyl-2-Pyrolidone (NMP) (OM Group, Inc., Cleveland, USA).
3. Acetone.
4. Methanol.
5. Isopropanol.

2.2 Fibrous Protein Sample

1. β-lactoglobulin (90% pure) containing a mixture of genetic variants A and B (Sigma, St. Louis, MO, USA, product no. L0130-5G).
2. Reverse osmosis filtered water.
3. Hydrochloric acid.
4. Sodium chloride.
5. Calcium chloride.

2.3 Substrates

1. Mica (see Note 5) (Agar Scientific Ltd., Stansted, UK).
2. Square glass cover slips, 22×22 mm^2, thickness #1 (Thermo Fisher Scientific New Zealand Ltd, North Shore City, New Zealand) or round glass cover slips, 15 mm, thickness #1 (26024, Ted Pella, Inc., Redding, USA) or round mica disks, 15 mm (Ted Pella, Inc., Redding, USA).
3. AFM Metal Specimen Discs, 15 mm (Ted Pella, Inc., Redding, USA).

Table 1
Brief overview of AFM-related methods and techniques applicable to protein nanotechnology

Problem	Solution	Method	Media
Visualization of distribution of functional group	Detection of areas with different coefficients of friction	Lateral force microscopy (7)	Various buffered pH values, absolute EtOH
Monitoring of dynamics of pH-dependent channel morphology	In vitro imaging	Contact mode AFM in liquid (8)	Various HEPES-buffered pH values
Dynamics of protein–protein interactions	Single-molecule recognition imaging	Chemical force microscopy (9, 10)	BSA/hSwi-Snf solution with ATP successfully added
Molecular recognition imaging	Adhesion force imaging	Chemical force microscopy (7)	Various buffered pH values, absolute EtOH
Single-molecule recognition	Adhesion frequency, rupture force, recognition time measurements	Dynamic force spectroscopy (11)	HEPES-buffered culture medium
Reconstruction of energetical profile of molecular interactions	Adhesion frequency, recognition time measurements	Dynamic force spectroscopy (11)	HEPES-buffered culture medium
Mechanochemistry of receptor–ligand bond under applied force	Bond lifetime measurement	Force Spectroscopy (12, 13)	Native environment
	Rupture force measurement	Single-molecule force spectroscopy (14)	PBS with free LexA successfully added
Measurement of pH-dependent adhesion forces	Electrostatic force measurements	Force spectroscopy (7)	Various buffered pH values, absolute EtOH
Correlation between structure, function, and mechanics of membrane proteins	Flexibility of individual membrane proteins	Indentation force spectroscopy and mapping (15)	Native environment (10 mM Tris–HCl, 150 mM KCl, pH 7.4)
Reconstruction of energetic landscape of protein folding	Temperature dependent (16) single-molecule unfolding and refolding (17)	Single-molecule force spectroscopy	Buffer solution
Bottom-up assembly of biomolecular structures	DNA hybridization based selective manipulation of single-protein molecules	Single-molecule lithography (18)	
Monitoring of kinetics of protein–DNA interaction	Measurement of protein–DNA dissociation rates	Single-molecule force spectroscopy (14)	PBS with free LexA successfully added
Label-free detection of proteolysis and proteolysis efficiency	In situ measurement of changes in mass of peptide layer	Nanomechanical resonance spectroscopy (19)	PBS buffer and air

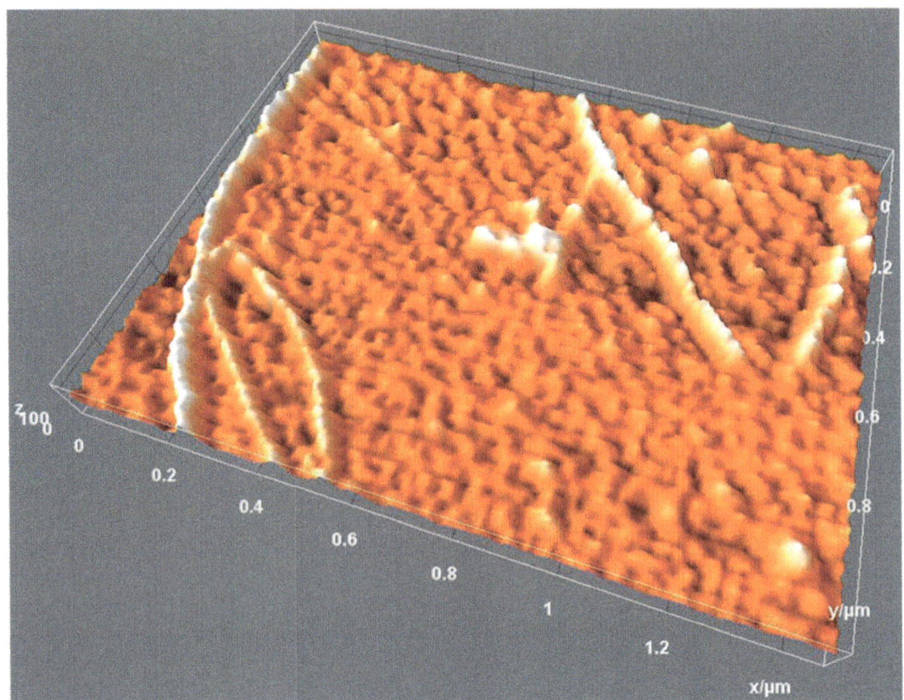

Fig. 1 3D rotated AFM images of protein nanofibrils taken at Nanosurf EasyScan 2 Flex-AFM in contact mode. Height of the fibers is about 5 nm. Image size is 1.5 × 1.1 mm. The image is processed with Fiji, a version of ImageJ image analysis software

2.4 Preparation of the Samples

1. 0.45 µm membrane filters PTFE J050A025A (Advantec MFS, Inc., Dublin, USA).
2. 0.2 µm syringe filters (Millex-GS, Millipore, Billerica, MA, USA).
3. Centrifugal filter with nominal cutoff 100 kDa (Amicon Ultra-4, Millipore, Billerica, MA, USA).
4. Centrifugal filter with nominal cutoff of 10 kDa (Amicon Ultra-15, Millipore, Billerica, MA, USA).
5. Ultrospec 2000 UV spectrophotometer (Pharmacia Biotech, Cambridge, UK).
6. Kimax glass screw capped glass tubes, 16 mm diameter (Schott, Elmsford, NY, USA).
7. Water bath (Lab Companion BS-11, Jeio Tech, Boston, MA, USA).
8. Power controlled ultrasonic bath DK102P (Monmouth Scientific, Bridgwater, UK).
9. Centrifuge Himac CR22G II super speed centrifuge (Hitachi Koki Co., Tokyo, Japan).
10. Spin-coater Laurell WS-400A-6NPP/Lite (Laurell Technologies Corporation, North Wales, USA).
11. Hot plate.

12. Double-sided tape 12.5 mm wide.
13. Tweezers for course work, e.g., for applying or removing double-sided tape, samples, etc.
14. Tweezers for cantilevers, provided with the microscope or purchased from Ted Pella, Inc., Redding, USA (*see* **Note 6**).
15. 20 μL micropipette.
16. 10 μL micropipette tips.
17. 10 μL micropipette tip rack.
18. 1.5 mL Eppendorf tubes.
19. Eppendorf tube rack.
20. Scissors.
21. Two glass Petri dishes.
22. KimWipes lint and fiber-free tissue.
23. Razor blades 05025-AB (SPI Supplies/Structure Probe, Inc., West Chester, USA).
24. Waste container.
25. Transfer pipette 3 mL.
26. Compound microscope CX21LED/CX21 (Olympus, Tokyo, Japan).
27. Binocular microscope SZX7 (Olympus, Tokyo, Japan).

2.5 Atomic Force Microscopy

1. Head Magnifier (SPI Supplies/Structure Probe, Inc., West Chester, USA).
2. NanoScope E (Bruker AXS, Santa Barbara, CA 93117, USA).
3. Standard 350 μm long silicon CSG11 cantilevers (NT-MDT Co., Zelenograd, Moscow, Russia) for contact mode with a nominal spring constant 0.03 N/m.
4. Standard 110 μm long silicon nitride CSC21 cantilevers (Mikromasch, Tallinn, Estonia) for contact mode with a nominal spring constant 2 N/m.
5. Specwell monocular microscope, magnification $10 \times 30x$, supplied with NanoScope E AFM or purchased from LS&S, Buffalo, USA.
6. Fiber optics illuminator EW-09745-00 supplied with AFM or purchased from Cole-Parmer, Vernon Hills, USA.

3 Methods

3.1 Preparation of 0.01 mg/mL Carbon Nanotubes in NMP

1. Filter 50 mL of NMP through a 0.45 μm PTFE membrane filter.
2. Add 10 mg of CNT to 10 mL of filtered NMP.
3. Perform high-powered sonication for 2 min.

4. Add 1 mL of 1 mg/mL CNT solution to 9 mL of filtered NMP.
5. Perform high-powered sonication for 2 min.
6. Add 1 mL of 0.1 mg/mL CNT solution to 9 mL of filtered NMP.
7. Perform high-powered sonication for 2 min.
8. Place all serial dilutions in a low-powered ultrasonic bath for 4 h.
9. Perform high-powered sonication for 1 min on all serial dilutions.
10. Centrifuge at 3,000 × *g* for 90 min.

3.2 Preparation of β-Lactoglobulin Solutions (20)

3.2.1 Preparation of the Solutions

1. Adjust the pH of reverse osmosis filtered water with HCl to 2 ± 0.05 to make "HCl buffer."
2. Dissolve β-lactoglobulin in HCl buffer to a final concentration of 1.2% (w/v).
3. Stir the solution overnight at 4°C.
4. Centrifuge the solution at 22,600 × *g* for 30 min and filter with 0.2 μm syringe filter.
5. Rinse the Amicon Ultra-15 centrifugal filter with HCl buffer prior to use.
6. Remove residual salts by ultrafiltration using an Amicon Ultra-15 centrifugal filter by centrifuging at 3,000 × *g* for 15 min.
7. After filtering three times, confirm that the conductivity of the protein solution is close to that of the buffer, 2 ± 0.05.
8. Determine the protein concentration in the desalted β-lactoglobulin solution by measuring the absorption at 278 nm with an Ultrospec 2000 UV spectrophotometer using a β-lactoglobulin standard curve and assuming 90% purity. A small proportion of protein may be lost during filtering, and an initial concentration of approximately 1.2% (w/v) before filtering can give a final concentration close to 1% (w/v) after filtering.
9. Store the solutions of β-lactoglobulin at 4°C. Use the solutions within 2 days of preparation.
10. Adjust the salt content of β-lactoglobulin solutions using the 1 M NaCl and 1 M $CaCl_2$ stock solutions.

3.2.2 Heating of β-Lactoglobulin Solutions

1. Heat 2 mL aliquots of β-lactoglobulin solution in screw-capped 16 mm diameter glass tubes in a water bath at 80 ± 0.1°C.
2. Following the requisite heating time, cool the tube in ice water for 10 min.

3.2.3 Preparation of β-Lactoglobulin Solution for Deposition on the Substrate

Purify the fibrils according to the ultrafiltration method of Bolder, Vasbinder, Sagis, and van der Linden (21) to reduce the concentration of impurities that could otherwise compromise the quality of the AFM images.

1. Wash the Amicon Ultra-4 centrifuge filter with 2 mL HCl buffer.
2. Add the heated protein solution (100 μL) to 2 mL HCl buffer in an Amicon Ultra-4 centrifuge filter.
3. Centrifuge the filter at 3,000 × *g* for 15 min.
4. Top up the retentate with 2 mL HCl buffer.
5. Perform the filtration three times in total.
6. Top up the final retentate with 1 mL HCl buffer.
7. Mix the retentate by inversion and transfer to a 1.5 mL Eppendorf tube. The stock solution of the heated and filtered protein is prepared. The final dilution of heated protein in the stock solution is approximately tenfold.
8. Add 10 μL of buffer solution to 10 μL of stock solution of the sample in 1.5 mL Eppendorf tube.
9. Mix the solution by pumping the micropipette three to four times (see Note 7).

3.3 Preparation of Substrate

Scanning probe microscopes may have a scanning-stage or a scanning-probe configuration. The NanoScope E microscope has a scanning-stage configuration. The stage of NanoScope E can accommodate samples up to 15 mm in diameter. 15 mm round cover slips and round or square mica chips 15 mm across cut with scissors from a sheet can be a good substrate for the sample preparation. The substrate can be mounted on the standard sample holder disks 15 mm in diameter; it can be glued in place with superglue or attached to the disk with double-sided tape (see Note 8). The central area of a sample or substrate must be kept intact in any circumstances to avoid cracks, contamination, and other possible defects.

3.3.1 Glass Cover Slips

Round or rectangular cover slips held with tweezers are prepared individually as follows:

1. Sonicate a cover slip in acetone for 10 s.
2. Immediately transfer the cover slip into a beaker of methanol and sonicate for 10 s.
3. Immediately transfer the cover slip into the beaker of isopropanol and sonicate for 10 s.
4. Immediately transfer the cover slip into the beaker of deionized water and sonicate for 10 s.

5. Put the cover slip into a Petri dish and dry under a partially closed lid to avoid dust contamination.
6. To avoid contamination from air, store the cleaned cover slips in the closed Petri dish for future use.

3.3.2 Mica

Remove upper layer of mica with double-sided (see Note 9) or conventional sticky tape.

1. Apply the tape to the surface of a mica chip and press the tape with fingers several times to ensure the strong and even adhesion of the tape across the surface of the chip.
2. Lift one side of the tape and remove the tape carefully in direction perpendicular to the surface of the chip. Observe the film of mica on the tape by reflecting light from it. Repeat the removal of the upper layer of mica with a freshly cut piece of double-sided tape if no shiny film is observed on the tape.
3. Store the cleaved pieces of mica in the closed Petri dish for future use to avoid contamination from air.

3.3.3 Spin Coating of Nanotubes (See Note 10)

1. Clean spin coater with acetone, methanol, and isopropanol.
2. Place cover slip or mica chip over the aperture of the spin coater.
3. Turn on the vacuum to fix the cover slip or chip.
4. Apply about 50 μL drop of MWCNT/NMP solution to the slide using a pipette.
5. Spin coat at 500 rpm for 10 s to make the initial spread.
6. Spin coat at 3,000 rpm (estimated g-force about 50 g) for 1 min (see Note 11).
7. Place the slides into Petri dishes with lids partially open on a hot plate at 200°C for 15 min to allow the NMP to evaporate (see Note 12).
8. Confirm even deposition of the nanotubes on the substrate using a compound microscope at 400× magnification in air.

3.3.4 Deposition of Fibrous Proteins

1. Form a 10 μL drop of the fibrous protein solution at the tip of a micropipette and touch the surface of mica with the bottom of the drop avoiding any direct contact between the tip and the mica surface.
2. Allow 5–10 s for the protein molecules to adhere to the mica surface.
3. Remove excess solution by shaking the vertically positioned substrate three to four times over a piece of tissue.
4. Rinse the surface of the substrate three to four times with a flush of buffer solution from the transfer pipette while holding the substrate above a waste container with the surface facing down at a 45° angle.

5. Confirm the uniformity of coating using a compound microscope at 400× magnification or with a head magnifier.
6. Allow the substrate to dry for 5 min on the top of a KimWipes tissue under the cover of the Petri dish to avoid contamination from the air.

3.3.5 Attachment of the Substrate to the Sample Holder

1. Put the double-sided tape evenly on the metal specimen disk. Remove excess tape with scissors.
2. Put a cover slip or a mica chip with the deposited nanotubes on the disk with the tape.
3. Ensure the firm attachment of the substrate to the disk by pushing two of the opposite corners or sides of the substrate (see Note 13) with both tips of the tweezers simultaneously.
4. Apply pressure to the other two corners or sides of the substrate.
5. Repeat the last step two to three times. Avoid touching the central regions of the substrate.

3.4 Imaging with Digital Instruments NanoScope E

3.4.1 Turning the Microscope ON

1. Assuming that all the connections and wiring are done in accordance with the manual for the microscope, turn the controller of the microscope ON.
2. Turn the power of the computer ON.
3. Load required user profile.
4. Start NanoScope v.6.0 software from the desktop of the computer. If the software is started into the previously saved workspace you may have a choice between starting a new and continuing the last saved session. Before starting a new session the old workspace must be closed first by clicking the cross icon of the left panel.
5. Click the "Real Time" icon at the toolbar for opening a new workspace.
6. Select "Scan Triple" checkbox (Fig. 2). The other windows can be activated later when needed. Three-view mode is the most flexible mode in providing the information of your choice: topography, deflection, and friction (lateral force mode); see Note 14.

3.4.2 Mounting Cantilever

Most AFM probes contain a cantilever protruding from a silicon chip (Fig. 3). A cantilever can be as long as 100–300 μm and as wide as 20–30 μm. On the free end of the cantilever a few micrometer high tip is fabricated. Conventional probes may have tip radius of about 10 nm. Radius of the tip of ultrasharp cantilevers can be as small as 2 nm. Radius and shape of the tip defines lateral resolution of the technique.

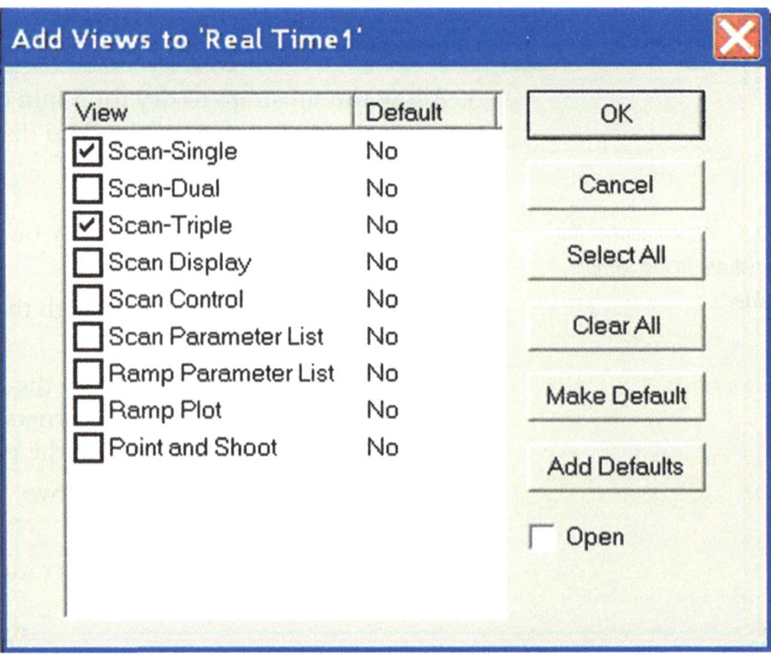

Fig. 2 Adding views to "Real Time" window. "Scan Triple" view is suggested to run the instrument. "Scan-Single" is invoked automatically and can be left unchecked. The other views can be added later when required

Fig. 3 Typical beam-type cantilever (Courtesy of Mikromasch, Tallinn, Estonia)

Fig. 4 Back view of NanoScope E. (**a**) Screw of vertical alignment of the photodiode. (**b**) Cantilever holder clamp. (c) Screw for horizontal alignment of photodiode. (**d**) Scanner connector. (e) AFM head connector. (**f**) Z-step motor trigger

AFM probes require microfabrication facilities to be produced with controlled geometrical and physical parameters. The tip radius, stiffness, and resonance frequency are the most critical parameters of the cantilever. Silicon (Si) and silicon nitride (SiNx) are the most commonly used materials of the probe. Si can be doped and be nearly as conductive as a metal. However Au, Pt, Ti, W, peptides, bovine serum albumin (BSA), PEGylated GlyPheLysGly (19), and a variety of other materials can be used for coating and modification of tips and entire surfaces of the probes. All the parameters above can define the character of interaction of the probe with substrate.

1. Free up about 0.5 m×0.5 m space on a bench near the microscope.
2. Wipe the surface of the bench of dust and contamination with fibreless cloth or KimWipes soaked with water or 70% alcohol.
3. Push the trigger of the Z-step motor into the "UP" position (Fig. 4f) for about 20 s to move the probe up from the sample surface. Observe the gap between the tip and the sample surface through the supplied monocular microscope (magnification of about 300×). The shadows or reflection image of the probe and cantilever are barely seen on the sample surface at this point.
4. Undo the clamp of the cantilever holder rotating the knob counterclockwise if seen from front (Fig. 4a).

Fig. 5 AFM head. Cantilever holder is sliding along the base attached to the scanner with the coiled springs. (**a**) Mounting spring of AFM head. (**b**) X and Y coarse positioning screw of AFM head. (**c**) Laser spot on the sample surface. (d) Cantilever holder. (**e**) Vertical alignment of photodiode. (**f**) Y-positioning screw of the laser. (**g**) X-positioning screw of the laser. (**h**) Window of the AFM head used for illumination or observation of the sample and cantilever from top. (**i**) Cantilever holder clamp. (**j**) Base of AFM head

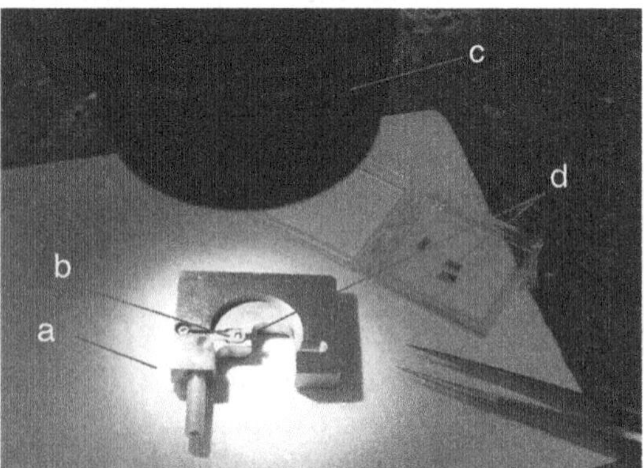

Fig. 6 Mounting cantilever. (**a**) Light spot from fiber optics illuminator. (**b**) Spring of cantilever holder. (**c**) Monocular microscope; (**d**) Cantilever

5. Remove the cantilever holder (Fig. 5a) from the AFM head under the angle to avoid contact of cantilever with the sample.
6. Put the holder on the bench upside down to provide access to the cantilever from the top (Fig. 6).
7. Rest both of your arms on the top of the bench.

8. Put the box with cantilevers near the cantilever holder. The cantilevers in the holder and in the box must be oriented in the same way (see Note 15).
9. Push the cantilever holder down with your left hand to release the cantilever spring (Fig. 6b). Clamp the probe chip with the tweezers from the sides and remove the chip carefully from the cantilever holder (see Note 16).
10. Place the removed old chip into a cantilever box; stick it in by pushing the center of the chip and avoiding any movement and contact in the cantilever area. Put a mark on the top of the box above the chip for possible future reference or examination.
11. Practice handling the probes with tweezers on old or used cantilever chips. Put a used chip on the silicone surface of the cantilever box with tweezers and press it gently to ensure adhesion of the chip to the silicone coating (see Note 17).
12. Grab the chip with the tweezers from the side, twist and move the chip carefully in plane with the silicone surface, and lift if slowly from the surface. A quick movement of the hand may cause the chip to slip or jump out of the tweezers.
13. Press the cantilever holder with your free hand down to lift the cantilever spring up and put the chip into the groove of the cantilever holder. Release the cantilever spring.
14. Align the cantilever precisely touching the chip sides but avoiding moving the tweezers tips into cantilever area (see Note 18).
15. When confidence in the replacement of the probes is developed repeat the procedure above with a new probe.
16. Place the cantilever holder with the mounted cantilever onto the ball mounts of the AFM head as shown in Fig. 5. Fix it with the clamp (Fig. 4b).

3.4.3 Positioning the Sample Holder

1. With the AFM head removed from the stage, increase the gap between the probe and the sample surface by pushing the lever of the head step motor (Fig. 7h) in the "UP" direction for about 10 s.
2. Take the disk with the sample mounted on it with your fingers from the sides and put the disk carefully on the magnetic stage of the scanner from the side and from as low a height as possible. Putting the disk in from a vertical direction may lead to the disk sliding off your fingers and crashing into the stage. Such hard contact may be destructive to both the scanner and the sample.
3. Confirm that the cantilever holder is removed from the AFM head. Take the head with one hand and put it on the ball mounts of the scanner. Twist and slide the head around the scanner to find the positions for the ball mounts.

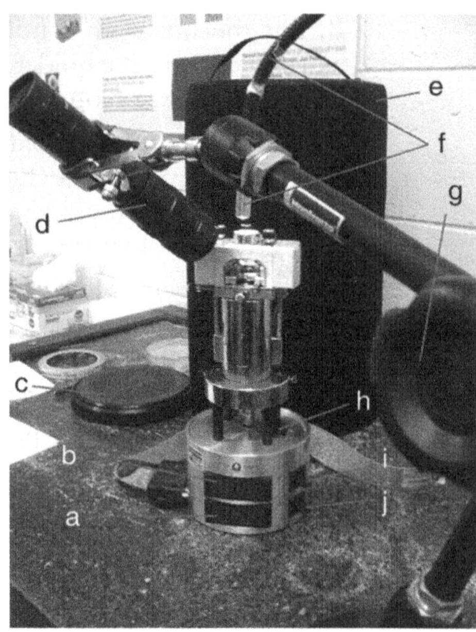

Fig. 7 Front view of NanoScope E with monocular microscope and top illumination aligned. (**a**) Lower indicator of photodiode for friction (lateral force) mode. (**b**) Upper indicator of photodiode for deflection signal in contact mode. (**c**) Silicone pad for vibration isolation. (**d**) Monocular microscope with 300× magnification for observation of cantilever. (**e**) Protection enclosure from acoustic noise and airflow. (**f**) Fiber optics illuminator. (**g**) Handler of the stand for monocular microscope. (**h**) Switch of AFM head step motor. (**i**) and (**j**) "SUM"/"DIFF" switches of top/bottom and left/right segments of photodetector

4. Pushing the center of the head with one hand near the upper window (Fig. 5h), stretch one of the scanner springs (Fig. 5a) and put it on the lever at the head at one side.
5. Change the hand holding the AFM head while keeping it pushed down and attach the other spring of the scanner to the lever on another side of the head.
6. Connect the cable of the head to the scanner socket (Fig. 4e). A laser spot should appear on the surface of the sample (Fig. 5c).
7. Put the cantilever holder into the AFM head and perform the laser alignment.

3.4.4 Laser Alignment

Laser alignment is not required for Nanosurf and other microscopes with pre-aligned probes. This makes the cantilever replacement and preparation prior to imaging quicker. One of the advantages of the microscopes like the NanoScope E is their capacity to work with nonstandard and custom-made cantilevers, which may be of interest for probing the properties of specific protein nanostructures.

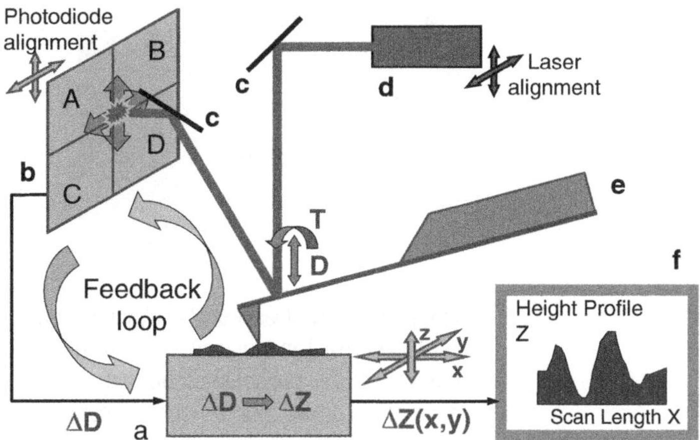

Fig. 8 Optical lever schematics. (**a**) XYZ piezo scanner with sample. (**b**) Four-quadrant photodiode. (**c**) Mirrors. (**d**) Laser. (**e**) Cantilever probe. (**f**) Computer screen. D—deflection of cantilever. T—torsion of cantilever. Feedback compensates changes in deflection of cantilever DD by moving piezo stage onto distance DZ while scanning in *XY* plane

Laser alignment is required for positioning of the laser spot on top of the probes with various geometrical characteristics. The laser beam reflected from the cantilever is used for the measurements of deflection of the cantilever D (Fig. 8), torsion of the cantilever T caused by the forces acting between the surface and cantilever. Bending of the cantilever results in the movement of the laser spot on a four-quadrant photodetector in the vertical direction, while torsion of the cantilever shifts the beam to the side. The difference in readings from the upper and lower half of the photodiode serves as a feedback signal for the piezo stage which is moved in the direction opposite to the deflection of the cantilever. The goal of the feedback system is to keep the deflection of the cantilever constant. The piezo stage compensates for the changes in height at every point of the sample according to the deflection signal from the photodetector. Movement of the stage in the *Z* direction is mapped as the variation in height across the image formed on the computer screen.

Laser beam deflection is proportional to the deflection of the cantilever. Laser displacement is measured by a four-quadrant photodetector. Four equal segments of a photodiode can detect deflection in the vertical direction as well as torsion of the cantilever.

Detection is more reliable and the signal from photodiodes is stronger at higher intensities of the light reflected from the cantilever. The cantilever surface is not uniform. It may have brighter and darker regions. A “brighter” area near the free end of the cantilever is the most favorable for the operation of the microscope. This area is the target of the procedure of laser alignment.

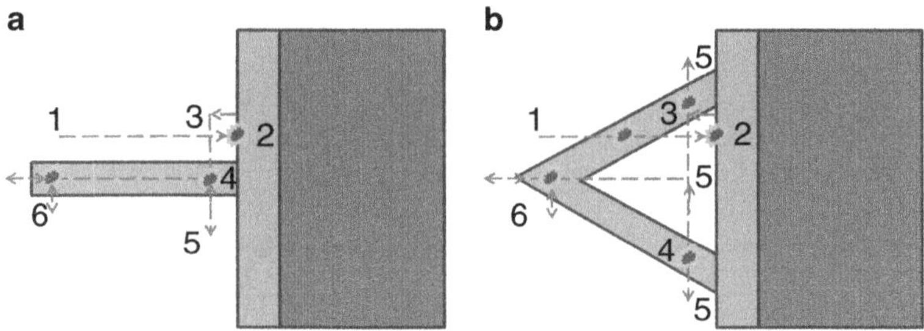

Fig. 9 Trajectory of laser spot at alignment on cantilever end. Top view. (**a**) Beam-type cantilever. (**b**) Triangular-type cantilever. 1—probable position of laser spot on the end of the previously mounted cantilever; 2—laser light scattered on the edge of the probe chip; 3—move laser along the chip edge; 4—position of maximum of "SUM" signal; 5—no signal position; 6—position of local maximum near the end of cantilever

1. Turn the switch of the upper indicator on the base of the stage (Fig. 7i) up. A sum from all four segments of the photodetector is displayed on the upper LCD display.
2. Remove the base of the microscope from the silicone vibration isolation pad and put it on the bench. The microscope on the silicone pad may make a wobbly motion under the pressure of hands and is difficult to control. Illuminate the cantilever from the top with the fiber optics supplied with the microscope. Adjust the monocular tube position to observe the cantilever under an angle of about 45° from the top and about 45° from the front, similar to what is shown in Fig. 7.
3. The goal of this step is to bring the laser spot from the initial position (position 1 in Fig. 9) to the free end of the mounted cantilever (position 6). Turn the rightmost knob on the top of the AFM head (Fig. 5g) to the right (clockwise) until the shiny spot on the edge of the chip is observed (position 2); see Note 19. Be careful to stop at this position. The laser beam scattered on the edge is easily observed but it may be not clearly seen when reflected from the smooth surface of the chip.
4. The next knob to the left on the top of the head (Fig. 5f) moves the laser beam along the edge of the chip. Make one turn to the right and back to see how the laser spot is moved along the edge of the chip. Step back away from the edge (Fig. 9, position 3). The laser spot is moved alongside the edge and eventually hits the cantilever. The optical microscope can visualize only the scattered light. However, the light reflected from the cantilever will be collected by the photodetector.
5. Maximize the value on the upper display on the base (about +3 to +4 V for a CSG11 probe) by finding position 4 on the cantilever. Move the laser spot towards the free end of the cantilever,

Fig. 10 Low-angle observation of gap between probe and the surface. (**a**) Monocular microscope tube. (**b**) Fiber optics illuminator. (**c**) AFM head. (**d**) Acoustic enclosure

rotating the right knob counter clockwise. Find the position of another maximum of the signal near the end of the cantilever (about 4 V for CSG11 probe). Change the direction of the movement of the laser beam and maximize the signal again. Repeat the movement in perpendicular directions to find the spot on the cantilever with the highest possible signal at position 6 (about 5.5 V for CSG11 probe); see Note 20.

6. With the laser beam aligned, the adjustment of the position of the photodiodes must be performed. Alignment of photodiodes is performed by turning knobs (a) and (c) shown at Fig. 4. Values on the photodiode indicators (a) and (b) (Fig. 7) must be maximized (at about 12 V) with the switches (i) and (j) (Fig. 7) in upper ("SUM") position and zeroed in the bottom (differential) position.
7. Laser alignment must be verified after photodiode alignment following the steps above.

3.4.5 Approaching the Surface

The gap between the cantilever tip and the surface can be observed most easily under the smallest possible angle between the microscope tube and the sample surface and about a 45° shift towards the cantilever direction as seen from Fig. 10.

1. Push the AFM head step motor switch to "DOWN" position (Fig. 7h) to bring the cantilever tip closer to the surface. The shiny end of the cantilever and its image should be seen as

close to each other as the length of the cantilever. The image of the illuminated end of the cantilever can be seen clearly or as a diffused shiny spot, depending on the properties of the sample surface.

2. Stop the manual approach at a distance larger than the length of the cantilever. In the case of imaging a new kind of sample, make a further approach in shorter steps, being ready to stop as soon as the cantilever has touched the surface and bent upwards. Bending the cantilever up should mean a hard contact of the probe with the surface. The very tip of the probe can be damaged and the probe may become unusable for quality imaging. Allow the gap between the cantilever end and its reflection of more than 10–20% of its length at all times (see Note 21). The manual approach stage is now finished. Automatic approach can be activated from the toolbar of the NanoScope software.
3. The tuning of the interface of the software should be performed at this stage. Any of the topography, deflection, and friction channels can be switched on for viewing images taken in forward (trace) and reverse (retrace) directions. Select one topography channel in trace and two deflection images in trace and retrace directions for this experiment (Fig. 11, "Data type" and "Line direction" fields, respectively).
4. Approach can be safe when the interaction between the probe and the sample is minimized. Put "Scan size" to 0 nm, "Deflection setpoint" (further Setpoint) to 0 V, "Integral gain" and "Proportional gain" to 5 (see Note 22). Click the "Engage" button (see Note 23). Travelling distance can be monitored at the lower right edge of the window, highlighted by the dashed oval in Fig. 12. With the photodiode switches in down position (differential readings from upper/bottom and left/right segments) the abrupt change of upper indicator reading (from −2 V to about 0 V in our case) is seen at the time of the contact (see Note 24).
5. Confirmation of the tip contacting the surface. Put the Setpoint to +1 V or higher at no scanning (0 nm scan area). The probe is in contact with the surface if no movement of the Z stage (slider with arrows) from the position of initial contact is observed (see Z-position panel at Fig. 13) and an image (see Note 25) is formed at a scanning area of 1×1 μm^2 or more. In other cases, when an increase in the Setpoint leads to a movement of the Z stage towards the "extended" limit of the scanner (see Note 26), the probe is out of stable contact. The laser may be out of alignment; a photodiode may detect light from the sample features or interference pattern from the surface. The sample surface may also be too soft or unstable, may follow or be deformed by the tip (see Note 27). In this case,

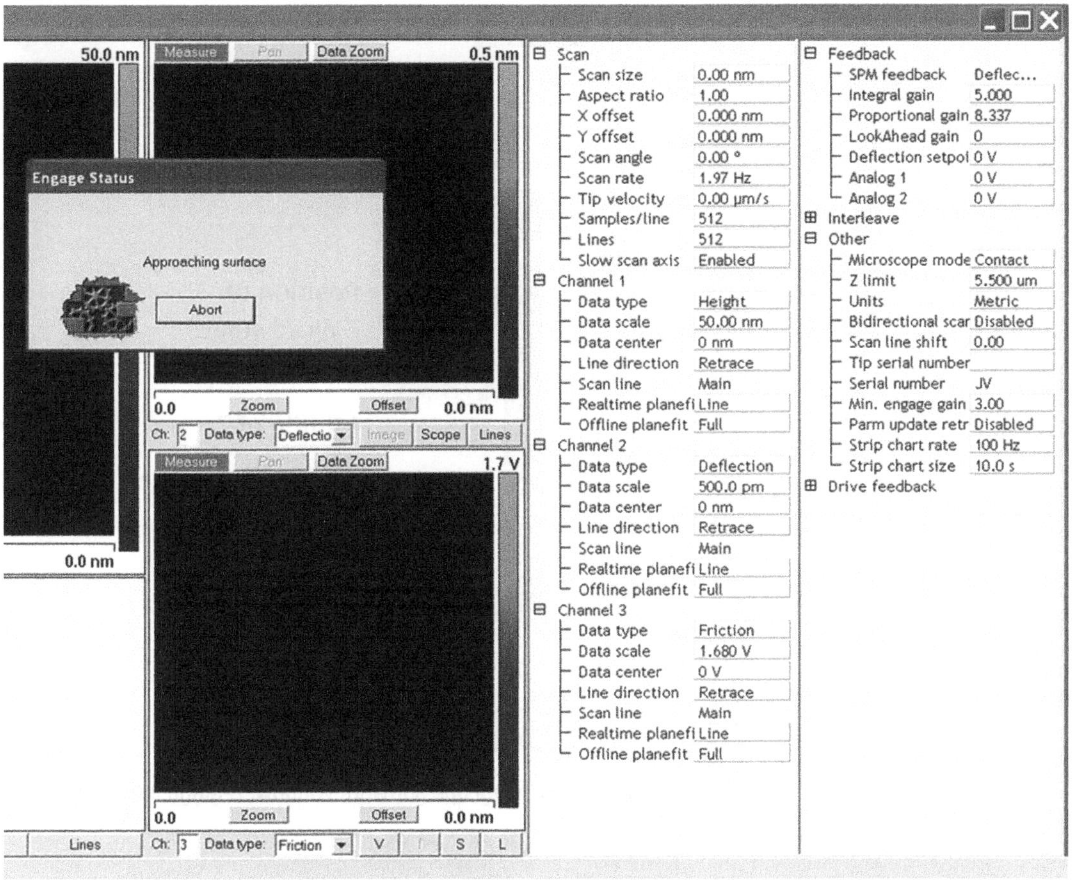

Fig. 11 Parameter window during approach of cantilever to the surface

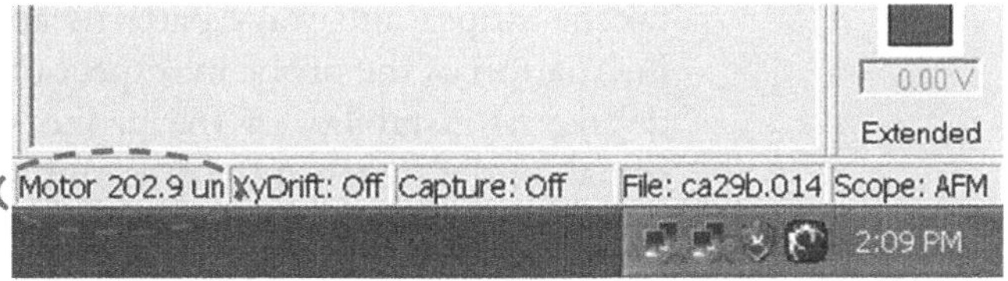

Fig. 12 Bottom-right corner of the NanoScope software window. Step motor position counter is highlighted with the *dashed oval*

either realignment of the laser or changes in the sample preparation should be performed before another attempt to approach the surface.

3.4.6 Adjustment of Imaging Parameters

Adjustment of the imaging parameters is required for capturing profiles free from artifacts. In AFM, two groups of artifacts can be distinguished, probe control related and the artifacts induced by surface condition of the sample. Probe control related artifacts are

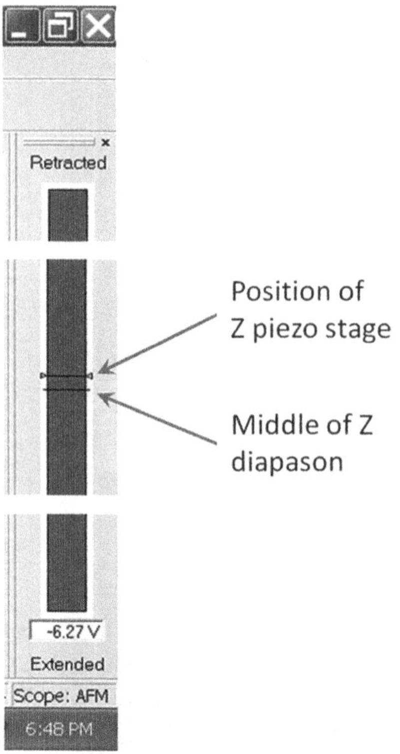

Fig. 13 Right side of the NanoScope software window. Z-stage indicator

described in this section. The sample related artifacts are listed and described in Subheading 3.4.8.

As soon as the probe is engaged, a 0 nm wide image is being taken. Stripes and wavy patterns appear on the screen. They show fluctuation of the probe in respect to the sample surface as well as the degree of instability of the probe–sample gap. Fluctuations of the signal on the screen at 0 nm scanning area are artifacts unless they are used for measurement of stability and calibration of the microscope.

Good tracking of the stable surface occurs when the topology profiles taken in trace and retrace directions are as close in shape and magnitude as possible (Fig. 14c). In the case of instability of contact between the tip and the surface the profiles may look different (Fig. 14d). The edge regions may have characteristic "waving up" and "falling down" profiles. At the edges of the scanning area the direction is changed. The cantilever is flipped into the opposite direction, which can change the effective length of the tip (Fig. 15d). The average difference in heights of profiles is about 20 nm in this example. The artifact can be eliminated with slower scan speeds and a higher Setpoint. The other possibility that creates this kind of artifact can be the nonlinearity of the piezo scanner in the *Z* direction.

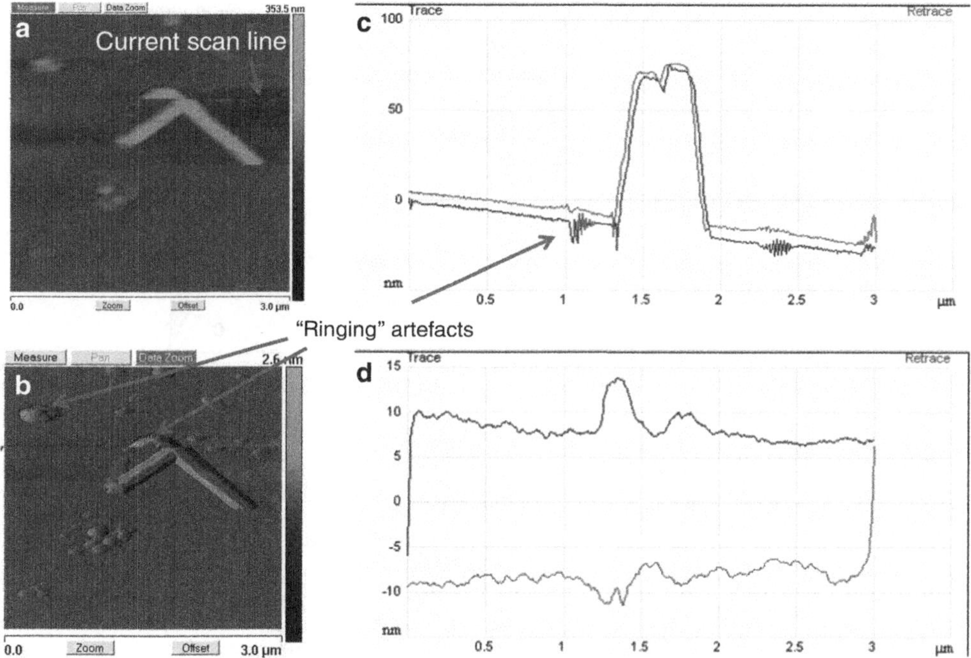

Fig. 14 Carbon nanotubes imaged in height (**a**) and deflection (**b**) modes. First bottom to top scan $5 \times 5\ \mu m^2$ followed by the second scan top to bottom $3 \times 3\ \mu m^2$ to visualize the oscillation artifacts on the top of the junction of nanotubes. (**c**) Trace and retrace profiles are nearly identical. (**d**) Trace and retrace profiles are different; edge artifacts are present. Not acceptable imaging conditions due to unstable interaction of the tip with the sample

"Ringing" artifacts (see Note 28) may look like oscillations in the topology profiles (Fig. 14). Oscillations can be seen as a wavy pattern on the image marked with arrows in profile (c) and images (a) and (b). "Ringing" is taking place when the system is set to be too sensitive. The deflection mode usually gives out more contrast images and is recommended for monitoring noise and oscillations and visualizing features otherwise not seen in other modes. Keep the deflection mode on at all times for better observation of oscillations and for capturing "better" images (see Note 29).

Widening of features on the image may appear because of improper tracking of the surface with the probe. While that may be not so critical at preview, a better image is usually required at the acquisition stage. Tracking of the surface can be controlled better by decreasing the scan rate or velocity of scanning, increasing integral and proportional gains, and increasing a contact force of the probe. Control of those parameters can be done through the corresponding fields in the control frame of the software (Fig. 11).

The experiment on adjustment of integral gain, proportional gain, setpoint, and scanning velocity is described below. The result of these adjustments is presented in Fig. 16.

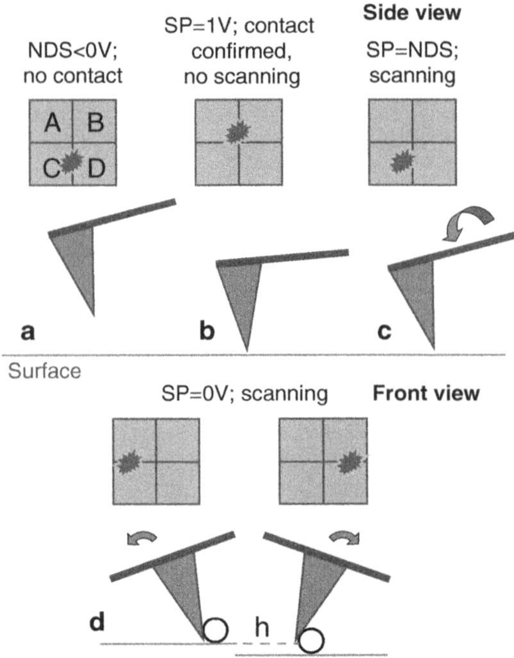

Fig. 15 Laser beam detection with four-quadrant photodiode. (**a**) No Deflection Signal (NDS) is set below Setpoint = 0 V at AFM head. Probe is approaching surface. (**b**) Deflection signal is equal to Setpoint. Probe is in contact with surface. (**c**) Setpoint is set equal to NDS at software. Scanning at contact forces is minimized. (**d**) Scanner compensates *h* nm high irregularity of the tip. Setpoint is 0 V

1. Put the "Data scale" of channel 1 (height) close to the expected height of the features (200 nm in this case) and to about 2 nm in the field of channels 2 and 3 (deflection). "Data scale" defines the brightness of the features on the screen. "Data center" defines the contrast of the images. Fine adjustment of brightness and contrast of images can be done later when imaging larger areas of samples with the features of interest before the final image is captured.
2. Position 1. Scan size is increased to 2 μm.
3. Position 2. The "Integral gain" was gradually increased to 7 until "ringing" lines appeared in the deflection image and in the height profile window (Fig. 16a). The "Integral gain" is successfully reduced to 5.4.
4. Position 3. The "Proportional gain" was gradually increased to 51, which resulted in the "ringing" in deflection channel and eventually in the topography channel too. The "Proportional gain" is gradually lowered down to the value 8.3 where almost no oscillations are observed in the deflection channel and imaging is returned to the normal conditions.

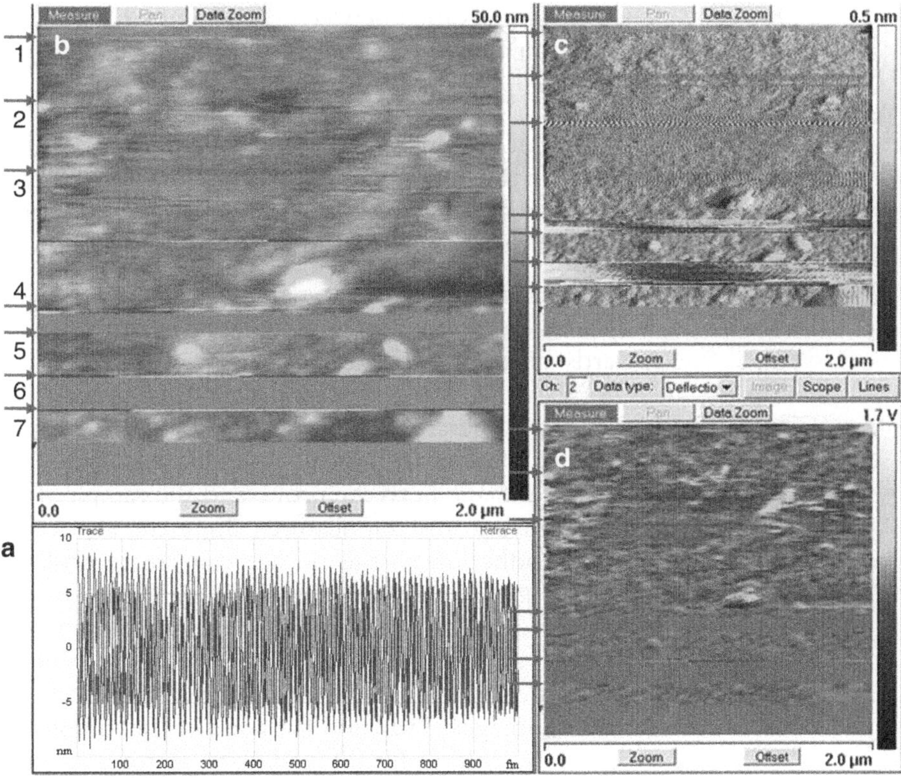

Fig. 16 Adjustment of scanning conditions. (**a**) Resonance oscillations ("ringing") in height profile. (**b**) Height image. (**c**) Deflection image. (**d**) Lateral forces (friction) image

5. Position 4. The Setpoint is varied from 0 V through −1 V to −2.5 V where the contact with the surface is lost. The Z piezo is fully retracted according to the position of the line with arrows near the "retracted" end of the scanner. This situation is not dangerous to the scanner or to the probe. At the Setpoint below the No Deflection Signal (NDS) value (see Note 30) the operation of the microscope may still be possible because the cantilever can be trapped by the capillary forces of a thin water layer and be bent towards the surface. This nonequilibrium condition however does eventually lead to the cantilever going out of contact with the surface. Returning the cantilever to the non-bending position generates a signal for feedback that cannot be compensated for by the scanner. The scanner has to move away from the probe to compensate for bending of the cantilever upwards. The scanner is retracted at the full range. The color of the Z scanner bar is turned to red.
6. Position 5. To have the contact restored the Setpoint is gradually increased to −2.1 V. The scanner is gradually moved towards the surface. The probe is brought into contact at this position.

7. The Setpoint is varied between –2.05 and –2.9 V. The contact with the surface is lost at position 6.
8. Position 7. The contact is restored again.

The stage of probing of the control factors above is important for establishing the optimum conditions of imaging particular samples with selected probes at the instrument available in the lab. The final tuning of the instrument can be described as follows:

1. Step back with the integral gain of the "ringing" conditions and/or increase the contact force by putting the Setpoint towards higher positive values. Swipe the Setpoint between NDS (about –2 V at our experiment) and up to 1–3 V. The bigger the difference between NDS and the Setpoint, the larger the force between the probe and the surface. Too high a force can damage both the probe and the sample. It may also lead to moving unstable features along the surface. Features with low adhesion to the surface can create artifacts in the images (Fig. 17). High and unstable particles can create not only linear artifacts but also create problems with visualizing, detection, and capturing of smaller features. Slopes of the edges of the structures are shown as bright and dark features in deflection images. Even and flat areas of the sample have little variation in colors. Bright and dark features in the height images show actual differences in the height of the structures. Deflection images are more immune to the absolute differences in the height of the structures. Deflection mode therefore can lead to higher contrast images and can be called a contrast enhancement technique, using terminology accepted in light microscopy. High contact forces can be used to our advantage too. A surface can be wiped free of the weakly adhered particles before capturing the actual image. High force is one of the requirements for the manipulation of nanostructures; see Subheading 3.6.
2. There is a chance of detecting no structures in the arbitrarily selected area of the sample at low concentration of the nanostructures. The scanning area should be gradually increased to 2 through 4 to 10 and 20 μm until characteristic structures appeared on the screen. A scanning area wider than 20 μm can already be too large for seeing structures as narrow as about 5 nm. If no structures can be seen at these conditions, try to reduce the scanning area to 10 μm and apply an offset (see Note 31) to probe the surrounding regions.
3. Scanning in AFM is usually done in trace and retrace in the *X* direction, followed by a step in the *Y* direction. Compared to a one-directional "raster" scanning where one scan in *X* direction is followed by a Y step, the double scanning can be much more informative and is very important for estimation of the

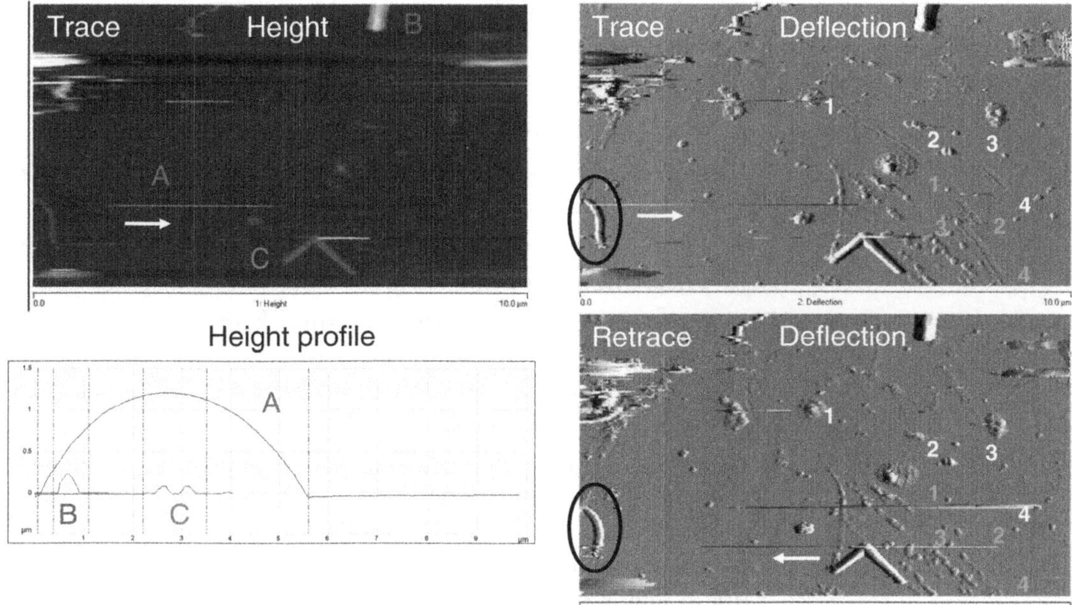

Fig. 17 Moving particle artifacts. Stable contrast in deflection images in trace and retrace directions indicate structures more stable mechanically. Features with contrast varied the most will have the shape changed or disappeared at the next scans. Trajectories of two particles pushed by the tip are shown by white and gray series of figures 1–2–3–4. Height profile A is formed by the tip driven into a sharp feature on the surface, thrown up into 1.25 μm height and flown a distance of 5.55 μm. Tip flying direction is shown with white arrows. Profiles of nanotubes B and C are shown for comparison. Note changes in shapes in the structures highlighted by *dashed ovals*. The image width is 10 μm. The height axis maximum value in the height profile is 1.5 μm

imaging conditions, below. While two-way scanning doubles the time of the acquisition, two images of the same mode can be achieved. They can be useful for more quantitative analysis, and the better image of those two would be selected for publication or presentation when required.

3.4.7 Preview Scanning

At the preview stage, the acquisition time is more important than the quality of the image. Scanning must be done at the highest possible speed while being nondestructive and providing information sufficient for decisions on further actions. Scanning can be done faster at a low number of lines and higher velocity. The number of pixels per line does not affect the acquisition time but creates a clearer image on the screen. With velocity exceeding 20 μm/s a widening of the structures in the images can be expected (see Note 32). Depending on the resonance characteristics of particular scanners, a 100 Hz or higher frequency of scanning may lead to distortions and reduction of the scanning area. Acquisition at 512 pixels/line and 128 lines will create a uniform image in the horizontal and a pixelated (see Note 33) image in the vertical direction on the screen (Subheading 3.6). This small number of lines will be sufficient however to distinguish the arbitrarily aligned linear structures on the flat surface.

1. Put 512 and 128 into "Samples/line" and "Lines" fields of the software, respectively.
2. Increase the scanning area to up to 10–20 μm if no features of interest are found at the first attempt.
3. Apply an offset to probe nearing regions (fields "X offset" and "Y offset" in Fig. 11).
4. If all the above steps lead to no features being observed on the surface, the probe should be withdrawn by clicking the "Withdraw" button two to three times or more if the sample is not levelled or flat enough.
5. Move the sample with the coarse positioning knobs of the head stage (Fig. 5b) in two perpendicular directions to probe another spot of the sample at a distance of a couple of millimeters away from the previous area. The combined microscopy option is not available on the NanoScope E. Therefore a quality optical preview of the sample surface and quick finding of features and spots of interest are problematic.
6. Approach the surface at the selected position and repeat the preview again.
7. If probing different regions of the sample at quick and slow scan speeds leads to no good results, change the sample or try to modify the sample preparation procedure.
8. Once the structures of interest are visualized in preview, either precision centering or zooming in around the features can be possible at this stage. Click the "Zoom" button at the bottom of the window of your interest ("Height" window in this case). The zooming rectangle will appear.
9. Move the rectangle around the field of view by clicking and dragging the mouse.
10. Adjust the size of the rectangle and place it around the features of interest. In this experiment the nanotube on the bottom of the image in Fig. 17 was found more interesting.
11. Click the "Offset" button to activate the lateral shift of the scanner towards the structures.
12. Put the cross mark at the center of the structure with the left-click of the mouse button.
13. Click the "Execute" button to perform the shift and start scanning around the selected features. Coarse zooming around the area of interest is complete.
14. Put even numbers (see Note 34) in the "Scan size" field for the precision zooming and for creating comparable images for future referencing, analysis, and publication.

15. At this point the features are put to the center and are scanned at high speed. If features of the sample are not seen clearly enough, perform the adjustment of the imaging conditions described in Subheading 3.4.6.

3.4.8 Capturing Images

If the appearance of the structures in preview is satisfactory, the actual images can be acquired.

1. Zoom in the region of the interest by clicking the "Zoom" button.
2. Put the zoom rectangle around the features of interest.
3. Click the "Execute" button to confirm the selection.
4. Adjust the scanning size from the corresponding field of the control panel.
5. Increase the number of lines to 512 or higher if required. The number of lines and pixels per line can be equal.
6. When all the acquisition parameters have been adjusted and no artifacts above are seen from the images the microscope is ready for imaging. Click the "Capture" button on the toolbar.
7. Confirm that the acquisition is being performed in the "Capture" field from the lower right corner of the frame of the main window (Fig. 12). The capture is switched to ON. The image will be captured during this run. "Next" is shown when the imaging parameters have been changed during the current scan. The capturing will proceed on the next run.
8. Press the buttons "Frame Up" and "Frame Down" when the acquisition is ready to start over from the bottom or from the top correspondingly. Alternatively, click the "Capture" button again to acquire the current screen. Confirm the capturing process by observing the "Forced" sign at the "Capture" field of the window.
9. Click the "Capture Now" button. Only part of the screen will be taken, starting from the edge of the frame and ending at the position where the button was pressed.
10. Alternatively, push the "PrtScr" (print screen) button from the keyboard to store the content of the computer screen in the clipboard of the computer. Push Ctrl-(Shift)-V or right-click "Paste" command from the contextual menu to have the content of the screen inserted into an image-editing program (e.g., Paint) for your reference. The image is processed by the computer in an unknown way and is not recommended for publications. However, it can be used as a quick reference tool that can save some time in the future.

3.4.9 Turning the Microscope OFF

1. Click the "Withdraw" button two to three times to remove the probe from the surface.
2. Push the lever of the scanner step motor (Fig. 7h) to the "UP" position for about 10 s to remove the tip from the surface into a safer position.
3. Disconnect the cable of the head from the scanner socket (Fig. 4e).
4. Pushing the center of the head with one hand near the upper window (Fig. 5h), stretch one of the scanner springs (Fig. 5a) and remove it from the lever at the head at one side.
5. Change the hand holding the AFM head, keep pushing it down, and remove the other spring of the scanner from the lever on another side of the head.
6. Put the AFM head on a flat and clean surface of the bench or in a specially prepared container to protect it from dust.
7. Save all your data including the workspace file with extension .wks to the local folder.
8. Copy the data files to a CD drive, USB pen, or through the network in accordance with the standard operating procedures in your lab.
9. Exit the software by clicking the cross icon in the right upper corner of the main window.
10. Turn off the computer.
11. Wait until the screen of the computer turns black.
12. Turn OFF the power switch at the back of AFM controller.

3.5 Image Processing and Analysis

Image processing and analysis are usually done during or shortly after the acquisition of a series of data. This practice helps to save time on sample preparations, searching and positioning of the AFM probe over the region of interest. Immediate data analysis is absolutely critical for unique samples with a short lifetime. The image analysis in NanoScope 6 software can be done during the acquisition. In this chapter the image processing and analysis is described for the example of the images of nanotubes (Fig. 18a) performed on a stand-alone NanoScope Analysis freeware package version 1.30, a NanoScope files viewer available from Bruker AXS, Santa Barbara, CA 93117, USA.

1. Download and install the software on your computer.
2. Backup all the source data in a separate directory or in another storage media before the data are processed.
3. Open a workspace file with .wks extension by selecting the "Workspace/Open Workspace..." command from the NanoScope Analysis menu. The workspace file contains links,

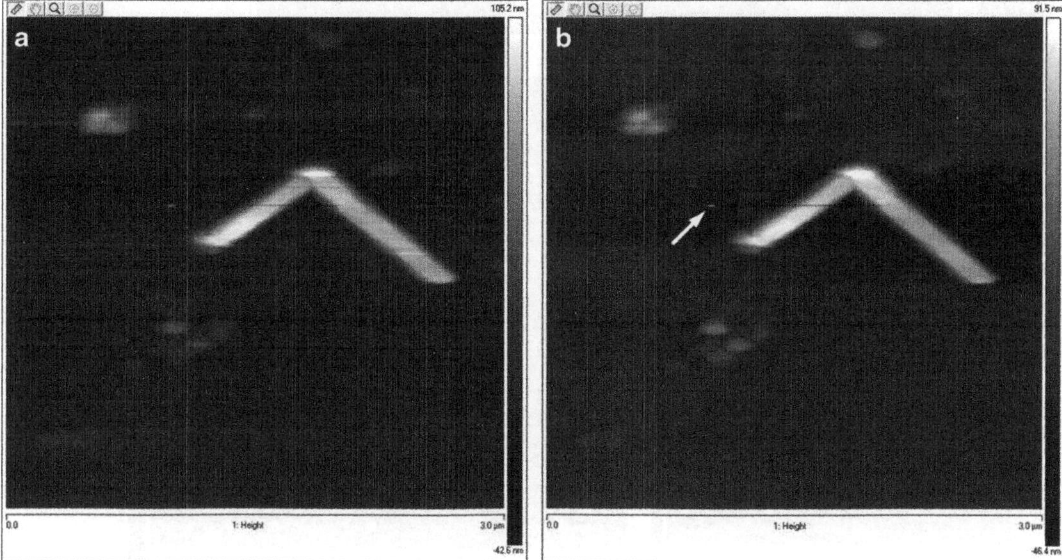

Fig. 18 Flattening: (**a**) initial image and (**b**) image after flattening. Scan lines are removed from the image. Particle removed from the tip can create an artifact pointed by arrow. The image sizes are 3 x 3 μm². The colour scale heights are 147.8 nm (**a**) and 137.9 nm (**b**)

views, and status information for all the files saved in the selected directory (see Note 35). The list of files will appear in the upper panel of the main window. Quick navigation between files can be done by pushing the "Ctrl-Tab" combination of buttons. Navigation between the data modes can be quickly done by pushing "Tab" and "Shift-Tab" buttons.

4. The individual files can also be accessed by clicking the "Directory…" icon in the "Browse Files" panel at the right corner of the window (see Note 36) (highlighted by the red oval in Fig. 19) and selecting the directory with the files. Single files can now be opened for viewing and editing.

3.5.1 Flattening

Flattening in AFM is a standard procedure where the scanned lines are brought into one plane. This procedure is required to compensate the low-frequency fluctuation of the position of the probe in respect to the sample surface.

1. Select an image from the list to process.
2. Select "Filters/Flatten" command from the menu.
3. Select "Flatten Order 1st" and "No Thresholding" in the pop-up window.
4. Click the "Execute" button. The initial and flattened images will look similar those shown in Fig. 18.

3.5.2 Removal of Faulty Scan Lines

Capturing ideal images with no artifacts and defects may require ideal sample preparation, carefully selected imaging parameters,

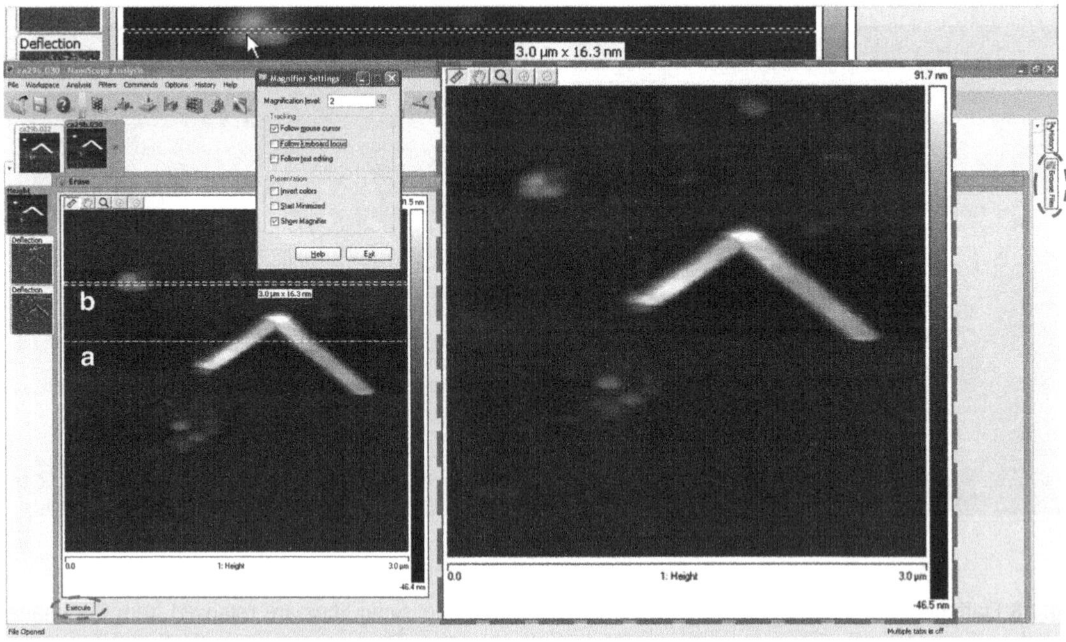

Fig. 19 Removal of faulty scan lines with Erase tool. Rectangular (**a**) or linear (**b**) regions selected at initial image (Fig. 18). Resulted image is shown in the insert with *dashed frame*. "Execute" button and "Browse Files" tab are selected with *dashed ovals*

thorough understanding of the mechanisms of probe–sample interaction, and the functioning of the microscope. Capturing ideal images can be time consuming. Frequently images may contain a single or a small number of lines with artifacts. These lines can be removed by averaging with surrounding lines.

1. Select "Filters/Erase" command from menu.
2. Run standard Windows program Magnifier for precision selection of lines and other regions with defects: "Start/All programs/Accessories/Accessibility/Magnifier". The magnification strip appears on the top of the screen (Fig. 19).
3. Click on the position of the lower artifact (Fig. 19a). A dashed line appears.
4. Click to select and move the line with the mouse, on the top of the scanned line with artifact.
5. Right-click in image and select "Area" for selection of a region.
6. Drag a rectangular area across the image (Fig. 19b) from one edge of the image to another edge, to leave no linear artifacts after the processing on either side.
7. Click the "Execute" button. The heights in selected regions are averaged with the surrounding areas. The resulting image can be seen in the insert with the dashed frame in Fig. 19.

3.5.3 Measurement of Geometrical Parameters of Acquired Structures

Compared to many other microscopic techniques capable of performing linear measurements only in plane AFM can produce 3-dimensional (3D) profiles with sub-nanometer accuracy.

Linear measurements in plane are readily available and can be done from the 2D view.

1. Click and drag the mouse along the nanotube. The resulting line shows distance in nanometers as well as the angle (Fig. 20a). The color of the line was changed to black for better visibility of the line on the top of white structures. (Right-click in the line and select "Set Colour/All". Select a color from palette.)

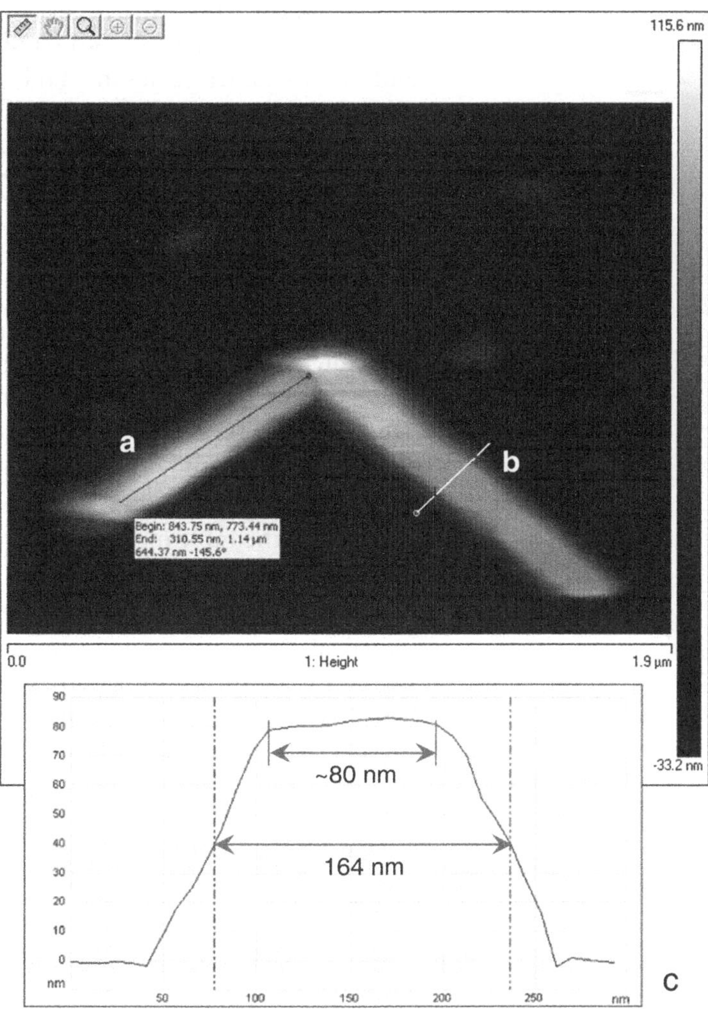

Fig. 20 Linear measurements and shape analysis. (**a**) Estimated length of nanotube is 644.37 nm, angle is −145°. (**b**) Position on the nanotube where profile (**c**) is taken. Measured height of nanotube is 84 nm with the width of 164 nm. Plateau on the top of the structure may originate from a tip artifact

2. Measurements of the width of 3D structures from the top view only are subjective and lead to errors as big as 30–50% depending on the structures measured. The objective measurements of width and height of the structures are usually done on the profile view or a cross-section.
3. Click "Analysis/Section" from the menu.
4. Click and drag a line across the features of interest. The profile of the structure appears at the "Section" frame of the "Section" window. Figure 20c shows a profile taken across the nanotube at the position b at the top view. Blue dashed markers on the profile graph are shown as blue dots on the white line on the top view.
5. The software gives out the horizontal and vertical distances, curvature and distance along the surface, angle, roughness, and other characteristics in a table. One of the objective methods of measurements is called FWHM (full width on half maximum). The method is illustrated in Fig. 20. A full length between profile points taken at the half of the height of the structure is measured.
6. Position one of the blue marker lines at the highest point of the profile.
7. Put another marker at the average position of the profile of the substrate near the nanotube profile. The vertical distance between the top of the structure and the base line is the height of the nanotube.
8. Move the marker line from the top of the structure into the middle position of one of the slopes. The "Vertical Distance" value from the table must show exactly the average value of the previously taken readings.
9. Move the other line onto another slope of the profile. The "Vertical Distance" value in the table must show about 0 nm and the marker lines must cross the profile at about the same level. The "Horizontal Distance" value seen from the table is the FWHM value and is about 164 nm in this example.

3.5.4 Flattening with Exclusion

Quality 3D images may require additional processing like flattening with exclusion of high regions (Fig. 21) followed by cropping of desired regions of the image. Conventional flattening is described in Subheading 3.5.1.

1. Drag the rectangles around bright features in the image (see Note 37) that correspond to elevated areas of the sample.
2. Push the "Execute" button. Height artifacts are removed from the image.

3.5.5 Cropping

In actual processing of images cropping is one of the first commands to be performed. Cropping can remove artifacts, eliminate

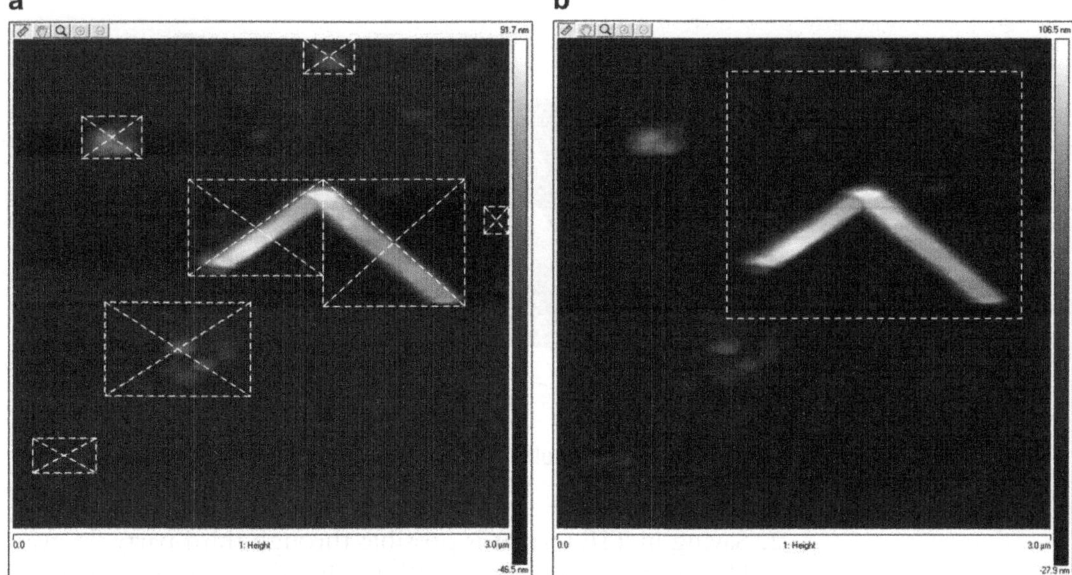

Fig. 21 Flattening with exclusion of regions. (**a**) High regions are selected with the *rectangles* to be removed from flattening. (**b**) Height artifacts are removed from the image followed by cropping into desired size shown by a *dashed rectangle.* The image sizes are 3 x 3 μm^2. The colour scale heights are 138.2 nm (**a**) and 134.4 nm (**b**)

a number of processing steps, and make the processing of the images easier and quicker.

1. Select "Filters/Crop" and "Split" command from the menu.
2. Click and drag a rectangle around the region of interest. Pushing "Create File(s)" button creates a file on the hard drive that can be accessed through the "Browse Files" panel to the right.

3.5.6 3D View

1. Open the flattened and cropped image by double clicking it at the "Browse Files" panel.
2. Select "Analysis/3D Image" command from the menu.
3. Click and drag the mouse to tilt the 3D model in different directions.
4. Use Ctrl-drag to zoom in and out of the image.
5. Use Shift-drag to move the image in the plane.
6. Right-click drag and Shift-drag the mouse to change the angles and intensity illumination. Stop in the position where all the features of interest are seen clearly but without glowing spots in the image (Fig. 22).

3.5.7 Saving Images

1. Right-click an image and select "Export/Screen Display". The images can be saved in either BMP or JPEG formats (see Note 38).

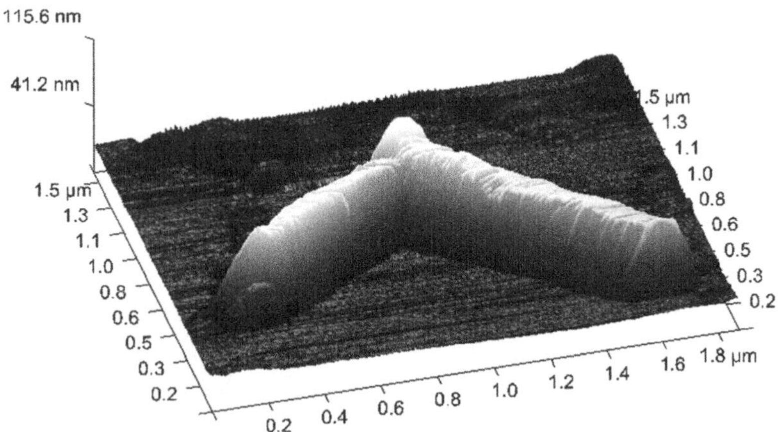

Fig. 22 3D view of nanotubes after flattening and cropping

2. Saving in TIF format is possible through third-party freeware like XnView (see Note 39). Right-click the image in NanoScope Analysis and select "Copy Clipboard" command.
3. Open XnView.
4. Push hotkey Ctrl-Shift-V or select "Edit/Paste" from XnView menu. The image exported into XnView can now be saved in a number of formats including TIFF.

3.5.8 Artifacts

Debris Removed from Tip

The artifact highlighted by an arrow in Fig. 18b is typical of a particle or debris removed from the tip at scanning. The quality of the image may be noticeably improved if the tip is cleaned up. Debris and particles may lead to formation of double-tip artifacts, to reduction of lateral resolution, and to instability of the contact of the tip with the surface and ideally should be avoided or firmly fixed to the sample surface.

Flattened or Double Tips

The measured FWHM of the structure in Fig. 20 is about two times its height. If a high rigidity of the nanotube is assumed then a tip artifact can be one of the reasons for this discrepancy in measurements. In this experiment the tip was heavily used to find a good arrangement of nanotubes that could lead to extensive wear of the tip. If the nanotubes are soft or loosely attached to the surface the tip may modify the nanotubes and produce similar artifacts. Adhesion of loose particles picked up by the tip from the surface could be another reason for this artifact. The measured plateau region of the profile is about 80 nm wide. Subtraction of 80 nm of tip width from 164 nm of structure width gives 84 nm, a value of the height of the structure.

3.6 Nanomanipulation

The same effect that created artifacts in Fig. 17 can be used to our advantage. A tip in hard contact with the surface can move the

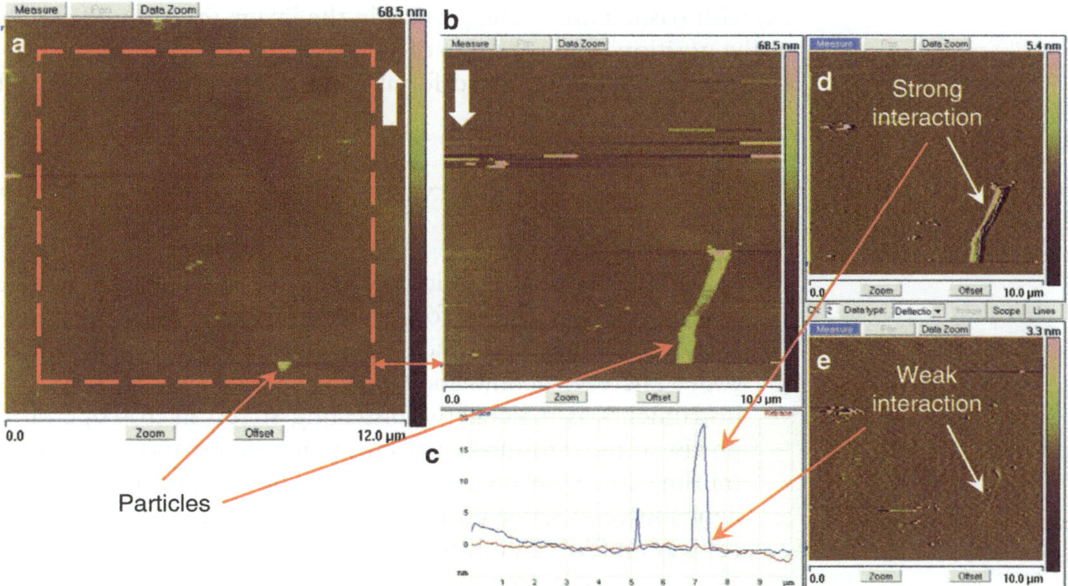

Fig. 23 Experiment on particle manipulation. Images (**a**) and (**b**) are taken in height mode. Images (**d**) and (**e**) represent the deflection of cantilever. Scan directions for 12 and 10 μm side images are indicated with white arrows. (**a**) Detection image, 12×12 μm^2 area scanned in bottom–top direction after modification. (**b**) Modification image, 10×10 μm^2 area scanned from top to bottom. Linear structure in the image is a trace of a particle found in detection image followed the modification. (**c**) Trace–retrace profiles taken at last modification run of the tip. Particle in image (**a**) is found in the position of the last modification run. (**d**) Deflection image in Trace direction. (**e**) Deflection image in retrace direction. Probe–particle interaction in Trace direction is much higher that can be seen from both profiles and deflection images

particles along the surface and eventually build a multicomponent nanostructure. Only the method of the particle manipulation is discussed in this chapter.

A cantilever with a 66 times higher spring constant (about 2.0 N/m) has been used for this experiment. High rigidity of the probe ensures higher interaction forces between the tip and the surface. In the first preview run through the surface with nanotubes, a loose particle was detected (Fig. 23b). The preview was done with 128 lines, with velocity 20 μm/s, and with a CSCS21 cantilever at its near-equilibrium position (Setpoint, 2.5 V); see Note 40. The loose particles have been distinguished from the fixed to the surface structures by the difference in height profiles in trace and retrace directions Fig. 23c and by a higher deflection signal in the Trace direction from deflection images (d) and (e). The next scan was performed over a bigger area of 12×12 μm^2 in the opposite direction, from bottom to top. The image presented in the Fig. 23 (a) is scaled to the dimensions of the image (b). A 10×10 μm^2 square corresponding to the initially scanned area is shown in image

(a) with dashed lines. The particle in the image (a) is found exactly at the position of the last line of scan of image (b).

The particle was moved for about 4 μm and intentionally left in the desired position.

4 Notes

1. In light and electron microscopy the particles can either be focused to form a probe for the scanning microscopy or distributed evenly across the whole field of view in wide-field microscopy. The even illumination of the field of view is critical for instruments as basic as the dissecting stereo microscope or as precise as the transmission electron microscope and as big as the transmission X-ray microscope that requires a synchrotron as the light source. A stable probe is as important for scanning microscopy as uniform illumination is for wide-field instruments.
2. Collection of data for spectroscopy requires no scanning across the sample.
3. Many of the aspects discussed in this chapter are applicable to the whole family of SPM instruments. Only a smaller number of the described features are specific to AFM. While terms "SPM" and "AFM" are not interchangeable, for simplicity's sake only "AFM" is used hereafter.
4. Feature-oriented scanning (FOS) and positioning (FOP) are relatively new techniques (22) that allow precision registration of relative positions of features in respect to each other. The method is potentially applicable to automatic tracking of movement of selected features in respect to other features on the surface and precise delivery of single atoms and molecules to the desired site of reaction recognized by the presence of other atoms or molecules (18). Scanning probe microscopes are made of a number of mechanical elements with different thermomechanical properties, and therefore sensitive to temperature fluctuation inside and around the microscope. FOS and FOP make the automatic compensation of drift and creeps of sample in small scale easier compared to other techniques (23) due to the fast feature-by-feature acquisition of topography.
5. Mica is an inert substrate used in this experiment for physical support of protein and nanotube samples.
6. These tweezers can be used for manipulating cantilevers only. Any damage of the tweezers may lead to frequent loss of cantilevers.
7. Micropipetting is done as follows:
   (a) Dial 10 μL on the micropipette.
   (b) Attach the 10 μL tip by pressing the opening of the tip at the tip rack with the end of the micropipette.

(c) Hold the Eppendorf tube in one hand and bring the tip of the micropipette close to the surface of the stock solution of fibrous proteins in the Eppendorf tube. Press the plunger button and immerse tip into the Eppendorf tube with the stock solution avoiding touching the walls of the tube. Release the button slowly. Quick release of the plunger may lead to errors in volumes taken.

(d) Remove the tip from the tube with the stock solution and introduce it into the Eppendorf tube prepared for the sample.

(e) Push the plunger button slowly with extra force at the end to have all the volume of the sample removed from the tip with excess of air.

(f) Contact the walls of the tube with the plunger fully pressed to have no sample left on the surface of the tip.

(g) Remove the tip into the waste container by pushing the release button of the micropipette.

(h) Add 10 μL of buffering solution to the Eppendorf tube following the steps above.

(i) The uniform solution is prepared by pumping the micropipette three to four times.

(j) Release the empty tip to the waste container.

8. Wiping the surface of the sample holder disk with KimWipes soaked with alcohol may be required after the experiment to remove all the residues of the sticking material from the tape and to keep the surface of the disk flat.

9. The removed layer will be as clean as the bulk mica substrate. The double-sided tape with the removed mica can be used for imaging too if needed. The mounting of a solid piece of mica is easier compared to a few layers of mica on a tape because of higher rigidity of the solid piece. The mounting of the mica on the disk is done by pushing each of two of the opposite corners of the chip with both tips of tweezers simultaneously.

10. Alternatively, the even distribution of the sample over the substrate can be achieved by dropping the sample on the substrate from the height of about 50 cm:

(a) Place a cover glass or a piece of mica on a top of a KimWipes tissue clean side up.

(b) Take a small amount of sample with 20 μL micropipette or a transfer pipette.

(c) Position the tip of the (micro)pipette at about 50 cm above the substrate (level of eyes if standing near the bench with the substrate on it).

(d) Press the (micro)pipette and allow one drop of solution (about 20–50 μL) fall on the center of the substrate and spread evenly over it. Fix the (micro)pipette to a laboratory stand if aiming the substrate by hand is difficult.

11. Estimated thickness of the coating is 2 μm.
12. Boiling point of NMP is 202–204°C.
13. Touching the central area may cause deformation of the mica, e.g., scratches on the surface, and make the substrate unusable.
14. Friction mode is topology sensitive and requires an absolutely flat surface for artifact-free friction images. Fibrous proteins or carbon nanotubes are not the ideal samples for imaging in friction mode. It may however visualize the features not seen in topography and deflection modes.
15. Mounting of the probes is a difficult procedure for beginners in scanning probe microscopy. Mounting of the cantilever to the holder can be done more easily when it is oriented in the same way as the cantilever in the box. Movement of the hand only in one plane without any twisting and turning is required. More complex movement of the hand may lead to the loss of the cantilever.
16. Try to work out an optimum pressure for the tweezers at this stage. Too low a force will cause slipping of the chip of the probe but too high force will cause the chip to jump off the tweezers.
17. Avoid moving the tweezers with chips above the areas with new cantilevers. Dropping the chips may cause damage to a number of the probes.
18. The alignment step may require a head magnifier or a binocular microscope for more reliable operation.
19. In most cases a laser beam is passing through the area near to the free end of the newly mounted cantilever. However, the laser beam is hitting air at this position and cannot be seen. The beam must be moved towards the edge of the chip to make it visible. The chances of finding the laser spot will be higher if looking at the scattering of the laser light from the edge of the chip.
20. The maximum can be found more easily when the values are observed increased and decreased on both sides from the optimum position. Pass carefully the position near the end of the cantilever and remember the highest "SUM" value from the photodiode. Try to restore this value moving in the opposite direction. Change the direction of the movement to perpendicular and try to maximize the signal again. Repeat the alignment in perpendicular directions to find the local maximum.
21. An additional idea about the proximity of the probe to the surface can be achieved while looking on the gap between the corners of the chip and the reflection images or shadows of the chip formed on the sample surface.

22. Proportional and integral gains define the reaction of the Z stage on changes in deflection signal received from photodetector. The difference between the signal from the photodetector and the Setpoint is called the Error Signal. The Error Signal is used to control the voltage applied to Z piezo stage and therefore the position of the stage in respect to the base of the cantilever. The voltage applied to Z stage is proportional to the Error Signal. The coefficient of proportion is set at the "Proportional gain" field of the software. The reaction of the stage controlled by the "Proportional gain" only is too slow for tracking sharp edges of the structures. The task of the integral gain is to push the stage towards the probe as soon as the contact is lost. This component of the Z-stage voltage is proportional to the time of the probe being out of the contact with the surface. Proportional and integral components of the voltage are summed and used to ensure the best performance of the system at highest possible velocities of the probe.
23. A travelling distance of about 100 μm before contact at automatic approach is an indication of a reasonably small initial gap between the tip and the sample surface. Glass or mica samples may require larger gaps in order to avoid hard contact with the surface.
24. In other cases, the reading from the upper indicator can gradually equalize with the Setpoint from the software. The reading can also fluctuate between NDS and the Setpoint values during or after approach. This is an indication of wrong laser alignment. The cantilever should be withdrawn from the surface by clicking the Withdraw button from the software toolbar two to three times followed by the alignment of the laser described at Subheading 3.4.4.
25. The image can be distinguished from noise by topographical contours connected in the direction of the slow scan (vertical or *Y* direction in the image). Noise is frequently seen as periodical and equally bright and contrast stripes running under the angle to the edges of the screen.
26. The contact with the sample is unstable when the Z-position indicator is moved more than about 1% of the range of the Z scanner at the Setpoint increased.
27. The imaging of about 1 μm thick freshly prepared protein film may result in the tip indenting the upper layer of the film creating unstable contact conditions. Attempted imaging under these conditions will lead to the removal of the film layer by layer and failure to achieve a reproducible image with every next frame scan. Quality imaging may require switching to tapping, noncontact, or other modes or preparation of a stable sample where thickness of the protein film is equal to or less than one monolayer of the material.

28. "Ringing" is seen as an oscillation pattern on the profile and as wavy areas on the images. "Ringing" usually takes place at the sharp edges of the structures and represents the resonance characteristics of the whole system including scanner, tip–sample gap, and feedback loop. It depends on cantilever rigidity, tip radius and shape, sample material properties, slope angle of the features, and other factors.

    The analogue with a racing car having jet engines in a vertical direction and robust suspension may help explain the behavior of the system that is "ringing." The car has a reliable contact with the ground in the plateau region. It can be driven at very high velocity as far as the surface is flat. But even a small step down from the plateau will drop the pressure to the wheels of the car. The drop of pressure is the signal to the feedback system for turning on the jets pushing the car down. The car starts accelerating towards the ground until its tires hit the surface. The car suspension is robust enough to withstand the shock. The pressure in the tires however becomes too high for the normal racing conditions. This high pressure is a signal to the feedback to switch the jet engines in the opposite direction, from pushing to lifting. If the impact to the surface was too big, the signal to the feedback system is large enough to through the car away from the surface. The contact with the wheels is lost and a pushing feedback signal is generated again. The feedback meets the resonance conditions and the car starts bouncing off the surface. The feedback system is more sensitive to the integral component of the amplification of the signal (an integral gain) from the wheels. Integral gain is usually the first of the parameters being dropped to stop the car from jumping at a given speed and the first to increase when the edges of the structures are not resolved well.

29. The image may appear "better" in different modes. Compared to microimaging and microphotography where visualization of features is the final goal, a direct comparison between the modes in microscopy is frequently not possible. Different modes are based on different physical principles. A mapping of physically different artifacts and features and their relative arrangement in time and space is the goal of microscopy and microanalysis. While deflection image can be seen with more contrast, the roughness and geometry of topographic features, for example, can be quantitatively obtained only from the height data.

30. No Deflection Signal is a photodetector voltage reading from the upper indicator of the scanner corresponding to the equilibrium position of the cantilever out of contact with the surface. The difference between the Setpoint and NDS defines the working deflection of the cantilever and acting forces

applied between the surface and cantilever. NDS can vary under thermal drifts, charges of the surface, and other conditions. The Setpoint can be controlled during the operation of the microscope to compensate the changes of the NDS and to maintain the desired imaging conditions.

31. A 10 and 20 μm scanning area covers just a fraction of the total area of the scanner. The Veeco JV scanner has a 125 μm full scanning range. The offset is a convenient tool for moving a smaller scanning area within the limits of the scanner. Zero nm offset means scanning the central area of the full range of the scanner. 70 μm X offset and −70 μm Y offset should mean shifting the 10 μm × 10 μm imaging area to the lower right corner of the scanner.

32. To the best knowledge of author, the world record in artifact-free scanning velocity is about 25 μm/s.

33. An image is made of pixels. The term pixels originated from the phrase "picture element." Images may have a small number of pixels (about 20–30 pixels in computer icons) across or about a hundred times bigger (up to about 3,000 pixels in a row of images taken from conventional photocamera matrixes). Modern computer monitors and data projectors have 1,000–2,000 hardware pixels across the screen. When "software" pixels of images are stretched over "hardware" pixels of monitors or data projectors two or more times, the pixels of images become visible on a screen as separate squares. The image becomes "pixelated." Similar considerations apply as those used for selecting image formats for printing. The resolution of the human eye is thought of being about 0.1 μm. An image with pixels at a density of 300 dpi (dots per inch) is believed to have no single pixels resolved by the naked eye. Sizes of images in scientific papers are about 2–3 in. each which puts the requirement for the quality images of having 1,000 pixels or more. One thousand pixels per side is a current de facto standard of presentation of images in digital and scanning microscopy. A higher number of pixels will lead to a slightly higher quality image at the expense of bigger image files and lower productivity due to the longer data acquisition and processing time.

34. Even numbers ideally should be selected from the sequence 1, 2, 5, 10, etc. for quicker and consistent referencing. One scale bar can be used for all images in the series. Features and structures in the images with an identical scan size can be easily compared.

35. If the workspace file cannot be opened select the files that require processing and analysis in Windows Explorer and drag and drop them onto the open window of the NanoScope Analysis software. Select the "Workspace/Save Workspace

As..." command from the menu and assign a relevant name and location for the newly created workspace.

36. Click the "Browse Files" tab at the right edge of the window. Click the Pin icon if the panel is required for permanent access.
37. The role of flattening is to level all the lines in the image and to bring them into one plane. Lines in raw SPM images are frequently clearly seen as bright or dark stripes. The tip is usually fluctuating and creeping with time, mostly due to the variations in temperature in various parts of the microscope. These fluctuations of height are seen as individual scanning lines.

    Bright features in the image represent high regions on the sample. Because the "center of mass" of all the lines is different, some of the lines become tilted with respect to others. One end of the line may become elevated and the other end lowered compared to the overall plane of the image. Conventional flattening is very convenient for smoothing the entire image. The flattening with exclusion is required for removing the height artifacts from the images.
38. BMP, TIF, and JPG are the most commonly used formats of the images. BMP images usually take a lot of space on a hard drive. JPG images may have artifacts due to the compression. TIF is a good compromise between BMP and JPG. TIF files are not too large but have quality of images acceptable for publication.
39. Available for download form http://xnview.com/.
40. When the cantilever is near the equilibrium position, the Setpoint is equal to the NDS seen from the upper photodiode indicator. In this position no cantilever bending induced forces are applied to the surface. The tip is trapped and kept in contact by mostly capillary and attractive intermolecular (Van der Waals) forces. The contact is close to unstable but the forces between the tip and the surface are minimized.

Acknowledgements

The author would like to thank Professor Richard Haverkamp, School of Engineering and Advanced Technology, and Dr. Mark Waterland, Institute for Fundamental Sciences, Massey University, for their support and opportunity to work with scanning probe microscopes. The author would also like to thank Dr. Mark Patchett, Institute of Molecular BioSciences, for his valuable suggestions, and Dr. Simon Loveday, Riddet Institute, and Mike Seawright, Institute for Fundamental Sciences, Massey University, for supplying protein and nanotubes samples and technical assistance with sample preparation for AFM.

References

1. MIAWiki page on interrelation between spectra, images and maps: http://confocal-manawatu.pbworks.com/w/page/16347061/Spectrum-Image-Map. Accessed 25 Dec 2011
2. Binnig G, Quate CF, Gerber C (1986) Atomic force microscope. Phys Rev Lett. doi:10.1103/PhysRevLett.56.930
3. Eaton P (2010) Atomic force microscopy. Oxford University Press, Oxford
4. Howland R, Benatar L, Symanski C (1998) A practical guide to scanning probe microscopy. DIANE publishing company, Darby
5. Morris VJ, Kirby AR, Gunning AP (2009) Atomic force microscopy for biologists. Imperial College Press, London
6. Li H, Oberhauser AF, Fowler SB et al (2000) Atomic force microscopy reveals the mechanical design of a modular protein. Proc Natl Acad Sci. doi:10.1073/pnas.120048697
7. Noy A, Frisbie CD, Rozsnyai LF et al (1995) Chemical force microscopy: exploiting chemically modified tips to quantify adhesion, friction, and functional group distributions in molecular assemblies. J Am Chem Soc. doi:10.1021/ja00135a012
8. Yu J, Bippes CA, Hand GM et al (2007) Aminosulfonate modulated pH-induced conformational changes in connexin26 hemichannels. J Biol Chem. doi:10.1074/jbc.M609317200
9. Stroh C, Wang H, Bash R et al (2004) Single-molecule recognition imaging microscopy. Proc Natl Acad Sci U S A. doi:10.1073/pnas.0403538101
10. Bash R, Wang H, Anderson C et al (2006) AFM imaging of protein movements: Histone H2A–H2B release during nucleosome remodelling. FEBS Lett. doi:10.1016/j.febslet.2006.06.101
11. Puech PH, Nevoltris D, Robert P et al (2011) Force measurements of TCR/pMHC recognition at T cell surface. PLoS One. doi:10.1371/journal.pone.0022344
12. Fuhrmann A, Ros R (2010) Single-molecule force spectroscopy: a method for quantitative analysis of ligand – receptor interactions. Nanomedicine. doi:10.2217/nnm.10.26
13. Rico F, Chu C, Moy VT (2011) Force-Clamp measurements of receptor–ligand interactions. In: Braga PC, Ricci D (eds) Atomic force microscopy in biomedical research: methods and protocols, methods in molecular biology. Humana Press, Totowa, NJ
14. Kuehner F, Costa LT, Bisch PM et al (2004) LexA-DNA bond strength by single molecule force spectroscopy. Biophys J. doi:10.1529/biophysj.104.048868
15. Rico F, Su C, Scheuring S (2011) Mechanical mapping of single membrane proteins at submolecular resolution. Nano Lett. doi:10.1021/nl202351t
16. Janovjak H, Kessler M, Oesterhelt D et al (2003) Unfolding pathways of native bacteriorhodopsin depend on temperature. EMBO J. doi:10.1093/emboj/cdg509
17. Rief M, Grubmüller H (2002) Force spectroscopy of single biomolecules. Chemphyschem. doi:10.1002/1439-7641(20020315)3:3<255::AID-CPHC255>3.0.CO;2-M
18. Kufer SK, Puchner EM, Gumpp H et al (2008) Single-molecule cut-and-paste surface assembly. Science. doi:10.1126/science.1151424
19. Kwon T, Park J, Yang J et al (2009) Nanomechanical in situ monitoring of proteolysis of peptide by Cathepsin B. PLoS One. doi:10.1371/journal.pone.0006248
20. Loveday SM, Wang XL, Rao MA et al (2010) Tuning the properties of β-lactoglobulin nanofibrils with pH, NaCl and CaCl2. Int Diary J. doi:10.1016/j.idairyj.2010.02.014
21. Bolder SG, Vasbinder AJ, Sagis LMC, van der Linden E (2007) Heat-induced whey protein isolate fibrils: conversion, hydrolysis, and disulphide bond formation. Int Diary J. doi:10.1016/j.idairyj.2006.10.002
22. Lapshin RV (2004) Feature-oriented scanning methodology for probe microscopy and nanotechnology. Nanotechnolgy. doi:10.1088/0957-4484/15/9/006
23. Sikora A, Sokolov DV, Danzebrink HU (2006) Scanning probe microscope setup with interferometric drift compensation. In: Wilkening G, Koenders L (eds) Nanoscale calibration standards and methods: dimensional and related measurements in the micro- and nanometer range. Wiley, Weinheim

Index

Juliet A. Gerrard (ed.), *Protein Nanotechnology: Protocols, Instrumentation, and Applications*, Methods in Molecular Biology, vol. 996, DOI 10.1007/978-1-62703-354-1, © Springer Science+Business Media New York 2013

Lightning Source UK Ltd.
Milton Keynes UK
UKOW06n0204140215

246171UK00013B/26/P